Undergraduate Lecture Notes in Physics

Undergraduate Lecture Notes in Physics (ULNP) publishes authoritative texts covering topics throughout pure and applied physics. Each title in the series is suitable as a basis for undergraduate instruction, typically containing practice problems, worked examples, chapter summaries, and suggestions for further reading.

ULNP titles must provide at least one of the following:

- An exceptionally clear and concise treatment of a standard undergraduate subject.
- A solid undergraduate-level introduction to a graduate, advanced, or non-standard subject.
- A novel perspective or an unusual approach to teaching a subject.

ULNP especially encourages new, original, and idiosyncratic approaches to physics teaching at the undergraduate level.

The purpose of ULNP is to provide intriguing, absorbing books that will continue to be the reader's preferred reference throughout their academic career.

Series editors

Neil Ashby
University of Colorado, Boulder, CO, USA

William Brantley
Furman University, Greenville, SC, USA

Matthew Deady
Bard College, Annandale-on-Hudson, NY, USA

Michael Fowler
University of Virginia, Charlottesville, VA, USA

Morten Hjorth-Jensen
University of Oslo, Oslo, Norway

Michael Inglis
SUNY Suffolk County Community College, Selden, NY, USA

More information about this series at http://www.springer.com/series/8917

Mark A. Cunningham

Beyond Classical Physics

 Springer

Mark A. Cunningham
Katy, TX, USA

Wolfram *Mathematica*® is a registered trademark of Wolfram Research, Inc

ISSN 2192-4791   ISSN 2192-4805   (electronic)
Undergraduate Lecture Notes in Physics
ISBN 978-3-319-63159-2   ISBN 978-3-319-63160-8   (eBook)
https://doi.org/10.1007/978-3-319-63160-8

Library of Congress Control Number: 2017946474

Printed on acid-free paper

This Springer imprint is published by Springer Nature

The registered company is Springer International Publishing AG

The registered company address is: Gewerbestrasse 11, 6330 Cham, Switzerland

For Elizabeth Ann

Preface

Beyond Classical Physics is intended to be a sequel to the first-year level *Neoclassical Physics*. I have followed the same pathway, emphasizing the experimental underpinnings of our modern mathematical representations of nature. The focus here is primarily on the microscopic world, where experiments indicate that interactions of the constituents of matter are vastly different from those we observe in macroscopic objects. This world is best described by what we have come to call quantum theory.

As we progress through the text, we shall endeavor to illustrate some of the historical contributions of the practitioners of the day. For example, Albert Einstein provoked a signal change in how physics is practiced with his development of a General Theory of Relativity but Einstein did not work completely in isolation. This text is not intended to be a history of the subject; indeed, we shall not attempt to provide attribution for every equation or experiment. Following such a path would lead to a publication too ponderous to contemplate. Instead, my intent is to provide an introduction to the state of physics at the beginning of the twenty-first century.

When I was a first-year graduate student, my electromagnetics instructor Feza Gürsey made a point of telling the class that Freeman Dyson had proved the renormalizability of quantum electrodynamics as a first-year student. Recently, Gerard t'Hooft had just proven the renormalizability of quantum chromodynamics at the one-loop level, also as a first-year student.

"What," he queried, "are you working on?"

As he scanned the room, no one was willing to make eye contact, so I suppose that he concluded that ours was another class of misfits, without a single distinguished intellect amongst the lot.

In retrospect and having sufficient time elapsed for Professor Gürsey's stinging rebuke to have softened, I have come to another interpretation: Gürsey was seeking to challenge us to become relevant, to step beyond the limitations of curriculum and seek the frontiers of knowledge. Students have a natural tendency to follow the curriculum, learning whatever is set forth in textbooks and required in the syllabus. Universities have become complacent in their curriculum development, instead focussing on administrative, accreditation goals that reward adherence to (least common denominator) standards, not innovation. The result is that most students do not see the frontiers of physics research until they are well into their graduate careers. It is a truly remarkable student who even knows what renormalization means at the first-year graduate level, much less have enough insight to contribute in a significant fashion. This is unfortunate, because renormalization became integrated into the physics literature by the 1950s. Physics has moved well beyond quantum electrodynamics.

I have chosen to include the use not only of *Mathematica* software but also the numerical codes NWChem, that performs electronic structure calculations, and NAMD/VMD that provide molecular dynamics and visualization/analysis capabilities. I recognize that these choices may seem to introduce an insurmountable obstacle to progress but tutorials in their use are available by their developers and the codes are freely accessible. In fact, given adequate computational resources, these codes can be used at the frontiers of research. These days, numerical simulation has risen to new importance. Experiment, of course, provides the defining result but often experiments need interpretation. Simulation, beyond ideal theoretical formulations, provides concrete, if flawed, results that can help to explain experimental results and guide further experimentation.

I have had undergraduate students utilize these codes to good effect. It takes discipline and hard work to become adept with these tools but students are generally quite enthralled with the results that they can obtain. They are brought to the precipice where they can begin to ask significant questions of their own. This is not a small achievement.

While not a lengthy textbook, I expect that there is sufficient material to fill two semesters, particularly if instructors require students to attempt reading the original publications listed in each section. I have found this to be an interesting exercise for students. Even if they do not fully (or partly) comprehend the work, in many cases one finds that the original paper does not mention explicitly the reason why it is cited. Attribution follows reputation, in some instances. In any case, consulting the original literature is still an important step along the pathway to becoming a scientist. I believe it never too early to begin.

That said, I recognize that in so short a work that there are many areas of physics that are not discussed. Again, my philosophy is that students will benefit more from investigating fewer topics in more detail than in pursuing a large quantity of topics, just to say that you've seen Bernoulli's law, in case it comes up in a subsequent class. I believe that if students are properly prepared, they will have the tools to follow any topic that subsequently fires their imagination.

As diligently as I have worked on this text, I admit that there are probably mistakes; I hope there are no blatant falsehoods. Nevertheless, all errors in the text are my responsibility and I apologize in advance for any particularly egregious examples. This work, of course, would never have been completed without the continual, and unflinching, support of my wife Liz. She has endured the creation of this book without complaint, even as I staggered through the effort. I cannot say thank you enough.

Katy, TX, USA   Mark A. Cunningham

Contents

List of Figures

List of Tables

List of Exercises

Introduction

Some 2500 years ago, Greek philosophers contemplated the nature of the universe. We know this from the manuscripts they wrote that have survived to the present day. Additionally, many of their works were also translated into Arabic by scholars at the great library in Alexandria and some of those translations survived the Dark Ages that befell Europe. One early philosopher, Democritus[1] and his mentor Leucippus were the founders of what is today termed the atomistic school. They were the earliest Western proponents of the concept that matter could not be repeatedly divided into smaller bits endlessly: there must exist some fundamental components that cannot be subdivided. This stance was subsequently vigorously opposed by Plato and his student Aristotle. As Aristotlian philosophy was extraordinarily influential in Western civilization, the atomistic view fell into disfavor for centuries.

Today, we know for certain that matter is composed of atoms. We can enumerate the list of all possible atoms and we know a great deal of the properties of atoms and the larger, composite structures known as molecules. We also know that atoms are themselves composed of parts, so use of the Greek word $\alpha\tau\omicron\mu\omicron\zeta$ for indivisible is somewhat unfortunate. These facts have been established over the last century, due to careful and extraordinary measurements combined with insightful mathematical representations of those experimental results, by a process that we call physics.

The Greek philosophers were guided by reason and intuition but generally sought theories that were in some sense ideal. For example, the model of planetary motion developed by Claudius Ptolemy incorporated the earth as the center of the universe, with the sun and planets occupying circular orbits around the earth.[2] Stars decorated a celestial sphere that

---

[1]Democritus ($\Delta\eta\mu\acute{o}\kappa\rho\iota\tau\omicron\zeta$) was born in Thrace in northern Greece around 460 BC. The historical records for Leucippus ($\Lambda\epsilon\acute{u}\kappa\iota\pi\pi\omicron\zeta$) are vague but what records survive generally refer to him and Democritus as master and pupil.

[2]Ptolemy ($\mathrm{K}\lambda\alpha\acute{u}\delta\iota\omicron\zeta$ $\Pi\tau\omicron\lambda\epsilon\mu\widehat{\iota}\omicron\zeta$) wrote his *Mathematical Treatise* ($\mathrm{M}\alpha\theta\eta\mu\alpha\tau\iota\kappa\acute{\eta}$ $\Sigma\acute{u}\nu\tau\alpha\xi\iota\zeta$) in Alexandria about 150 AD. It was later translated into Arabic and is known today as the *Almagest*, from a translational corruption of the Greek word ($\mu\epsilon\gamma\alpha\lambda\acute{u}\tau\epsilon\rho\eta$) meaning greatest.

© Mark A. Cunningham 2018
M.A. Cunningham, *Beyond Classical Physics*,
Undergraduate Lecture Notes in Physics,
https://doi.org/10.1007/978-3-319-63160-8_1

also revolved around the earth. Circles and spheres are, of course, ideal shapes.

The Greek traditions were, in large measure, swept aside by Isaac Newton in 1687 in his *Philosophiae Naturalis Principia Mathematica*, in which he argued for a scientific methodology that was based on measurement, not speculation based on some perceived ideal. Our modern scientific methodology has evolved largely from Newton's ideas and has provoked significant progress in our understanding of the natural world. Indeed, physicists are no longer called natural philosophers as they are no longer content with speculative debates about ideal principles. They have, instead, become realists, basing their theories on experimental observation.

More precise measurements of planetary motion by Tycho Brahe and his students in the late 1500s led Johannes Kepler to conclude in 1609 that the orbit of Mars was, in fact, an ellipse centered on the sun. This highly inconvenient fact signalled the death knell for the Ptolemaic universe. Coupled with Galileo's observations of four moons of Jupiter in 1610, it was then clear that the geocentric universe and perfectly circular planetary orbits proposed by the Greeks did not agree with observational data. Newton ultimately resolved the issue with his gravitational theory described in the *Principia*. Newton's demonstration that an inverse square force law produced elliptical orbits provided a relatively simple explanation of the observed planetary motions.

In this text, we shall follow the development of physics from Newton onward, focussing in particular on the most recent developments. Like Aristotle before him, though, Newton cast a large shadow upon the scientific enterprise. For example, the enormous success of Newtonian mechanics lent gravity to Newton's opinions about other aspects of the natural sciences. Newton considered himself an astronomer and spent a sizable portion of his time trying to understand the nature of light. Newton's *Opticks*, published in 1704, expressed his ideas on the corpuscular theory of light. This model held sway until James Clerk Maxwell's synthesis of a complete theory of electromagnetism in 1865 demonstrated that light was governed by a wave equation.

The nineteenth and twentieth centuries saw the development of ever more precise instruments, enabling scientists to investigate phenomena at extraordinary length scales: detecting light emitted by stars thousands (and millions) of light years away and uncovering the structures within the cells that make up living organisms. In particular, what scientists found as they investigated the microscopic world was that the Newtonian ideas about forces and trajectories could not explain the phenomena they observed: the microscopic world does not behave in a manner consistent with our experiences with macroscopic objects. A new framework was developed that we call quantum mechanics that enabled a quantitative description of the microscopic world and we shall discuss this new theory in some detail.

1.1. Perception as Reality

A reasonable place to begin our discussion is to assess what we mean by an observation. Of course, historically, we would have meant using our eyes to look at some phenomenon. Tycho Brahe's original data, compiled with the use of a number of instruments of his own design, used Brahe's eyes, or those of an assistant, as the sensor. Galileo was the first to observe celestial objects with a magnifying device but he also used his own eyes as sensors. Many years later, in Ernest Rutherford's darkened laboratory, his assistants measured the behavior of α particle scattering by observing dim flashes of light through a microscope as the α particles impacted a zinc selenide crystal. Apparently, some of his assistants were not particularly reliable as detectors.

Today, of course, we can construct highly sophisticated machinery and operate it under the control of computers that do not require lunch breaks or complain about working nights and weekends. Our new experiments utilize sensors that have sensitivities far exceeding those of the human eye. We can resolve time differences at this writing of about 1 as (10^{-18} s), roughly the time it takes light to traverse the width of a single atom. Such precision has provided us with extraordinary information about the universe around us but we still must rely on converting these data into some sort of visual representations that we can observe with our eyes. This strategy is a sensible one, as our visual faculties are quite advanced. Technically, the information bandwidth that our brain is capable of processing is vastly greater for visual inputs than aural or tactile ones.

So, a significant portion of the discussion in this text will be dedicated to the development of different representations of our findings. Physicists are inveterate drawers of pictures, which represent (vastly) simplified representations of complex ideas and formulas. The challenge for students is to recognize the meanings of these squiggles and scrawls and provide the appropriate interpretation in terms of mathematical formulas. The pictures are always intended as an aid to understanding, although that is always a dubious prospect upon first encounter. Nevertheless, we are going to discuss a number of phenomena that do not possess any physical form.

Consider this: suppose that you are producing an epic cinematic masterpiece. How do you convey to your audience that the room you are filming contains a very strong magnetic field? The field itself is invisible, silent, odorless and tasteless. Humans have no senses that are directly affected by magnetic fields. So, if the presence of the field is somehow key to the story, how do you make it apparent to the audience that the field is present?

In physics class, we will most often depict a magnetic field by sketching files lines on the board, affixing some arrowheads on the lines to indicate the local field direction, as is indicated in figure 1.1. The field lines are a **representation** of the magnetic field that we might find useful. Here the student should recognize that the arrows associated with the magnetic field lines are different physical quantities than the arrow associated with the velocity vector. In print, we have emphasized this difference with different gray shadings. The student should also recognize that real magnetic fields do not manifest themselves with convenient little arrows.

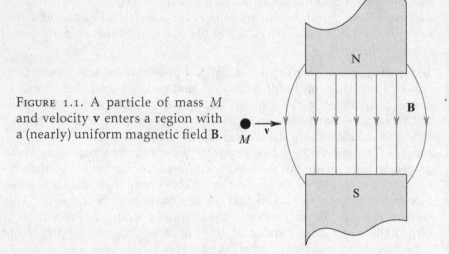

FIGURE 1.1. A particle of mass M and velocity $\mathbf{v}$ enters a region with a (nearly) uniform magnetic field $\mathbf{B}$.

Such an illustration does allow us to visualize the configuration of the experimental apparatus. In a real experiment, of course, all of the equipment would most likely be enclosed within a vacuum chamber and sealed behind walls to mitigate any radiation hazards. Visitors to any physics laboratory will undoubtedly be disappointed; there are few outward signs of progress visible to the experimenters. Such is not the case for other human endeavors such as construction projects, where structures rise from the ground and progress is readily apparent.

In 1972, Osheroff, Richardson and Lee were investigating the low-temperature behavior of ^{3}He with a device that slowly increased the pressure on the liquid ^{3}He. (Here we use the standard notation that an atom X, with a nucleus that contains Z protons and N neutrons and, thereby, an atomic mass $A = Z + N$ is denoted $_Z^A X$. The Z value is implicitly defined by the name of the element but is sometimes supplied as a convenience.) In figure 1.2, we see the results of their experiment.

The pressure on the ^{3}He sample was slowly increased, reaching a maximum at point C, which is arbitrarily defined as the zero point in the figure. At point D, the drive was reversed and the sample depressurized. Kinks in the slope of the pressure vs. time curve at points A and B while the sample is being pressurized indicate that there are two changes of phase within the sample.[3] There are equivalent kinks in the depressurization path at points A' and B' that appear at the same pressures and temperatures as for points A and B. This is the experimental evidence for superfluid phases in liquid ^{3}He.

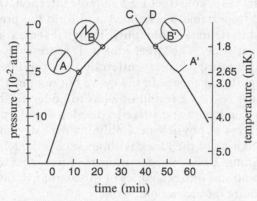

FIGURE 1.2. Compressional cooling of liquid ^{3}He reveals two phase changes that have subsequently been interpreted as evidence for superfluid phases. The kinks in the curve of pressure vs. time yielded a Nobel prize for Osheroff, Richardson and Lee. Reprinted figure with permission from: *Phys. Rev. Lett.* **78** (1972) 885. Copyright (1972) American Physical Society.

Arguably, if you were to tell someone that you performed research in superfluids, I believe that most people's imaginings would evoke much more spectacular results than kinks in a curve. Surely superfluids must somehow coalesce into some sort of solid forms that would allow their remote manipulation by alien beings. Perhaps drinking superfluids would convey super powers to mere humans, enabling them to twist, contort and stretch themselves into any desirable shape. Unfortunately, superfluid research is much more mundane; superfluidity is an emergent phenomenon that characterizes the collective behavior of the quantum system. Osheroff,

---

[3]Osheroff, Richardson and Lee were awarded the 1996 Nobel Prize in Physics "for their discovery of superfluidity in helium-3."

Richardson and Lee did not report any violent shaking of their apparatus at points A and B. Neither did they mention any mysterious glow from their apparatus that heralded their discovery. Publicly, none of the three claimed any sort of super powers.

In figure 1.2, the pressure that is plotted as a function of time is actually the (calibrated and) appropriately scaled voltage produced by some sensor. The temperature indicated on the right-hand side of the figure is also not that read from a thermometer but another voltage that has also been scaled appropriately; at temperatures of a few milliKelvin, one cannot rely on the mechanical expansion of metals to provide a suitable temperature measurement. Devising measurement instrumentation is key to progress.

Physics is the process of constructing a suitable interpretation of the measurements. The (repeatable) kinks (at A and B) in the pressure *vs.* time curve illustrated in figure 1.2 undoubtedly signal some sort of change in the material under study. The larger kink at D represents a change in the measurement apparatus; an intervention by the experimenters. It's certainly a more significant change in the curve but not one that arises from the sample properties. As a result, one has to understand in detail how experiments are conducted and strive to develop an interpretation of the results. The process of physics inevitably involves developing a mathematical representation of the ideas; without such a model, figure 1.2 has no intrinsic meaning. Unlike images of tigers or trees that we can immediately recognize and sort into categories of threatening or non-threatening, plots of data require interpretation.

Figure 1.2 notwithstanding, it can be said that everything we know about the universe is due to scattering and spectroscopy. This is a bit of hyperbole but not altogether insensible. Light falls on the retinas of our eyes and we interpret the resulting electrical signals in our brains. The light may arrive at our eyes directly from some distant source like the sun or stars but often reflects from some other object, like a house or a neighbor, before reaching our eyes. Reflection, as we shall see, results from the integration of a vast number of individual scattering events. It is an average property of the surface.

Spectroscopy is the energy-dependent measurement of scattering intensity. In our initial discussions, we shall mean the intensity of light but spectroscopic experiments can also be conducted with elementary particles. In that case, we would mean the intensity of electron or proton fluxes at the position of the detector. The human eye has a modest spectroscopic capability: we can distinguish many different colors but the wavelength resolution of our optical sensory system is not very precise. In figure 1.3, we plot the (normalized) spectral sensitivities of the three types of cone cells found in the retinas of our eyes that are responsible for bright light/color vision, normally called *photopic* vision. There are also rod cells present in the retina that enable us to see in low light conditions (*scotopic* vision). There is only one kind of rod cell, with a sensitivity similar to that of the M cells but with the peak shifted to 498 nm.

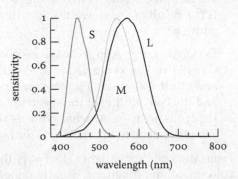

FIGURE 1.3. Human eyes contain three types of cone cells. Each type responds differently as a function of wavelength and are termed short (S), medium (M) and long (L).

The spectral sensitivities are determined by color-matching experiments that were initially pioneered by Hermann Graßmann in the 1850s. Subjects were seated in front of a white screen onto which was projected a pure color from a test lamp and a composite color obtained by adjusting the intensities of light from three standard sources, nominally red (700 nm), green (546.1 nm) and blue (435.8 nm), until the colors were perceived to be identical.[4] Interestingly, not all colors can be reconstructed in this fashion. In some cases, the red lamp was required to be placed on the same side of the screen as the test lamp; this situation arises from the significant overlap in the medium- and long-cone sensitivities observed in figure 1.3.

The sensitivity functions provide an example of a typical sensor response, which can be characterized as a convolution of what is called the instrument factor and an input signal. Mathematically, we can write this as follows:

$$(1.1) \qquad V(x) = \int d\xi \, F(x - \xi) I(\xi),$$

where x is the independent variable, such as wavelength or energy, F is the instrument factor and I is the input signal. The convolution V represents the output signal of the sensor, most often a voltage as a function of x.

---

[4]The *Commission Internationale de l'Eclairage* (CIE) was convened originally in 1913 to standardize all things relating to illumination. The 1931 convention in Cambridge promulgated the color standards still in use today.

EXERCISE 1.1.   We can demonstrate a useful property of Gaussian functions: the convolution of a Gaussian distribution with another Gaussian yields a Gaussian. Define a *Mathematica* function

$$F_1[x\_] := PDF[NormalDistribution[xi,wi],x].$$

We can produce the convolution with the `Convolve` function. What is the result of the convolution with a Gaussian with mean zero and width `W`?

EXERCISE 1.2.   Approximate a spectrum with a series of narrow Gaussians. Use `xi`=(2,5,5.5,7,9.2) with amplitudes (1,0.6,0.7,0.3,0.8) and widths `wi`=0.02. Plot the spectrum over the domain $0 \leq t \leq 10$ and the convolution of the spectrum with Gaussians of widths ranging from 0.1 to 1. At what point is the doublet no longer resolved? What happens to the spectrum for large values of `W`?

From the cone sensitivities, in 1931, the CIE determined corresponding color matching functions, usually denoted as $\bar{b}$, $\bar{g}$ and $\bar{r}$. The RGB values of any color are defined as

$$(1.2) \quad R = \int d\lambda\, \bar{r}(\lambda)I(\lambda), \quad G = \int d\lambda\, \bar{g}(\lambda)I(\lambda) \quad \text{and} \quad B = \int d\lambda\, \bar{b}(\lambda)I(\lambda).$$

We can think of the RGB values as the components of a three-dimensional vector in color space. Owing to the overlap of the cone sensitivity functions, the red, green and blue vectors are not an orthogonal basis set and the color-matching functions are not positive-definite, as illustrated in figure 1.4.

FIGURE 1.4.   CIE color matching functions for the RGB color scheme. Note that the red (r) curve is significantly (and unphysically) negative in the region below 520 nm.

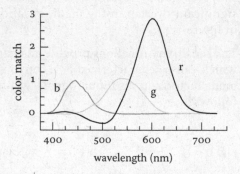

For colors corresponding to wavelengths of light below about 520 nm, it was necessary to move the red source to the same side of the screen as the test lamp, corresponding to negative values of red (700 nm) intensity. One cannot, of course, have negative values of intensity but the color matching functions do explain why you perceive yellow ($\approx$580 nm) when red and green sources are mixed.

EXERCISE 1.3. One can view the RGB color space with the *Mathematica* Graphics3D command:

```
Graphics3D[{Opacity[0.7],
    {RGBColor[#],Sphere[#,0.05]}&/@Tuples[Range[0, 1,.2],3]},
    Axes → True,AxesLabel → {"Red","Green","Blue"},
    Lighting → "Neutral"]
```

The 1931 CIE convention defined an alternative color space, called XYZ that does not require negative intensities in the color matching functions. The XYZ coordinates are obtained via linear transformations of the RGB coordinates. They can be viewed with the following command:

```
ChromaticityPlot3D[RGBColor[#]&/@Tuples[Range[0, 1,.2],3]]
```

Replace the RGBColor directive in the chromaticity plot with the directive XYZColor. How does this change the plot?

The science of spectroscopy was pioneered in 1859 by Robert Bunsen and Gustav Kirchoff, who utilized Bunsen's newly developed flame source to heat samples to high temperature, where the different samples produced distinctly different colored flames. This phenomenon had been observed previously but Bunsen and Kirchoff passed the light from their flame through prisms, whereupon the two discovered that each element had a spectrum that consisted of a series of specific colors, usually referred to as lines, owing to the fact that the light from the source was initially passed through a narrow slit to improve the precision of the measurements. After passing through the prism, the incident light separated into a series of narrow lines. Prisms rely on the refraction of light at interfaces and the fact that the refractive indices n are wavelength-dependent, a phenomenon known as *dispersion*. At an interface between two different media, such as air and glass or air and water, the path of a light ray is governed by the following equation:

$$(1.3) \qquad n_1 \sin\theta_1 = n_2 \sin\theta_2,$$

where the n_i are the refractive indices of the two media and the θ_i are the angles measured to the normal to the surface of the interface. Equation 1.3 is generally called Snell's law, after the Dutch astronomer Willebrord van Roijen Snell who derived it in 1621 but didn't publish the result, although the phenomenon was studied by Claudius Ptolemy in the middle of the second century.[5] We shall revisit Snell's law subsequently but the dependence of the refractive indices on wavelengths is small, limiting the

---

[5]Equation 1.3 first appears in print in Christiaan Huygens' *Dipotrica*, published in 1703. Only parts of a *ca.* 1154 Latin translation of an Arabic copy of Ptolemy's *Optics* exists today.

precision with which Bunsen and Kirchoff could determine spectra. Nevertheless, in 1861, the two discovered the elements cæsium and rubidium from their emission spectra.

The practice of spectroscopy was greatly improved by the development of diffraction gratings.[6] For normally incident light, with a grating spacing of d, maxima in the diffracted pattern will occur when the following relation is satisfied:

$$(1.4) \qquad\qquad d \sin\theta = n\lambda.$$

Here, θ is the angle between the normal to the grating and a (distant) viewing screen, n is an integer (known as the order) and λ is the wavelength.[7] As a result, the precision of a spectroscope can be controlled directly by adjusting the grating spacing d, whereas years of investigations into various glasses had not yielded any significant control over the weak dispersion that gave prisms their analytic power. The Australian scientist Henry Joseph Grayson developed a series of precise ruling engines to scribe lines with micrometer spacings on glass microscope slides. Grayson's intent was to provide precise length measurements to biologists studying cells but his technology was readily adapted to spectral analysis. By 1910, Grayson had achieved gratings with roughly 4700 lines/mm, producing an extraordinary new tool for spectroscopists.

> EXERCISE 1.4. If you possess a diffraction grating ruled with 4700 lines/cm and can measure angles to a precision of 1°, to what precision can you measure the wavelength of light around 600 nm? How does your measurement improve if your angular precision is 1 arc-minute or 1 arc-second?

With these new instruments, physicists were able to establish that, when excited by electrical discharges or heated by flames, each element emitted light at a specific set of wavelengths unique to that element. The relative intensities at each of the wavelengths were also characteristic. For example, in the visible spectrum of neon, illustrated in figure 1.5, there is a strong peak at 692.94 nm and other strong peaks at 703.24, 717.39 and 724.51 nm. There are many other peaks in the visible range from 400–800 nm, but the intensities of these four peaks account for the strong red color that we associate with neon lamps.

---

[6]In May of 1673, the Scottish mathematician James Gregory wrote a letter to his publisher John Collins describing his observation that light passing through a bird's feather split into different colored spots. Gregory did not pursue his investigations, deferring the field of optics to Isaac Newton, who was very active in researching optical phenomena and very protective of his domain.

[7]Note that the order n in equation 1.4 is not related to the index of refraction n discussed earlier. This will not be the last time we will encounter notational difficulties. *Caveat emptor.*

FIGURE 1.5. Emission spectrum of neon gas. Neon has a characteristic red color, owing to the cluster of strong lines around 700 nm. The units of the intensity scale are arbitrary.

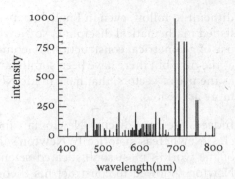

wavelength(nm)

EXERCISE 1.5. The data for figure 1.5 were obtained from the Atomic Spectra Database (ASD) maintained by the US National Institute of Standards and Technology (NIST). The database can be found by searching for "strong lines neon" in a web browser. As of this date, the database is found on the www.nist.gov site but searching should find the database even at some date in the future when the agency is renamed.

Collect the spectral data for Ne I (neutral neon) from the database for the wavelength range 400–800 nm and reproduce figure 1.5. Obtain spectral data for neutral helium and neutral argon. Plot their spectra.

It is now possible to retrieve vast amounts of experimental information about the universe we inhabit from carefully cultivated databases maintained by physicists in different repositories. In the previous exercise, we demonstrated the existence of the Atomic Spectra Database, a repository that is not a static catalog of experimental data but, instead, a dynamic entity that is continually updated and collated. At present, it is to be expected that students will not understand the majority of notations and remarks embedded within the pages; that is, of course, the purpose of this text. Nevertheless, students should appreciate that physics has been driven by the accumulation of vast quantities of data of ever-increasing precision.

1.2. Classical Physics

We often date the onset of classical physics with Newton's publication of his *Principia* in 1687. This is a somewhat artificial boundary but reflects the importance of the philosophical principles Newton espoused. The modern scientific method has evolved from those principles. For the intrepid student, sifting through the pages of the *Principia* itself may not prove as satisfying as one might initially hope. Newton's language is quite difficult to follow, even in English translation. Calculus was not an established mathematical discipline, so Newton argued his points through the use of geometrical constructions. Geometry was an established discipline at the time but there have been significant improvements in notation, such as the use of vectors, that have made Newton's gravitational theory much more accessible.

Indeed, we now discuss Newtonian mechanics in introductory physics classes; whereas historically, Newton's ideas were only accessible to a few of the world's most sophisticated scientists. Over time, the procedures Newton followed to construct his theory of gravitating bodies were extended to include many other physical phenomena. By the beginning of the nineteenth century, Lagrange and Hamilton were providing new insights into how one might generalize the process of deriving the equations of motion for any system. A brief timeline of some notable scientists is provided in figure 1.6. We note that physicists use the generic term equations of motion to describe the set of equations that define the time evolution of the state of a system. This can include many phenomena and is not restricted solely to systems in which physical objects are moving.

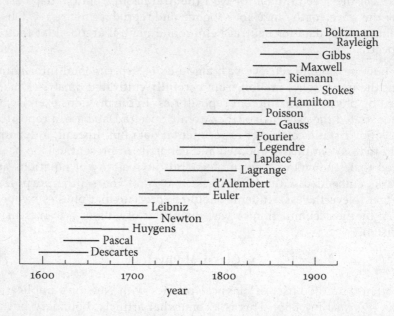

FIGURE 1.6. Lifetimes of selected mathematicians and physicists. Newton and Leibniz are the founders of the calculus. Over the succeeding generations their initial ideas were expanded and systematized by others.

An intrinsic difference between physics and mathematics is that physical models are implicitly inexact; they represent an approximation of reality. In geometry, two triangles are either congruent or they are not; there are absolutes. In physics, we attempt to develop models that capture the essence of a phenomenon and then we compare our results to (imprecise) experimental data. For example, in fluid systems, we know ultimately that the fluid is composed of small molecules. Yet, we talk about the properties of a bulk fluid, mass density, for example, as if it were a continuous function. We take derivatives of the density even though, at some microscopic level, we reach the atomic scale and the fluid is a jumble of individual molecules, not a continous soup of matter. The mathematical limit of letting some small parameter go to zero really isn't defined for real fluids. Nevertheless, treating fluids as continuous media seems to work reasonably well for macroscopic amounts of fluid.

There are, of course, some physical systems where the granularity of matter cannot be ignored. Consider the sand dunes pictured below.[8] Sand grains are small, so one might be tempted to treat sand as a fluid and, in some sense, sand does behave like a fluid: it can flow through an hourglass, for example. In many details, though, sand is not a fluid: it can form stable mounds (dunes) that persist after the driving force (wind) is removed. By contrast, water waves quickly dissipate without a driving force. Fluids do not possess internal structure and thereby seek a minimal (locally flat) level in a gravitational field.

FIGURE 1.7. The Curiosity rover took this image of dunes on Mars in late 2015 that features many striking examples of the physics of granular materials. Image courtesy of NASA.

So, how can we treat sand? One can envision building a model that treats each grain individually. One would then need to determine the force on

---

[8]For the faint-hearted students, steel yourselves. This is not going to be a lengthy essay on dirt. It is, instead, a cautionary tale on emergent behavior.

that grain due to its neighbors and, from there, determine how that grain would subsequently move. Unfortunately, such a strategy leads to enormous calculations, due to the vast number of grains in a dune and the fact that the computational work will scale poorly, like some power of the number of grains (N^2 or N^3 or worse). Such calculations are really intractable, so we shall have to devise an alternative approach.

As we are not particularly interested in individual sand grains but in the behavior of many grains, we can utilize statistical methods, pioneered by Gibbs and Boltzmann, to assess the properties of ensembles of grains. This means that we will attempt to devise models that can reproduce the major features that are visible in figure 1.7. For example, there are large mounds separated by a relatively large distance and numerous ripples with smaller amplitudes that are more closely spaced. Moreover, the tops of the ripples and mounds form cusps, where the surface is smooth and then changes shape abruptly. These features almost certainly depend upon the wind velocity, particle size, and size distribution, and other physical parameters associated with the grains. We will have to posit some sort of attractive, short-range force acting between grains and the exact form of this force will be empirically defined. That is, we will not be able to derive it from some set of fundamental principles; we will have to adjust the form of the force in order to reproduce experimental results.

We can apply the process of model construction, comparison with experiment, conduct of additional experiments and model refinement repeatedly until we are satisfied. Satisfaction here is determined by various measures: goodness of fit, for example, is a quantitative measure but there are often non-quantitative assessments that come into play. We may, for example, expend a good deal of effort on various modifications to the mathematical model without significant improvement in the results and simply accept that further progress will require more effort than is reasonable. Knowing when to quit is an important skill that physicists need to develop.

Classical physics is not defined solely by phenomena that are visible to the human eye. Electromagnetic phenomena can be described classically, at least in the realm where currents are composed of many charge carriers, not single electrons. Maxwell's equations provide us with a means for determining the fields that arise as a result of charge distributions and currents that is extraordinarily successful. Using Maxwell's equations, we can design and construct devices that transmit and receive radio waves, thereby enabling communication over great distances. We can predict the behavior of electrical circuits before they are fabricated, ensuring that they will ultimately perform as desired. So, even though we cannot personally detect the presence of a current flowing in a wire, we can understand the consequences of its existence through the Maxwell equations.

Fortunately, we can use our experiences dealing with the portion of the electromagnetic spectrum to which our eyes are sensitive, to help us understand the remainder. For example, when waves encounter an interface between two different media, simple rules of behavior have emerged, such as the angle of incidence (measured from the normal to the surface) is equal to the angle of reflection. Mathematically, this phenomenon arises from the requirement that the fields be continuous at the interface. Indeed, this particular result is also true for acoustic waves and surface waves on water.

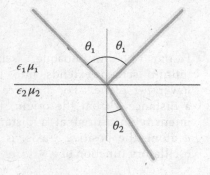

FIGURE 1.8. A thin beam of light, entering from the upper left, impinges upon a surface between two regions with different electromagnetic properties. The reflected beam (upper right) makes the same angle θ_1 with respect to the normal as the incident beam. The transmitted (refracted) beam makes a different angle θ_2.

The experiment sketched in figure 1.8 forms the foundation of geometrical optics and is one that is conducted in every introductory course. This behavior supported Newton's corpuscular theory of light for many years. Because light has a very short wavelength compared to the beam size, the ray approximation indicated by the figure provides an effective means of understanding how light propagates through optical systems composed of lenses and mirrors. For longer wavelengths, diffraction effects can be significant and the ray approximation is less useful. Indeed, even for visible light, one can see diffraction effects if you look closely.

The French physicist Augustin-Jean Fresnel considered a simple example of light emanating from a point source that encountered an opaque, semi-infinite screen. For convenience, we will locate the source on the negative x axis at a distance s away from the origin. An opaque, semi-infinite screen is placed in the y-z plane at the origin, blocking a direct view of the source for points in the negative z direction, as indicated in figure 1.9. The relative intensity of a beam measured some distance d beyond the screen is given by the following expression:

$$(1.5) \qquad I(\zeta) = \frac{I_0}{2}\left[\left(S(\zeta) + \tfrac{1}{2}\right)^2 + \left(C(\zeta) + \tfrac{1}{2}\right)^2\right],$$

where the Fresnel functions $S(z)$ and $C(z)$ are defined as follows:

$$(1.6) \qquad S(\zeta) = \int_0^\zeta dt\,\sin(\pi t^2/2) \quad \text{and} \quad C(\zeta) = \int_0^\zeta dt\,\cos(\pi t^2/2).$$

Here, the variable ζ is proportional to the distance from the edge:

$$(1.7) \qquad \zeta = z\sqrt{\frac{ks}{2d(s+d)}},$$

where the wavenumber k is inversely related to the wavelength λ: $k = 2\pi/\lambda$.

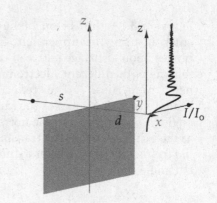

FIGURE 1.9. An opaque, semi-infinite screen extends into the lower y-z plane. For a source at a distance s from the origin, the intensity measured at a distance d along the positive x-axis, is an oscillatory function of z.

The oscillatory behavior observed in figure 1.9 is not what one might naïvely expect. One might have anticipated that there would be a sharp shadow line at the edge but the oscillations in the intensity are readily observed. Indeed, observations of diffraction signalled the demise of the corpuscular theory of light. This was a phenomena that required that light be a wave.

EXERCISE 1.6. Plot the functions FresnelS and FresnelC over the range $0 \le \zeta \le 10$. Plot the relative intensity $I(\zeta)/I_0$ over the range $-2 \le \zeta \le 10$. How does this behavior differ from what you would expect of a corpuscular description of light?

EXERCISE 1.7. Suppose that you try to reproduce Fresnel's experiment with a laser pointer light source ($\lambda = 600$ nm) and a razor blade as the opaque screen. If you place the laser pointer a few cm from the blade and allow the light to fall on a blank sheet of paper a few cm distant, what would be the spacing of the bright fringes in the diffraction pattern? How does the spacing change as you adjust the distances s and d from equation 1.7?

The phenomena is more readily discerned if one places the viewing screen at a large distance from the opaque screen. In the limit where the source is also placed far from the opaque screen, we have a situation known as Frauenhofer diffraction.[9] If we consider a narrow slit in an infinite opaque screen, where the slit width a is comparable to the wavelength λ ($a/\lambda \approx 1$), then the intensity can be determined as a function of θ, the angle from the initial beam direction (x-axis). For normal incidence and a single wavelength source, we find the following result:

$$(1.8) \qquad I(\theta) = I_0 \, \text{sinc}^2\left[\frac{ka}{2}\sin\theta\right],$$

where the wavenumber k is given by $k = 2\pi/\lambda$ and the sinc function is defined as follows:

$$\text{sinc}\, x = \frac{\sin x}{x}.$$

EXERCISE 1.8. The intensity of Frauenhofer diffraction is sharply peaked at $\theta = 0$, with nulls in the intensity at specific values of ka. Plot $I(\theta)$ over the domain $-\pi/2 \le \theta \le \pi/2$. Study the behavior for the domain $2 \le a/\lambda \le 10$.

1.3. Quantum Physics

Waves, in most instances, do not interact directly. As a result, the patterns produced by arrays of slots can be readily determined. A modest complication is that the intensity we observe is proportional to the square of the field amplitude. Nonetheless, these calculations are not exceedingly difficult and the characteristic patterns generated by multi-slot interference are further evidence of the wave nature of light. For two slits of width a, with centers separated by a distance b, the intensity as a function of angle is given as follows:

$$(1.9) \qquad I(\theta) = I_0 \cos^2\left[\frac{kb}{2}\sin\theta\right]\text{sinc}^2\left[\frac{ka}{2}\sin\theta\right].$$

---

[9]Joseph von Frauenhofer was born into a family of glassmakers in 1787 and orphaned at age 11. Apprenticed to an optician, Frauenhofer came to the notice of Prince Elector Maximilian Joseph IV of Bavaria when his master's house burned to the ground. Under Maximilian's aegis, Frauenhofer was eventually able to conduct his own optical research, ultimately discovering the dark absorption lines in the solar spectrum. Frauenhofer invented the diffraction grating spectrometer to explore the spectral lines in more detail but never published his work, in order to avoid disclosing the trade secrets of glassmaking. He never contributed to the theory underlying Frauenhofer diffraction and was not a part of the scientific community during his lifetime.

Note that this equation indicates that the Fresnel diffraction effects simply multiply the interference effects.

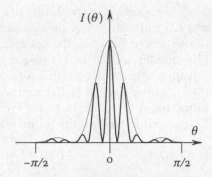

FIGURE 1.10. The intensity of light interfering from two slits (black curve) is modulated by the diffraction envelope (gray curve) of a single slit. Nulls in the intensity allow the determination of the slit width and separation in terms of the wavelength λ, or *vice versa*.

The intensity observed in a two-slit diffraction pattern is illustrated in figure 1.10. The patterns that arise depend on the ratios a/λ and b/λ, where a and b are the slit width and separation, respectively, and $\lambda = 2\pi/k$ is the wavelength.

EXERCISE 1.9. Plot the two slit diffraction intensity over the domain $-\pi/2 \leq \theta \leq \pi/2$ for different values of a/λ and b/λ. How do the patterns change as, for example, b/λ changes from 100 to 10? How do the patterns change as a/λ changes from 3 to 10?

The initial use of quantization in physics can be traced to Max Planck's strategy for constructing a sensible probability distribution for electromagnetic energy in a box. Planck proposed that light came in discrete units, so that instead of integrating over all frequencies, which led to divergent integrals, one summed over the number of discrete units within the box. The summation converges to a finite result and really amounts to enforcing a high-frequency cutoff on the integration over frequency. In any event, physicists began thinking about what came to be called **photons**.

One of the first experiments that tried to illuminate some of the properties of these photons was conducted by G. I. Taylor, a student of J. J. Thomson.[10] Under Thomson's direction, Taylor studied the patterns produced by light diffracting around a thin needle. The light source was a high-temperature gas flame collimated through a thin slit. Light passing through the slit fell on the needle and then upon a photographic plate. Taylor found that an exposure time of about 10 s achieved a particular level of blackness in the developed plates. He then placed a piece of dark

---

[10]Taylor's "Interference Fringes With Feeble Light" was published in the *Proceedings of the Cambridge Philosophical Society* in 1909.

glass in front of the slit, to reduce the light intensity, and experimented with exposure times to recover the same level of blackness in the developed plates. Taylor assumed that the exposure time was inversely proportional to the intensity of the beam and then conducted a series of additional experiments, adding additional dark glass at each step. In the final step, an exposure of over 2000 hours was required. Ultimately, Taylor found no difference between any of the exposed plates. They all produced a pattern akin to that seen in figure 1.10, with a few minor variations in overall blackness that could be attributed to not correctly estimating the exposure times.

We can utilize modern technology to update the Taylor/Thomson experiment, with surprising results. In particular, we can replace the photographic emulsion with an image intensifier that can detect single photons. The image intensifier utilizes high voltages to convert production of an initial photoelectron by an initial photon capture event into an exponential cascade of electrons. Thus, a single photon striking the surface may result in 10^{12} electrons at the end of the amplification stage. Modern devices encapsulate the cascade events within micrometer-scale channels, preserving high spatial resolution of the initial detection event. The microchannel electrons are dumped into a charged-coupled device (CCD) camera and the camera integrates the signals. There is a fair amount of interesting physics taking place within the detection system but we shall not stop to investigate it in detail at this moment.

FIGURE 1.11. Single photons diffracting through a three-slit experiment display particle-like properties, scattering in seemingly random fashion (a). Counting more events (b) and (c) results in the emergence of the expected diffraction pattern. Image courtesy of Department of Physics, Princeton University.

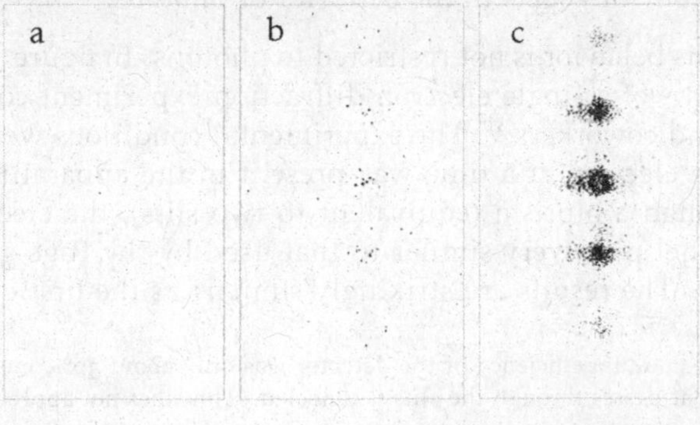

The images depicted in figure 1.11 represent the results of single-photon diffraction (in this case, through three slits). That is, the light intensity within the apparatus was reduced such that, on average, only a single photon's worth of energy was incident on the slit at any one time. As a result, the image produced is the result of a single photon interfering with itself, not other photons that would make up a classical wavefront.

> EXERCISE 1.10. The initial laser power used in the experiment portrayed in figure 1.11 was 1 mW. What is the energy contained in a single photon with wavelength $\lambda = 632.8$ nm? How many photons per second are required to produce that initial power level? The initial laser beam was attenuated by a factor of 5×10^{-11} before striking the slits. What is the average time between photons in the attenuated beam? What is the average distance between photons in the attenuated beam?

What we see is that, after passing the slits, each photon hits the detector in some apparently random position.[11] Only after many photons reach the detector does a diffraction pattern begin to emerge. Apparently, diffraction is an emergent property of a statistically large number of individual photons.

The essence of the new quantum theories that were being developed in the early 1900s was that microscopic particles were described by wave functions that determined their state. Squaring the wave functions produced the probability distribution of the particle. Such theories were remarkably distinct from the completely deterministic theories pioneered by Newton; they had more in common with the statistical mechanics theory pioneered by Gibbs and Boltzmann. Indeed, the theoretical framework developed by Gibbs needed no essential modifications to incorporate quantum ideas. There were some physicists that were troubled by the reliance on a probabilistic formulation of nature; Einstein, in particular, felt that determinism should not be abandoned. Yet, in figure 1.11, we see direct evidence of that probabilistic aspect in the behavior of photons.

Moreover, this behavior is not restricted to photons. In figure 1.12, we depict the results of a single electron diffraction experiment conducted by Tonomura and coworkers.[12] The experimental conditions were such that only a single electron at a time was present in the apparatus and, after passing through a biprism (equivalent to two slits), the electron struck a microchannel plate very similar to that used by the Page group to detect photons. The results are strikingly similar: as the first electrons are

---

[11]Actually, the quantum efficiency of the detector was only about 30%, meaning that not every photon that passes through the slits is detected. This does not appreciably alter the conclusions.

[12]"Demonstration of single-electron buildup of an interference pattern" was published in the *American Journal of Physics* in 1989.

FIGURE 1.12. Time se-
quence of single electron
scattering through a dou-
ble slit. (a) Initial electrons
scatter seemingly ran-
domly. (b–c) As more
electrons are counted,
the (d) classical double
slit diffraction pattern
arises. Image courtesy of
the American Institute of
Physics.

counted, there is an apparently random distribution. Then, as more electrons are counted, a diffraction pattern emerges. The electrons interfere
with themselves in a fashion just like photons, with identical results.

The consequences of this observation are profound. Electrons are not particles in the sense of exquisitely small balls of some form or another; they
diffract, meaning that they are waves. As each entity, electron or photon, made its way through the apparatus, the encounter with the slits
resulted in a modification to the probability density function, such that
when the entities reached the detectors, there was a probability that the
entity would be detected at each point on the detector. That probability is
defined by the intensity plots that we have described above. We have seen
this sort of emergent behavior previously when discussing random walks.
The location of any individual walkers is essentially unknowable but the
distribution of many walkers becomes very predictable as the number of
walkers becomes large.

EXERCISE 1.11. Use the RandomReal function to generate random
numbers in the range (0,1). If the number is greater than one half,
take a step in the positive direction. Otherwise, take a step in the
negative direction. For each walker, take twenty steps. Plot the distribution for 100 walkers and 10,000 walkers.

This behavioral equivalence between electrons and photons is usually described as *wave-particle duality*. The usual interpretation is that electrons
and protons display characteristics of both waves and particles, depending upon the circumstances. We shall have much more to discuss on the
matter going forward, but the results pictured in figures 1.11 and 1.12
demonstrate the futility of using macroscopic ideas to explain microscopic
phenomena. As we shall see, the attempts to describe atoms with some
sort of planetary model, which is an appealing analogy, cannot succeed.

The microscopic world is vastly different from our personal perceptions of the universe.

1.4. Mathematical Insights

In 1905, Albert Einstein published four papers that proved subsequently to be of major importance in defining the direction of modern physics. This accomplishment is even more remarkable when one realizes that Einstein was the only member of his graduating class (1900) at the Eidgenössische Technische Hochschule (ETH) in Zürich who was not offered a position to continue his studies as a research assistant.[13] Einstein managed to eventually find teaching positions in private schools before the intervention of his friend Marcel Grossmann's father secured Einstein a position at the Swiss Patent Office in 1902. (The Patent Office had advertised for a patent examiner, second class, but only offered Einstein a third class appointment.) Despite the dubious beginning of his career, Einstein was greatly enthusiastic about the post: it meant that he had the financial means to marry Milena Maric, a fellow physics student he had met at the ETH. Einstein's father objected to the match, so the couple delayed their marriage until after his death in 1903. Einstein continued at the Patent Office until 1909, being promoted to examiner, second class in 1906.

Einstein's 1905 submissions to *Annalen der Physik* included an explanation of the photoelectric effect, provided a theoretical basis for Brownian motion, defined what we today call special relativity and remarked on mass/energy equivalence.[14]   Also in 1905, Einstein was awarded a doctorate degree from the University of Zürich for his dissertation *On a new determination of molecular dimensions* that he dedicated to Grossmann. It is no small wonder that 1905 has come to be known as Einstein's *annus mirabilis*.

In 1908, Einstein's *Habilitation* thesis *Consequences for the constitution of radiation following from the energy distribution law of black bodies* was accepted by the University of Bern and Einstein became a lecturer there briefly before accepting a professorship at the University of Zürich in 1909. Einstein's growing reputation as a theoretical physicist finally began to bring him more opportunities; he moved several more times before moving to the University of Berlin in 1914, where he became the founding director of the Kaiser Wilhelm Institute of Physics.

The early 1900s were heady years for physicists. The majority spent their time working toward developing a consistent description of what we call

---

[13]Wayward students take heart!

[14]Einstein was awarded the Nobel Prize in Physics in 1921 "for his services to Theoretical Physics, and especially for his discovery of the law of the photoelectric effect."

quantum phenomena. Einstein was notably silent in the ongoing discussions. He was philosophically opposed to the idea that a successful physical theory could or should be built upon a statistical formulation. The wave functions that had been introduced to describe quantum states were interpreted (when squared) to be the probability of a state possessing a particular property.

Instead, Einstein focussed on continuing the work that he started in 1905 with the Theory of Special Relativity. Suppose that we have two coordinate systems: $\mathbf{x} = (ct,x,y,z)$ and $\mathbf{x}' = (ct',x',y',z')$; the Special Theory of Relativity requires that the transformations between the two systems be linear:

$$(1.10) \qquad 0 = \frac{\partial^2 x'}{\partial x^2} = \frac{\partial^2 y'}{\partial x^2} \;\; \dots \;\; et\ cetera.$$

Einstein was interested in understanding what happened if you relaxed that restriction: what happens if the transformations are nonlinear? There was no compelling reason to study this particular question and, in truth, there were many interesting experimental results being produced in the quantum arena. Nevertheless, even as a student Einstein kept his own counsel, much to the consternation of his instructors, and he demonstrated a fierce independence that initially cost him the opportunity to pursue physics professionally. So, it is not surprising that, while others made names for themselves in the new quantum theories, Einstein continued along his singular path.

We know from dimensional analysis that in the following linear transformation:

$$x' = at + bx,$$

the coefficient a must have the units of a velocity and b is dimensionless. Hence, the Special Theory of Relativity concerns observers in frames that are, at most, moving with constant relative velocities. If we are to consider transformations that are not linear, then we could have transformations like the following:

$$x' = at^2 + bt + cx,$$

where here the a coefficient has the units of an acceleration. In his studies of a more complex set of transformations, Einstein understood that he would have to cope with accelerating systems and he also understood that this would yield a number of additional complications.

Take, for example, the simple situation of a rotating coordinate system, as depicted in figure 1.13, as depicted in figure 1.13. If the riders on a merry-go-round roll a ball back and forth between one another, the ball takes what appears to be a curving trajectory from one rider to the next. Viewed from a stationary spot above the merry-go-round, another observer will

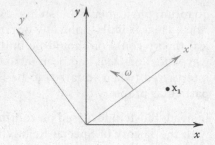

FIGURE 1.13. In a rotating frame of reference ($\mathbf{x}'$), a point that is fixed in space ($\mathbf{x}_1$) will change its position over time.

see that the ball actually follows a straight-line path, just as one should expect for an object that is not being accelerated.

To see how this occurs, let us choose two coordinate systems with the same origin. A point in one system is defined by $\mathbf{x} = (x, y)$. In the rotating system, a point is defined by $\mathbf{x}' = (x', y')$. The angle θ between the two systems is a function of time: $\theta = \omega t$, where ω is the angular velocity. The relationship between the two coordinate systems is provided by the rotation matrix:

$$(1.11) \qquad \begin{bmatrix} x' \\ y' \end{bmatrix} = \begin{bmatrix} \cos \omega t & \sin \omega t \\ -\sin \omega t & \cos \omega t \end{bmatrix} \begin{bmatrix} x \\ y \end{bmatrix}.$$

Here, the coordinate systems coincide at time $t = 0$ and at integer multiples of $T = 2\pi/\omega$ (known as the period) thereafter.

Suppose now that x is itself a (simple) function of time: $x = x_o + vt$. It is not difficult to show that trajectories in the rotating frame are curved. (We do so in the following exercise.) Newton's laws of motion require that an object that is deviating from straight line motion is doing so as the result of an acceleration created by an external force. As a result, observers in an accelerating frame of reference will infer the presence of forces that do not actually exist.

EXERCISE 1.12. Define the following functions:

```
x₁[t_]:=(xo + v*t) Cos[w t] + yo Sin[w t]
y₁[t_]:=-(xo + v*t) Sin[w t] + yo Cos[w t].
```

Use the *Mathematica* function ParametricPlot to examine the trajectories.

```
Manipulate[
  ParametricPlot[{x₁[t],y₁[t]}/.{xo->1.5,yo->0.3,w->w₁,
    v->v₁},{t,0,1},PlotRange->{{-4,4},{-4,4}}],
```

{w1,0,1},{v1,-1,1}]

What happens when $\omega = 0$ and you change the value of v? What is the shape of trajectories when ω is no longer zero? How does altering the initial values x_0 and y_0 affect the results?

Einstein made a key discovery in 1907, when he published his notions on an equivalence principle. One can observe, for example, that the force F on a charged particle in an electric field E is given by the following equation:

$$(1.12) \qquad \mathbf{F} = q\mathbf{E},$$

where q is the charge of the particle, and the electric field can be, for example, the field produced by other charged particles. This will give rise to an acceleration of the particle:

$$(1.13) \qquad \mathbf{a} = \frac{\mathbf{F}}{m} = \frac{q\mathbf{E}}{m}.$$

One can also recast Newton's gravitational equations into the following form:

$$(1.14) \qquad \mathbf{F} = M\mathbf{G},$$

where M is the gravitational mass of an object and G is the gravitational field of, say, other masses. The gravitational force will also give rise to an acceleration:

$$(1.15) \qquad \mathbf{a} = \frac{\mathbf{F}}{m} = \frac{M\mathbf{G}}{m} \equiv \mathbf{G},$$

where in the last step, we have made the assumption that the gravitational mass M of an object is the same as the inertial mass m.

Einstein reasoned that if you were placed inside of a box that was far from any other masses and accelerated in some fashion, say with a rocket motor, then you would be unable to differentiate between that acceleration and the acceleration due to the gravitational force that arises when standing on the earth's surface. Einstein believed that this equivalence between gravitational and inertial masses provided a key insight. As a result, he became convinced that his search for a more general theory of relativity would give rise to a new approach to describing gravity.

For the next several years, Einstein worked diligently on his theory of gravity, but without making much progress. The situation changed when, in 1912, Einstein moved back to Zürich and began working with his old classmate Marcel Grossmann. Part of Einstein's struggle with the development of this new theory was in trying to find a means to describe his ideas in mathematical terms. Unfortunately, mathematicians are content to live in an exceedingly abstract world, patterning their writings on those of Carl Friedrich Gauss, whose personal motto was *pauca sed matura* (few, but ripe). When questioned (somewhat chastisingly) for the paucity of words in his writings, especially any documentation of his motivating ideas, Gauss answered that architects do not leave their scaffolding in place when construction is complete; neither would he. As a result, mathematical papers are hard reading for physicists, even Einstein.

Such was Einstein's situation in 1912: How does one extrapolate the ideas of special relativity into a more general form and is there some sort of mathematics that can be used? Fortunately, Marcel Grossmann was a professor of mathematics at the ETH and willing to serve as Einstein's guide through the mathematical literature. Einstein, for his part, was mathematically adept, a quality that should not be misinterpreted to mean that he was a mathematician, for he was not.[15]

This is a perpetual problem for physicists, particularly those engaged in theoretical research at the frontiers. Mathematicians engage in absolutes: one either proves a theorem or one does not. Physicists, on the other hand, live in a world of approximation: fluid density is treated as a continuous function, even though at some level the fluids are composed of individual molecules. Often, it is a problem of language. As physicists, we utilize words like momentum and energy to mean specific things. So too do mathematicians, who use words like group and algebra to mean specific things.[16] In the initial stages of discovery, though, there is often some ambiguity about the choice of words and notation that should be used to represent a concept.

In any case, Grossmann was willing to serve as Einstein's translator in his struggles to find the appropriate mathematical language to represent his ideas. Grossmann recognized that the tensor calculus proposed by Elwin Bruno Christoffel in 1864 might provide the mathematical structure that Einstein needed. There were subsequent works by Tullio Levi-Civita and Gregorio Ricci-Curbastro, who further developed Christoffel's ideas but their writings in the mathematical literature were too abstract for Einstein to follow, so Grossmann spent a fair portion of 1912 educating Einstein in the use of tensor calculus and the ideas of the differential geometry developed by Gauss and Bernhard Riemann.

---

[15]The eminent mathematician David Hilbert told his biographer that the average schoolboy on the streets of Göttingen knew more about four-dimensional geometry than Einstein but it was Einstein who put together the new theory of gravity.

[16]One of the goals of the influential Twentieth Century French mathematicians who founded the *Association des collaborateurs de Nicolas Bourbaki* was to systematize the language of modern mathematics across all of its subdomains. Their efforts led to the discovery of surprising relationships between what were previously considered disparate fields.

Einstein and Grossmann wrote a series of papers in 1912 that almost captured what Einstein sought: a coherent, self-consistent theory of gravity that arose from the geometry of spacetime. In fact, they even considered what was to become the final theory but discarded it as being not quite correct. Now, a hundred years or so after the fact, students may be puzzled by the fact that it took another three years for Einstein to finally complete his mission. There is a Sidney Harris cartoon published some years ago that depicts a puzzled Einstein standing before a chalkboard with equations $E = ma^2$ and $E = mb^2$ crossed out. All that he had to do, one supposes, is to take the next logical step.

In truth, Einstein's struggles with his theory were more complex than the cartoon would indicate but a visit he made to Göttingen, where he had been invited to lecture on his ideas by David Hilbert, finally provided Einstein with the impetus to complete his work. Einstein presented six two-hour lectures on his theory of gravity in late June and early July before returning to Berlin. In recounting the results of his expedition to friends, Einstein was cheerful: "I have convinced Hilbert and (Felix) Klein." Among the difficulties that Einstein had encountered was an apparent failure of his theory to conserve energy. Einstein recognized this to be a potentially fatal flaw.

As it happened, Hilbert had hired a young mathematician who also possessed some knowledge of physics and could therefore help him in his own efforts to understand the subject: Emmy Noether.[17] Upon her arrival in Göttingen, Noether quickly solved two important problems: first, identifying all of the invariants of an arbitrary vector or tensor field in a Riemannian space and, second, proving that infinitesimal transformations of the Lorentz group give rise to conservation theorems. It is this second result, known to physicists as Noether's theorem, that symmetries in the equations of motion give rise to conserved quantities, that stands as one of the most important contributions to modern theoretical physics.

Noether's work demonstrated that Einstein's theory was consistent with global conservation of energy, eliminating a significant barrier to his program. For his own part, Hilbert became fascinated with Einstein's use of advanced mathematics to describe physical phenomena and set out to see what he could contribute to the discussion. Hilbert corresponded throughout November, 1915 with Einstein. Einstein, undoubtedly feeling some urgency in his work now that Hilbert was also working on the problem, submitted four papers (one each Thursday) to the Prussian Academy

---

[17]Noether's appointment at Göttingen was contentious, the faculty refused to allow a woman to lecture. Hilbert famously replied, "I do not see that the sex of the candidate is an argument against her admission as *Privatdozent*. After all, we are a university, not a bath house." Noether was finally permitted to submit her *Habilitation* in 1919.

refining his ideas on the gravitational theory and stating the complete set of field equations in the November 25th submission. Interestingly, on November 20, 1915 Hilbert submitted a paper of his own to the Göttingen Royal Society in which he *derived* Einstein's gravitational theory from the principle of least action. Hilbert's paper was not printed until March of 1916 and it seems likely that he modified it after Einstein's final theory was published in December. Hilbert never claimed priority for the discovery of the field equations. His interactions with Einstein certainly contributed to Einstein's eventual formulation of the theory but it is quite astonishing to discover that Einstein's great work can be derived from a higher principle.

Einstein's principle of relativity is a guiding force in subsequent theoretical developments in physics. All fundamental theories must be consistent with this principle. Noether's contribution that there is a deep connection between symmetries in the equations of motion and experimentally observed conserved quantities has also fundamentally changed how we view the universe. There are certainly areas of physics where physics is conducted in a manner that would be familiar to Newton and Laplace, such as our earlier discussion of sand grains but fundamental laws of nature at the microscopic scale appear to be governed by the deeper principles uncovered by Einstein and Noether. We shall discover more of the impact of these ideas in the subsequent chapters.

On the Nature of the Photon

In 1864, the Scottish physicist James Clerk Maxwell undertook the task of combining all that was then known about electric and magnetic phenomena into a single, encompassing theory. Maxwell used as a template the mathematical structure of the well-known theory of elastic media. In this regard, Maxwell followed in the long tradition of applying established mathematical tools to a new problem. Maxwell's original papers are somewhat difficult to read because they do not incorporate the modern vector notation invented by Oliver Heaviside. In this regard, it makes Maxwell's achievement even more compelling as he was able to understand the mathematical structure present in the equations even without the notational support that makes it more evident.

The principal modification that Maxwell made to the existing theories of electric and magnetic phenomena was to add a term that he perceived was necessary due to the symmetry of the equations. It is known as the displacement current, for historical reasons, but the real impact of Maxwell's contribution was that it converted a system of equations that described electrical and magnetic phenomena as separate entities into a single, comprehensive theory of electromagnetism. Indeed, as we shall encounter shortly, the division of phenomena into electric and magnetic components depends on the frame of reference. This makes the formulation inconsistent with the principle of relativity and we shall see subsequently how to remedy that problem.

In large measure, Maxwell's equations form the most successful physical theory ever developed. To be sure, thermodynamics provided the underpinnings for the industrial revolution and quantum electrodynamics has demonstrated unprecedented agreement with high-precision measurements. Nonetheless, electromagnetic phenomena form the heart of all of our modern technology. We generate electrical power and distribute it across the world. We build electrical motors that drive ubiquitous machinery and have mastered the generation of electromagnetic radiation that cooks our food in microwave ovens and allows us to communicate

© Mark A. Cunningham 2018
M.A. Cunningham, *Beyond Classical Physics*,
Undergraduate Lecture Notes in Physics,
https://doi.org/10.1007/978-3-319-63160-8_2

over vast distances. We use lasers to scan groceries at checkout counters and send down images from satellites orbiting over the planet that enable us to more accurately predict weather patterns. All of these disparate phenomena can be modelled quite precisely with Maxwell's equations.

Additionally, Maxwell's equations have served as a template for essentially all further developments in theoretical physics. As a theory of abstract fields, Maxwell's equations have been extended, generalized and scrutinized in an effort to explain other physical phenomena, leading to the development of the Standard Model of high-energy particle physics and string theory beyond.

2.1. Maxwell's Equations

Maxwell's equations represent a classical field theory that links the existence and time variation of the field quantities to sources: charge distributions and currents. We shall begin by stating Maxwell's equations in a modern, integral form, using the SI system of units. Units are always problematic in dealing with electromagnetic phenomena. As a practical matter, force has the dimension of $(M \cdot L \cdot T^{-2})$, where M is a mass, L is a length and T represents time. The Coulomb force $(Q\mathbf{E})$ requires a dimension for the electric field of $(M \cdot L \cdot T^{-2} \cdot Q^{-1})$. There are alternative unit systems that can potentially simplify the equations by suppressing some of the constants that arise but their use is not compelling.

Using vector notation, the equations can be written rather compactly. In their integral form, Maxwell's equations can be written as follows:

$$(2.1) \quad \int_{\partial \mathcal{V}} d^2 \mathbf{r}_2 \, \mathbf{n}(\mathbf{r}_2) \cdot \mathbf{D}(t_2, \mathbf{r}_2) = \int_{\mathcal{V}} d^3 \mathbf{r}_1 \, \rho(t_1, \mathbf{r}_1),$$

$$(2.2) \quad \int_{\partial \mathcal{V}} d^2 \mathbf{r}_2 \, \mathbf{n}(\mathbf{r}_2) \cdot \mathbf{B}(t_2, \mathbf{r}_2) = 0,$$

$$(2.3) \quad \int_{\partial \mathcal{S}} d\mathbf{l} \cdot \mathbf{E}(t_2, \mathbf{r}_2) = -\frac{d}{dt} \int_{\mathcal{S}} d^2 \mathbf{r}_1 \, \mathbf{n}(\mathbf{r}_1) \cdot \mathbf{B}(t_1, \mathbf{r}_1) \quad \text{and}$$

$$(2.4) \quad \int_{\partial \mathcal{S}} d\mathbf{l} \cdot \mathbf{H}(t_2, \mathbf{r}_2) = \int_{\mathcal{S}} d^2 \mathbf{r}_1 \, \mathbf{n}(\mathbf{r}_1) \cdot \mathbf{J}(t_1, \mathbf{r}_1) + \frac{d}{dt} \int_{\mathcal{S}} d^2 \mathbf{r}_1 \, \mathbf{n}(\mathbf{r}_1) \cdot \mathbf{D}(t_1, \mathbf{r}_1).$$

Here, we have implicitly assumed that all vectors are three-dimensional. The fields are functions of both time and space. These are quite formidable expressions, so let us spend some time discussing their meaning.

The electric displacement $\mathbf{D}$ is related to the electric field $\mathbf{E}$ by what are known as constitutive relations. In physical media, the presence of an external electric field causes a number of different effects: charge separation in metals and polarization in insulators. For modest field strengths in most media, the two are simply proportional: $\mathbf{D} = \epsilon\mathbf{E}$, where ϵ is known as the dielectric permittivity. In general, the relationship can be more complex, ϵ may be a tensor, wherein any single component of the electric displacement will depend upon all of the components of the electric field:

$$D_i = \epsilon_{ix}E_x + \epsilon_{iy}E_y + \epsilon_{iz}E_z,$$

Indeed, it is not necessary for $\mathbf{D}$ and $\mathbf{E}$ to be related linearly but we shall defer discussions of that nature to later in the text. Similarly, the magnetic induction $\mathbf{B}$ is often treated as having a linear dependence on the magnetic field $\mathbf{H}$: $\mathbf{B} = \mu\mathbf{H}$, where μ is called the magnetic permeability. In vacuum, ϵ and μ are just constants, usually denoted as ϵ_o and μ_o.

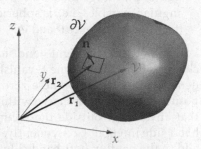

FIGURE 2.1. A charge distribution is located within the volume V. Points $\mathbf{r}_1$ lie within the volume. The integral extends over the surface of the volume ∂V. Points $\mathbf{r}_2$ lie on the surface, with local normal $\mathbf{n}$.

In the first equation 2.1, known as Gauss's law, we find that, if we integrate $(d^3\mathbf{r}_1)$ over a (three-dimensional) volume V that contains a distribution of charge ρ (known as the charge density), this will be equal to the two-dimensional $(d^2\mathbf{r}_2)$ integral of the component of the electric displacement $\mathbf{D}$ that is everywhere normal to the surface (∂V) that bounds V. An illustration of such a volume is depicted in figure 2.1. At the point $\mathbf{r}_2$ on the surface of the volume, there is the (outwardly directed) normal $\mathbf{n}$. The component of the displacement parallel to the normal that emerges through the infinitesimal surface element (denoted by the small rectangle) is then summed (integrated) over all such surface elements to obtain the total flux.

There are several important assumptions captured in equation 2.1. First, we have defined a continuous function ρ that defines the charge density within the volume. As all charge ultimately arises from the charges on individual electrons or nuclei, this is clearly an approximation but one which serves generally well for macroscopic systems. We have also assumed that the volume is finite, or mathematically, that the charge density has finite support. If we were to consider a universe with a constant charge extending to infinity, then the charge density integral would diverge, leading to the loss of our ability to make any use of equation 2.1. For the mathematically aware, there is also the implicit assumption that the surface ∂V is orientable, meaning that we can define a normal every-where. Not all manifolds have this property and we are additionally limit-ing ourselves to surfaces that are reasonably smooth, or at least piecewise smooth. Indeed, most examples found in introductory texts restrict the surfaces to spheres and cylinders.

> EXERCISE 2.1. The volume element in spherical coordinates is given by the following:
> $$d^3\mathbf{r} = dr\, d\theta\, d\varphi\, r^2 \sin\theta,$$
> where r is the radial coordinate, θ is the polar angle measured from the z-axis and φ is the azimuthal angle measured from the x-axis. If we have a charge distribution $\rho(\mathbf{r}) = \rho(r)$ that is independent of the angular variables, i.e., is spherically symmetric, what is the result of the integrating over θ and φ?

We can note that the right-hand side of equation 2.1 will vanish if the total charge within the volume V vanishes. This does not mean that the charge density must be everywhere zero, only that there are equal amounts of positive and negative charge within the volume. In any case, it is also true that any charge outside the volume V does not contribute to the right-hand side integral. Consequently, any charge outside the volume does not add to the net flux (left-hand side) through the surface of the volume.

In many instances, we will be interested in the fields due to just a single charge or a few charges. In those cases, we can still utilize equation 2.1 as a statement of Gauss's law by utilizing the delta function $\delta(x)$ introduced by Dirac. The delta function has the following property:

$$(2.5) \qquad \int_a^b dx\, f(x)\,\delta(x-c) = f(c),$$

provided that $a \le c \le b$. Strictly speaking, the delta function is known to mathematicians as a distribution or a generalized function. Actually, Dirac's contrivance provoked a large amount of work by mathematicians to demonstrate that such a function could even be defined in any sort of sensible fashion. For students, the appearance of a delta function is generally a godsend, evaluating a complex integral is reduced to writing down the value of the integrand at the point $x = c$.

In the case of a single charge, located at the point $\mathbf{r}_o = (x_o, y_o, z_o)$, the charge density will be given by the following expression:

$$\rho(\mathbf{r}_1) = Q\,\delta(x_1 - x_o)\delta(y_1 - y_o)\delta(z_1 - z_o) = Q\,\delta^3(\mathbf{r}_1 - \mathbf{r}_o).$$

Inserting this expression into the charge density integral, we obtain:

$$\int_V d^3\mathbf{r}_1\, \rho(\mathbf{r}_1) = Q,$$

provided that r_o lies within the volume $\mathcal{V}$, otherwise, the integral vanishes.

Turning now to the second Maxwell equation 2.2, we find a similar expression for the magnetic induction **B** but, in this case, the integral always vanishes. This result can be taken to mean that there is no magnetic charge.

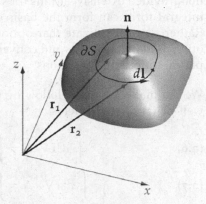

FIGURE 2.2. An oriented surface is bounded by a closed loop ∂S. The positive direction around the loop is defined by the right-hand rule. Points r_1 lie on the surface and points r_2 lie on the boundary ∂S.

The next equation 2.3 is known as Faraday's law. It relates the time rate of change of the magnetic flux through a surface S to the line integral of the electric field around the perimeter ∂S. At some level, equation 2.3 is the basis of our modern technology: the changing magnetic flux through a surface defined by a metal loop gives rise to an electric field (and a current by Ohm's law) within the loop. This is the definition of an electrical generator or a transformer, depending upon the application. If we, instead, consider the induced current to be the source term, equation 2.3 provides the definition of an electric motor. These facts may not be explicit in the rather curious glyphs contained within the mathematics but those outcomes can be obtained nonetheless.

In figure 2.2, we illustrate the relationship between the surface S and its boundary ∂S. Points r_1 within the boundary of an oriented surface, with local normal $n(r_1)$. Points r_2 on the boundary of the surface trace a one dimensional path, where locally the direction dl is positive in a right-handed sense.

The final equation 2.4 was originally proposed by Ampére but subsequently modified by Maxwell, who included the last term on the right-hand side out of concerns of symmetry with the corresponding Faraday's law. In its final form, the Ampére-Maxwell law provides that the current flow and the time rate of change of the electric flux through a surface S is proportional to the line integral of the magnetic field around the boundary of the surface. It is actually the last term, known historically as the displacement current, that provides the final clue as to the nature of light.

The student, at this point, will most likely not be suddenly awakened to the secrets of the universe. How these equations depict the nature of electromagnetic phenomena is still opaque. In truth, mathematicians have not provided extensive means for solving integral equations, although the integral form can form the basis of numerical, finite element methods. Rather than investigate that option here, we shall follow the path most often used by physicists and convert the integral equations into a series of partial differential equations.

We can make use of the following vector identities, that hold for suitably well-behaved vector fields:

$$(2.6) \qquad \int_V d^3\mathbf{r}\, \nabla \cdot \mathbf{A} = \int_{\partial V} d^2\mathbf{r}\, \mathbf{n(r)} \cdot \mathbf{A} \quad \text{and}$$

$$(2.7) \qquad \int_S d^2\mathbf{r}\, \nabla \times \mathbf{A} = \int_{\partial S} d\mathbf{l} \cdot \mathbf{A}.$$

Here, $\mathbf{A}$ represents some vector field and ∇ is a differential operator. The vector $\mathbf{n}$ is the unit normal to the surface ∂V. In Cartesian coordinates, the operator ∇ has the following form:

$$\nabla \equiv \left[\frac{\partial}{\partial x}, \frac{\partial}{\partial y}, \frac{\partial}{\partial z} \right].$$

Note that we use the term operator and not vector. Because ∇ is composed of differentials, it really only has meaning when applied to functions. Technically, mathematicians will call the differential operator a 1-form but we will defer this discussion until later. As an example, ∇ applied to a scalar function f results in a vector that we call the *gradient*:

$$(2.8) \qquad \nabla f(x,y,z) = \left[\frac{\partial f}{\partial x}, \frac{\partial f}{\partial y}, \frac{\partial f}{\partial z} \right].$$

It is possible to construct two other mathematical objects from the differential operator. The term $\nabla \cdot \mathbf{A}$ is called the *divergence* of the field $\mathbf{A}$ and is a scalar function. The object $\nabla \times \mathbf{A}$ is called the *curl* of $\mathbf{A}$ and is a vector. We'll discuss more about the mathematical underpinnings of the operator subsequently but, for the present, we can simply state that we interpret these operations to mean the following:

(2.9) $\nabla \cdot \mathbf{A} = \dfrac{\partial A_x}{\partial x} + \dfrac{\partial A_y}{\partial y} + \dfrac{\partial A_z}{\partial z}$ and

(2.10) $\nabla \times \mathbf{A} = \hat{\mathbf{x}}\left[\dfrac{\partial A_z}{\partial y} - \dfrac{\partial A_y}{\partial z}\right] + \hat{\mathbf{y}}\left[\dfrac{\partial A_x}{\partial z} - \dfrac{\partial A_z}{\partial x}\right] + \hat{\mathbf{z}}\left[\dfrac{\partial A_y}{\partial x} - \dfrac{\partial A_x}{\partial y}\right].$

Using the vector identities 2.6 and 2.7, we can rewrite Maxwell's equations as a series of differential equations:

(2.11) $\nabla \cdot \mathbf{D} = \rho,$

(2.12) $\nabla \cdot \mathbf{B} = 0,$

(2.13) $\nabla \times \mathbf{E} = -\dfrac{\partial \mathbf{B}}{\partial t}$ and

(2.14) $\nabla \times \mathbf{H} = \mathbf{J} + \dfrac{\partial \mathbf{D}}{\partial t}.$

Here, we have suppressed an integration over an unspecified volume $\mathcal{V}$, assuming that equations 2.11–2.14 are true pointwise, precisely because the volume is not specified. In truth, there is a formidable amount of mathematics underlying our ability to switch back and forth between the differential and integral forms of the Maxwell equations. Undoubtedly, few students will have much appreciation for that fact at the moment and few students will observe that equations 2.11–2.14 represent any sort of real progress in understanding the meaning of Maxwell's equations. Indeed, solving coupled partial differential equations remains a formidable challenge. Nevertheless, a good deal of progress has been made in constructing solutions and, with the advent of modern computers, numerical approaches have extended the applicability to quite complex geometries.

It is nonetheless remarkable that we can derive two separate representations of the Maxwell equations. We have chosen to write them in a fashion that emphasizes the operator nature of both the integration and differentiation operators. We will see ultimately that the two are inverses of one another but the fact that we have two representations means that, if we cannot find a way to solve the integral equation, perhaps we can find a way forward by considering the differential equation. This sort of strategy has been used quite successfully, as we shall see.

EXERCISE 2.2.   Use Gauss's law for magnetic fields, equation 2.2, and the divergence identity, equation 2.6, to convert the separate integrations into a single volumetric integration. Show that you can recover equation 2.12.

We can show an important result with modest effort, though. Consider taking the curl of equation 2.13. We can make use of a vector identity:

$$\nabla \times \nabla \times \mathbf{A} = \nabla(\nabla \cdot \mathbf{A}) - (\nabla \cdot \nabla)\mathbf{A} = \nabla(\nabla \cdot \mathbf{A}) - \nabla^2 \mathbf{A}.$$

We then find the following result:

$$\nabla \times (\nabla \times \mathbf{E}) = -\nabla \times \frac{\partial \mathbf{B}}{\partial t}$$

$$\nabla(\nabla \cdot \mathbf{E}) - \nabla^2 \mathbf{E} = -\frac{\partial}{\partial t}(\nabla \times \mathbf{B}),$$

where in the second step we have assumed that the order of differentiation can be reversed. Let us now make the simplification that we are considering free space and that the constitutive relations between the fields are simply proportional. That is, we assume that $\mathbf{D} = \epsilon_o \mathbf{E}$ and $\mathbf{B} = \mu_o \mathbf{H}$. Then we can use equation 2.14 to write

$$\nabla \frac{\rho}{\epsilon_o} - \nabla^2 \mathbf{E} = -\frac{\partial}{\partial t}\left(\mu_o \mathbf{J} + \epsilon_o \mu_o \frac{\partial \mathbf{E}}{\partial t}\right)$$

(2.15) $$\epsilon_o \mu_o \frac{\partial^2 \mathbf{E}}{\partial t^2} - \nabla^2 \mathbf{E} = -\nabla \frac{\rho}{\epsilon_o} - \mu_o \frac{\partial \mathbf{J}}{\partial t}.$$

Now in the absence of local sources ρ and $\mathbf{J}$, equation 2.15 is just the homogeneous wave equation for each of the components of the electric field, provided that we identify $\epsilon_o \mu_o = 1/c^2$, where c is the velocity of light. Note the particular importance of the displacement current in the establishment of the wave equation. In any case, this was a significant prediction of Maxwell's equations, one which was verified by Heinrich Hertz in a series of experiments that he conducted during the years 1887–1889.[1]

> Exercise 2.3. Expand the term $\nabla \times \nabla \times \mathbf{A}$ into Cartesian components and show that the vector identity discussed above is correct.

> Exercise 2.4. Take the curl of the Ampére-Maxwell equation 2.14 and derive a similar wave equation for the magnetic field, using the same constitutive relations for the field.

Hertz's apparatus was quite ingenious, making repeated use of Faraday's law; a basic representation of his early devices is illustrated in figure 2.3. A battery was connected via a switch to a circuit composed of a transformer and capacitor, as depicted in figure 2.4. When the switch is closed, a transient voltage pulse is created in what amounts to an LC circuit. That pulse is amplified to high voltage through the transformer and applied to the two arms of the antenna. If the voltage difference is large enough, a spark leaps the gap between the two arms of the antenna, resulting in a

---

[1] Hertz published "Über sehr schnelle electrische Schwingungen" in the *Annalen der Physik* in 1887.

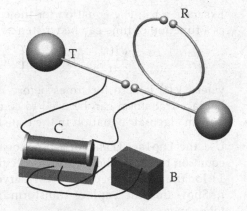

FIGURE 2.3. Hertz's apparatus included a battery (B) connected to an induction coil (C) and thence to a transmitting antenna (T). The receiver (R) was constructed of a metal loop.

large current pulse along the antenna. This current pulse results in the radiation of electromagnetic energy. If a changing magnetic field impinges through the receiver, a current will be generated in the loop. For sufficiently large currents, a spark will leap the gap.

Hertz found that, if he worked in a darkened room and allowed his eyes to adapt sufficiently, that he could readily detect the small sparks emanating from his receiver. He found that the waves could be blocked by metal screens, that the waves bounced from metal screens with an angle of reflection equal to the angle of incidence and, by orienting the receiver in different directions, that the waves were polarized. That is, the electromagnetic fields were well described by the wave equation. Hertz was gratified to confirm the predictions made by Maxwell but saw no particular use for the phenomenon. Commercialization would require the intervention of others like Guglielmo Marconi and Alexander Graham Bell.

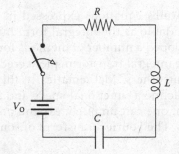

FIGURE 2.4. Hertz's transmitter circuit can be modelled as a simple RLC circuit, as depicted at right, where the resistance arises from the finite resistivity of the wires.

EXERCISE 2.5. The equation for the current I flowing in the circuit as a function of time can be written as follows:

$$-L\frac{dI(t)}{dt} - RI(t) - \frac{1}{C}\int_0^t d\tau\, I(\tau) + V(t) = 0,$$

where $V(t)$ is zero for times before $t = 0$ and V_0 thereafter. The Laplace transform can be used to convert the differential equation into an algebraic equation in the transform variable s.

Use the `LaplaceTransform` function on each of the terms in the equation to derive the transformed equation. Note, use the variable I1 for the current because I is reserved for the imaginary number i. Show that the Laplace transform of the current is given by the following:

$$\tilde{I}(s) = \frac{V_0}{Ls^2 + Rs + 1/C}.$$

Use the `InverseLaplaceTransform` function to demonstrate that if $I(0) = 0$, then

$$I(t) = \frac{V_0}{\sqrt{R^2 - 4L/C}}\, e^{-\alpha t}\left[e^{\beta t} - e^{-\beta t}\right],$$

where

$$\alpha = \frac{R}{2L} \quad \text{and} \quad \beta = \left[\frac{R^2}{4L} - \frac{1}{LC}\right]^{1/2}.$$

The behavior of the circuit depends upon the value of β. If $1/LC$ is larger than $R^2/4L$, then the square root becomes imaginary and the exponentials become oscillatory.

Plot the behavior of the system if $V_0 = 20$ V, $R = 1\,\Omega$, $L = 3$ mH and $C = 47$ nF for $0 \leq t \leq 1$ ms. What happens if you increase R?

2.2. Fields and Potentials

Maxwell's equations expressed in their differential form remain just as formidable as the integral form but physicists and mathematicians have developed a number of methods for their solution. One of the most useful is the integral transform, pioneered by Joseph Fourier and Simon Laplace, among others. Mathematically, this strategy is based on the fact that one can define a functional space and a basis of functions within that space. Then, one can uniquely expand *any* function in terms of the basis functions. The Fourier transform of some function $F(x)$ is defined as follows:

$$(2.16) \qquad \tilde{F}(k) = \frac{1}{\sqrt{2\pi}}\int_{-\infty}^{\infty} dx\, F(x)e^{-ikx}$$

and the inverse transform is defined as follows:

$$(2.17) \qquad F(x) = \frac{1}{\sqrt{2\pi}} \int_{-\infty}^{\infty} dk\, \tilde{F}(k) e^{ikx}.$$

Here we have adopted the physicist's standard notation, wherein an over-all factor of $(1/2\pi)$ that is required to normalize the outcome is shared between the forward and inverse transforms. Mathematicians often adopt alternative definitions that avoid the square root and associate the factor with either the forward or inverse transform. Students should be aware that different texts may be using different definitions.

Fourier's inversion theorem can be thus stated as follows:

$$F(x) = \frac{1}{2\pi} \int_{-\infty}^{\infty} dk\, e^{ikx} \int_{-\infty}^{\infty} d\zeta\, F(\zeta) e^{-ik\zeta}$$

$$(2.18) \qquad = \frac{1}{2\pi} \int_{-\infty}^{\infty} d\zeta \int_{-\infty}^{\infty} dk\, e^{ik(x-\zeta)} F(\zeta).$$

We note that this can be true only if

$$(2.19) \qquad \delta(x-\zeta) = \frac{1}{2\pi} \int_{-\infty}^{\infty} dk\, e^{ik(x-\zeta)}.$$

We have discovered an integral representation of the Dirac delta function, which will prove useful subsequently.

The utility of applying Fourier transforms to differential equations stems from the following observation:

$$\frac{dF(x)}{dx} = \frac{1}{\sqrt{2\pi}} \frac{d}{dx} \int_{-\infty}^{\infty} dk\, \tilde{F}(k) e^{ikx}$$

$$(2.20) \qquad = \frac{1}{\sqrt{2\pi}} \int_{-\infty}^{\infty} dk\, (ik)\tilde{F}(k) e^{ikx}.$$

That is, the x-dependence of the Fourier transform lies solely within the exponential. The transformed function $\tilde{F}$ is a function of the transform variable k and not a function of x. So, taking the derivative, so long as it is possible to exchange the order of integration and differentiation, simply brings down a factor of (ik). Higher order derivatives are multiplied by powers of (ik). As a result, differential equations become algebraic equations in the transform domain. Those can be readily solved and all that remains is an (often formidable) integral back into the spatial domain. Physicists often make use of Fourier transforms in the time domain, to remove the time-dependence of a system of equations. In this case, the transform variable ω has the dimension of inverse time, or frequency. One then talks of working in the frequency domain and a specific time-dependence can be recovered by performing the inverse transform integration.

The transforms can also be applied to vector equations. Suppose now that we have a function of the vector $\mathbf{r} = (x, y, z)$. Then, we can transform each component separately, with a (vector) transform variable $\mathbf{k} = (k_x, k_y, k_z)$:

$$F(\mathbf{r}) = \frac{1}{(2\pi)^{3/2}} \int_{-\infty}^{\infty} dk_x \int_{-\infty}^{\infty} dk_y \int_{-\infty}^{\infty} dk_z \tilde{F}(\mathbf{k}) e^{i(k_x x + k_y y + k_z z)}$$

(2.21)
$$\equiv \frac{1}{(2\pi)^{3/2}} \int d^3 \mathbf{k}\, \tilde{F}(\mathbf{k}) e^{i \mathbf{k} \cdot \mathbf{r}}.$$

Here, we have illustrated the three dimensional Fourier transform.

EXERCISE 2.6. Consider the function $F(\mathbf{r})$ as defined in equation 2.21. What is the partial derivative with respect to x? Compute the gradient of the function: $\nabla F(\mathbf{r})$, where

$$\nabla = \left(\frac{\partial}{\partial x}, \frac{\partial}{\partial y}, \frac{\partial}{\partial z} \right).$$

Write your answer in terms of the vector $\mathbf{k}$.

In vacuum, the transformed Maxwell equations (suppressing the integrals over $d\omega\, d^3\mathbf{k}$) become the following:

(2.22) $$i\epsilon_0 \mathbf{k} \cdot \tilde{\mathbf{E}} = \tilde{\rho},$$

(2.23) $$i\mu_0 \mathbf{k} \cdot \tilde{\mathbf{H}} = 0,$$

(2.24) $$i\mathbf{k} \times \tilde{\mathbf{E}} = i\mu\omega\tilde{\mathbf{H}}, \quad \text{and}$$

(2.25) $$i\mathbf{k} \times \tilde{\mathbf{H}} = \tilde{\mathbf{J}} - i\omega\epsilon_0\tilde{\mathbf{E}}.$$

In source-free regions, the first two equations 2.22 and 2.23 tell us that the wave vector $\mathbf{k}$ is perpendicular to both the (transformed) electric $\tilde{\mathbf{E}}$ and magnetic $\tilde{\mathbf{H}}$ fields, as depicted in figure 2.5. The second two equations tell us that $\tilde{\mathbf{E}}$ and $\tilde{\mathbf{H}}$ are themselves perpendicular. The wave vector $\mathbf{k}$ defines the direction of propagation of the wave. By convention, the polarization of the wave is defined by the direction of the electric field.

EXERCISE 2.7. For a vector function $\mathbf{F}(\mathbf{r}) = (F_x(\mathbf{r}), F_y(\mathbf{r}), F_z(\mathbf{r}))$ show that, in the transform domain, the divergence and curl operators yield the following results:

$$\nabla \cdot \mathbf{F} \Rightarrow (i\mathbf{k}) \cdot \tilde{\mathbf{F}} \quad \text{and} \quad \nabla \times \mathbf{F} \Rightarrow (i\mathbf{k}) \times \tilde{\mathbf{F}}.$$

The transformed fields are often called plane wave solutions of Maxwell's equations. These fields have infinite spatial extent and so are not physically realizable but many sources can be treated approximately as plane waves, particularly monochromatic (single frequency or small bandwidth) sources. Indeed, many treatments of electromagnetic phenomena will simply posit that there are solutions that have the spatial dependence

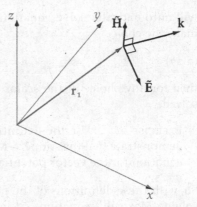

FIGURE 2.5. For plane waves, the
direction of propagation of the
electromagnetic wave is defined by
the wave vector **k**. This direction
is orthogonal to both the electric $\tilde{\mathbf{E}}$
and magnetic $\tilde{\mathbf{H}}$ fields.

$\exp(i\mathbf{k}\cdot\mathbf{r})$ and proceed from there. In fact, these treatments are skipping
over the derivation from the Fourier transform. While the exponential
terms are technically infinite, we anticipate that real fields will have finite support and the integrals will be finite due to the properties of the
transformed field $\tilde{\mathbf{E}}(\mathbf{k})$. We will come back to this point subsequently.

Even with the simplifications associated with integral transforms, we are
still left with a series of vector equations. One possible approach to dealing with Maxwell's equations utilizes an alternative formulation in terms
of potentials. One might ask the question, as regards equation 2.12, "What
sort of function has no divergence?" It is possible to construct one by simply taking the curl of *any* vector function. Similarly, one can use the fact
that the curl of the gradient of any scalar function also vanishes. In the
static case ($\omega = 0$), we can use this result to solve directly for the electric
field in equation 2.13.

EXERCISE 2.8.   For some vector function $\mathbf{F} = (F_x, F_y, F_z)$ show explicitly that $\nabla\cdot(\nabla\times\mathbf{F}) = 0$. Show also that for a scalar function F that
$\nabla\times(\nabla F) = 0$.

More generally, the time-dependent electric field cannot be just the gradient of a scalar function. If we make the decision that we shall define the
magnetic induction **B** in terms of a vector potential **A**, then the choice

(2.26) $\mathbf{B} = \nabla\times\mathbf{A}$

will automatically solve equation 2.12, for any choice of **A**. If we also make the choice[2]

$$(2.27) \qquad \mathbf{E} = -\nabla V - \frac{\partial \mathbf{A}}{\partial t},$$

then for any choice of the scalar function V, equation 2.13 will also be solved.

> EXERCISE 2.9. Use the definitions of equations 2.26 and 2.27 and demonstrate that the Ampére-Maxwell equation 2.14 leads to a wave equation for the vector potential.

So, with these definitions of the potentials, we have a broad capability of solving Maxwell's equations, at least in source-free regions. We can look for functions V and **A** that satisfy boundary conditions for any particular problem and construct the electric and magnetic fields accordingly. We note, though, that equations 2.26 and 2.27 admit some ambiguity in the definitions of the potentials **A** and V. If these potentials provide solutions **E** and **B** to Maxwell's equations, then so do the modified potentials

$$(2.28) \qquad \mathbf{A}' = \mathbf{A} + \nabla\phi \quad \text{and} \quad V' = V - \frac{\partial\phi}{\partial t},$$

where ϕ is an arbitrary scalar field $\phi = \phi(t,\mathbf{r})$. This transformation is known as a **gauge** transformation[3] and represents a non-obvious symmetry of Maxwell's equations. Modern theories of elementary particles are all gauge field theories; we shall discuss this property in more detail subsequently.

> EXERCISE 2.10. Show that the modified potentials **A**' and V' produce the same electric and magnetic fields as the potentials **A** and V.

What we would like to do next is study how to devise solutions that involve sources. In this case, the potentials provide a means for doing just that. The mathematics is rather involved, so we will skip over the bulk of the derivations. A general strategy for dealing with source terms involves computing the Green's function. Suppose that we have some differential equation involving the function $F(x)$. We can write this generically as follows:

$$\mathcal{D}F(x) = s(x),$$

---

[2]We use the symbol V for the electric potential, as is traditional and distiguish it from the volume $\mathcal{V}$ by using a different font. Students will have to be constantly vigilant, as such subtleties can be obscured when copying equations from the chalkboard.
[3]The term *gauge* is due to Hermann Weyl, who was studying the effects of scaling transformations on Lagrangian systems.

where $\mathcal{D}$ represents all of the derivative operators and other functions multiplying F and s is the source term. The Green's function is defined as the function that satisfies the differential equation with a delta function source:

$$\mathcal{D}G(x,x_1) = \delta(x-x_1).$$

The function F now can be recovered by performing the following integral[4]:

(2.29) $$F(x) = \int dx_1\, G(x,x_1)s(x_1).$$

Now one might question how one goes about solving an equation where the source term is singular but, recall, we have already derived an integral representation for the delta function. If we work in the transform domain, things are not as bleak as it might seem at first.

In any case, the potentials for Maxwell's equations can be solved using this strategy. We find that the potentials have a surprisingly simple form:

(2.30) $$V(t,\mathbf{r}) = \frac{1}{4\pi\epsilon_o} \int d^3\mathbf{r}_1 \frac{\rho(t_1,\mathbf{r}_1)}{|\mathbf{r}-\mathbf{r}_1|}$$

(2.31) $$\mathbf{A}(t,\mathbf{r}) = \frac{\mu_o}{4\pi} \int d^3\mathbf{r}_1 \frac{\mathbf{J}(t_1,\mathbf{r}_1)}{|\mathbf{r}-\mathbf{r}_1|},$$

where $t_1 = t \pm |\mathbf{r}-\mathbf{r}_1|/c$ is the time at the source location. There are two possible solutions, known as the advanced (+) and retarded (−) solutions. We think of the retarded solutions as being the physical ones and shall subsequently ignore the advanced solutions. Here we are invoking a causality argument: the fields at some distance $d = |\mathbf{r}-\mathbf{r}_1|$ from the source cannot know about changes in the source until the time that it would take light to travel that distance. Using the advanced solution would give us knowledge of the source motion prior to its actual motion.

We have utilized a fair amount of complex analysis to this point, by which we mean complex numbers not just complicated formulas. It will take time before students can become accustomed to talking about imaginary numbers with a straight face. It is an equally disheartening experience to try to explain to one's colleagues that we are dealing with retarded solutions; there is inevitable gleeful commentary that all of physics is retarded and that all involved in the study of such an inane subject must have some form of mental handicap. One should display forebearance.

---

[4]We're skipping over some details of other terms that generally can be forced to vanish by assuming that the Green's function and the source term vanish sufficiently fast at large distances.

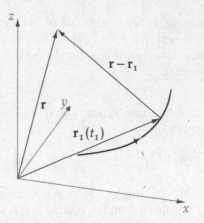

FIGURE 2.6. A charged particle follows the path $r_1(t_1)$ as indicated by
the dark curve. The fields are measured at some distant point r.

2.3. Point sources

The results presented in equations 2.30 and 2.31 are deceptively simple. The integrands are actually singular at the source point $r = r_1$ and,
thus, actually performing the integrals requires some mathematical fi-
nesse. Even for the case of a single point source, where $\rho = Q\delta^3(r - r_1)$, it
is not as simple as just collapsing the integrals, owing to the fact that we
have to take into consideration the retarded time. Nevertheless, these in-
tricacies can be overcome, with the result that the potentials for a moving
point source can be written as follows:

$$(2.32) \qquad V(t,r) = \frac{1}{4\pi\epsilon_0} \frac{q}{|r - r_1| - (r - r_1)\cdot v_1/c} \quad \text{and}$$

$$(2.33) \qquad A(t,r) = \frac{\mu_0}{4\pi} \frac{qv_1/c}{|r - r_1| - (r - r_1)\cdot v_1/c},$$

where q is the charge and $v_1 = dr_1/dt_1$ is the velocity of the point charge at
the retarded time. These are known as the Liénhard-Wiechert potentials.

EXERCISE 2.11.   A complicating factor in the analysis of the Liénhard-
Wiechert potentials is that the time t_1 at the source point depends on
the position of the field point r. Consequently, derivatives of t_1 with
respect to the spatial components of r do not vanish. We have

$$t_1 = t - \frac{1}{c}\sqrt{(x - x_1)^2 + (y - y_1)^2 + (z - z_1)^2}.$$

What is the derivative with respect to x:

$$\frac{\partial t_1}{\partial x} = ?$$

What then is the result of applying the gradient operator ∇t_1?

We can now recover the fields through equations 2.27 and 2.26. The algebra is tedious but manageable. We obtain the following:

(2.34)

$$E(t,\mathbf{r}) = \frac{q}{4\pi\epsilon_0} \frac{|\mathbf{r}-\mathbf{r}_1|}{[|\mathbf{r}-\mathbf{r}_1| - (\mathbf{r}-\mathbf{r}_1)\cdot\mathbf{v}_1/c]^3}$$

$$\left\{ \left(1 - \frac{v_1^2}{c^2}\right)\left(\frac{\mathbf{r}-\mathbf{r}_1}{|\mathbf{r}-\mathbf{r}_1|} - \frac{\mathbf{v}_1}{c}\right) + (\mathbf{r}-\mathbf{r}_1)\times\left[\left(\frac{\mathbf{r}-\mathbf{r}_1}{|\mathbf{r}-\mathbf{r}_1|} - \frac{\mathbf{v}_1}{c}\right)\times\frac{\mathbf{a}_1}{c^2}\right] \right\}$$

(2.35)

$$B(t,\mathbf{r}) = \frac{\mathbf{r}-\mathbf{r}_1}{c|\mathbf{r}-\mathbf{r}_1|} \times E(t,\mathbf{r}).$$

Here, $\mathbf{a}_1 = d^2\mathbf{r}_1/dt_1^2$ is the acceleration of the charge.

EXERCISE 2.12. As a sanity check, set $\mathbf{v}_1 = 0$ and $\mathbf{a}_1 = 0$ in equation 2.34 and see if you recover Coulomb's law.

These results are reasonably gruesome but we can make some immediate observations. First, as we can see from figure 2.6, the vector $\mathbf{r} - \mathbf{r}_1$ provides the direction from the point charge to the field point at the retarded time. Second, from equation 2.33, we see that the magnetic field is perpendicular to both that direction and the electric field. Recall that the plane wave solutions we discussed previously also had the magnetic field perpendicular to the electric field and the direction of propagation.

In deriving complicated results like those found in equations 2.34 and 2.35, it is a very good strategy to conduct dimensional analysis throughout the derivation, to avoid unfortunate algebraic mistakes. Upon completion, it is also a good idea to check the dimensionality of your result. Dimensionality is a bit difficult to assess in electromagnetics but we can recall from Coulomb's law that the electric field of a point charge has a factor of $(4\pi\epsilon_0)^{-1}$ multiplying the charge divided by distance squared. We see just such a term in the first part of equation 2.34. This means that the remaining factor in the curly brackets must be dimensionless. In fact we have organized the result in a fashion that makes this reasonably apparent.

EXERCISE 2.13. Express equation 2.34 dimensionally and verify that the terms in the curly brackets are dimensionless and that the remaining term involving $\mathbf{r}$ scales like (L^{-2}).

Moreover, we can observe that the first term in the curly braces has no dependence on the distance $d = |\mathbf{r} - \mathbf{r}_1|$, only on the direction from the source. As a result, this first term will have an overall inverse square dependence on d. It is, thus, the modification of the Coulomb force law that incorporates a moving (constant velocity) charge. The second term in the curly braces is proportional to the acceleration of the charge and contains an additional factor of d, so that asymptotically the field will scale inversely with d. This component is termed the radiation component of the field.

To understand what we mean by the term radiation, we should say that the Poynting vector **S** defines the energy flux of the electromagnetic field. We'll go into a bit more detail subsequently but the definition of the Poynting vector is as follows:

$$(2.36) \qquad\qquad\qquad \mathbf{S} = \mathbf{E} \times \mathbf{H}.$$

Energy conservation of the electromagnetic fields will be shown shortly to take the following form:

$$(2.37) \qquad \frac{d\mathcal{E}}{dt} = \frac{d}{dt} \int_V d^3\mathbf{r}\, \frac{1}{2}\left[\epsilon_0 E^2 + \mu_0 H^2\right] + \int_{\partial V} d^2\mathbf{r}_1\, \mathbf{n}(\mathbf{r}_1)\cdot\mathbf{S},$$

where $\mathcal{E}$ is the total energy within the volume V and the last term in the equation represents the energy flux through the boundary ∂V. Radiation will be defined as the fields that propagate to infinity, that is only those components that remain finite as the bounding surface ∂V is removed to infinity.

If we take the bounding surface to be a large sphere, then the differential elements for the surface integral will be

$$d^2\mathbf{r}_1 = r_1^2 \sin\theta\, d\theta\, d\varphi,$$

which clearly scales like the distance squared. If we consider actually computing the product $\mathbf{E} \times \mathbf{H}$ from equations 2.34 and 2.35, we should note that we will obtain, from the product of the first term in the curly braces with itself, a term that scales like d^{-4}. Overall, this will scale like d^{-2} (due to the d^2 coming from the differential element) and will vanish as $d \to \infty$. We will also obtain terms, from the product of the first term in the curly braces with the second term, that scale like d^{-3} and, overall, like d^{-1} and will likewise vanish. The only terms that can possibly remain will be the one obtained from product of the second term in the curly braces with itself, which is independent of d and, thus, can remain finite as $d \to \infty$.

> EXERCISE 2.14. Examine the terms in the Poynting vector dimensionally and verify the assertions made in the discussion above.

We should note that our somewhat haphazard analysis of the previous paragraph can be made rigorous, so the mathematically aware need not be disheartened. Nevertheless, electromagnetics is the discipline where all of the equations become difficult. Rather than mindlessly trying to perform all of the integrals one encounters, it is generally better to try to understand something about the integrals before trying to solve them.

In this particular instance, we needn't even perform the bulk of the integrations as we are only seeking solutions that will remain finite at large distances.

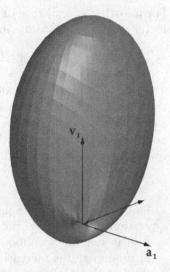

FIGURE 2.7. When the acceleration is perpendicular to the direction of motion, the radiation is predominantly in the forward direction. Here, the velocity is $0.4c$.

Using these results for the fields generated by an accelerated charge and after another long algebraic effort, we can arrive at the following result for the radiation field of a charged particle that is being accelerated perpendicular to its velocity:

$$(2.38) \quad |\mathbf{S}_{\mathrm{rad}}| = \frac{\mu_0 q^2 a_1^2}{16\pi^2 c} \frac{[1-(v_1/c)^2\cos\theta]^2 - [1-(v_1/c)^2]\sin^2\theta\cos^2\varphi}{[1-(v_1/c)\cos\theta]^5},$$

where here a_1 is the magnitude of the acceleration, θ measures the angle between the field point and the velocity $\mathbf{v}_1$ and φ is the angle between the field point and the acceleration $\mathbf{a}_1$. The energy flux is illustrated in figure 2.7.

This radiation is known as synchrotron radiation and is a serious problem that arises when constructing particle accelerators, particularly for electron beams. If we utilize magnetic fields to bend the electron beam into a circle, a substantial portion of the energy injected into the electrons will be radiated in the form of x-rays. From figure 2.7, we see that the radiation is emitted in the direction of propagation, perpendicular to the (centripetal) acceleration. Initially considered a parasitic problem, there are now synchrotrons built explicitly for the purpose of generating intense x-ray beams for crystallography. In these machines, no one is doing physics with the electron beams; they are, instead, studying the three-dimensional structures of protein crystals and other novel materials.

EXERCISE 2.15. Use the `Manipulate` function to study the behavior of the radiation flux from equation 2.38 as a function of the velocity (v_1/c). Use the `SphericalPlot3D` function to plot S_{rad}.

For the other special case of an acceleration that is collinear with the velocity, we obtain the following result:

$$(2.39) \qquad |S_{rad}| = \frac{\mu_0 q^2 a_1^2}{16\pi^2 c} \frac{\sin^2\theta}{[1-(v_1/c)\cos\theta]^5}.$$

The radiation in this case, depicted in figure 2.8, is symmetrical around the direction of travel (no φ dependence) and is zero at $\theta = 0$; it is known as *bremsstrahlung*. As the velocity increases, the angle of maximum flux continues to decrease, such that at extreme relativistic velocities, the maximum flux is directed at an angle that approaches zero.

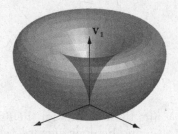

FIGURE 2.8. When the velocity and acceleration are collinear, the radiation pattern is symmetrical about the direction of motion and is increasingly tilted into the forward direction as the velocity increases. Here the velocity is $0.4c$.

EXERCISE 2.16. Use the `Manipulate` function to study the behavior of the radiation flux from equation 2.39 as a function of the velocity (v/c). Use the `SphericalPlot3D` function to plot S_{rad}. What happens as $v/c \to 1$?

2.4. Relativistic Formulation

We've skipped over a number of details to get to this point but it is time to confront some crucial issues directly. For example, we can ask what happens when we apply a Lorentz transformation to Maxwell's equations and the answer is rather disappointing: the electric and magnetic fields transform in a very complicated manner. Assigning students the task of deriving those transformations is commonplace in introductory texts but not terribly illuminating. We know that the fields are solutions to a wave equation and that the wave equation is invariant under a Lorentz transformation. So, let us instead make use of Einstein's adventures into the mathematical hinterlands and ask an alternative question: Is there some way to write Maxwell's equations that makes the question of Lorentz invariance obvious and straightforward.

As Einstein discovered, the answer is yes. Yes, it is possible to do so but it took Einstein a decade to fully comprehend the required mathematical infrastructure. In Einstein's defense, the mathematics was new and not yet widely understood by anyone outside the mathematical community. So, Einstein had to translate mathematics into physics, learning new vocabulary and constructing useful examples. Mathematicians are concise writers and often satisfied with existence and uniqueness proofs. Constructive proofs and nontrivial examples are often in short supply.

In any event, the appropriate language is provided by tensor calculus, where tensors are the mathematical extension of vectors. Unfortunately, the additional benefits of tensor calculus comes with a significant amount of mathematical baggage that can be perplexing to physicists. We shall endeavor to persevere. We are interested in using tensor calculus, which is the extension of calculus to multidimensional objects and non-Cartesian geometries. We will find the notation to be nearly impenetrable and the motivation completely opaque but, if we hold fast to a few touchstones, we will emerge with at least a modest appreciation for the underlying mathematical structure.

If we look at the definition of some vector $\mathbf{v}$ in a space of N dimensions, we usually begin by defining a basis $\mathbf{e}_1, \ldots, \mathbf{e}_N$, where the $\mathbf{e}_i$ are linearly independent. Actually, impressive works in modern geometry demonstrate that we do not need to define a basis at all; everything can be done in a fashion that is basis-independent. For the moment, we shall proceed in a more or less straightforward fashion, one that should be reasonably familiar. In previous examples studied in physics class, we would normally construct the $\mathbf{e}_i$ to be orthonormal but this is not a requirement and, in some cases, is not desirable. For example, in dealing with crystals, we would want to choose a basis that aligns with the crystal axes. In other cases, using a curvilinear coordinate system may prove to be advantageous.

With respect to the basis formed by the $\mathbf{e}_i$, the vector $\mathbf{v}$ can be resolved into components:

$$\mathbf{v} = \sum_{i=1}^{N} v_i \mathbf{e}_i = (v_1, \ldots, v_N).$$

In order to determine the values of the components v_i, we need to use the dual vectors $\tilde{\mathbf{e}}_i$. In crystallography (and in three dimensions), these are known as the reciprocal vectors and can be constructed as follows:

$$(2.40) \quad \tilde{\mathbf{e}}_1 = \frac{\mathbf{e}_2 \times \mathbf{e}_3}{\mathbf{e}_1 \cdot (\mathbf{e}_2 \times \mathbf{e}_3)}, \quad \tilde{\mathbf{e}}_2 = \frac{\mathbf{e}_3 \times \mathbf{e}_1}{\mathbf{e}_1 \cdot (\mathbf{e}_2 \times \mathbf{e}_3)} \quad \text{and} \quad \tilde{\mathbf{e}}_3 = \frac{\mathbf{e}_1 \times \mathbf{e}_2}{\mathbf{e}_1 \cdot (\mathbf{e}_2 \times \mathbf{e}_3)}.$$

These dual vectors have the following property:

(2.41) $\tilde{e}_i \cdot e_j = \delta_{ij}$,

where the δ function here is zero if $i \neq j$ and one if $i = j$. Note that, in Cartesian coordinates, the dual vectors are equivalent to the original basis vectors but this is not generally true.

> EXERCISE 2.17.   Use the definitions from equation 2.40 and show that equation 2.41 is true.

Suppose now that we define a different set of basis elements e_i', where we have scaled the original e_i by some factor. This would occur, for example, if we switch between using meters and centimeters as our units of measure. What happens then to the values of the vector components? If our original coordinate values x_i were provided in terms of meters, then the new coordinate values x_i' will be one hundred times larger, reflecting the change to units of centimeters. Mathematicians have termed this property **contravariant**. That is, the values of the components v_i scale inversely to the change of scale of the basis vectors.

Consider now transforming a dual vector

$$\tilde{v} = \sum_{i=1}^{N} \tilde{v}_i \tilde{e}_i.$$

If each of the basis elements is scaled by a factor, then the dual basis elements will scale inversely to that factor. The components of the dual vector will scale proportionally. That is, if our units change from meters to centimeters, the components $\tilde{v}_i'$ of the dual vector will also be one hundred times smaller than the components $\tilde{v}_i$ of the original vector. This property is termed **covariant**.

> EXERCISE 2.18.   Consider the following basis vectors:
>
> $$e_1 = (1,0,0), \quad e_2 = (0.2,1,0) \quad \text{and} \quad e_3 = (0,1.3,1.8).$$
>
> What are the reciprocal vectors $\tilde{e}_i$? What are the components of the vector $\mathbf{v} = 3\hat{x} + 4\hat{y} + 2.5\hat{z}$ in the e_i basis? What are the components of $\mathbf{v}$ in the reciprocal basis?
>
> Repeat the analysis for the new basis $e_i' = e_i/100$.

In moving forward, we face a serious notational problem. We have heretofore utilized a bold typeface to identify vectors but we don't have a convenient typographical mechanism for identifying multidimensional tensors. In many matrix algebra classes, one will find that upper-case variables will be restricted to matrices and lower-case variables will be vectors. Consequently, the following equation:

$$\mathbf{A}\mathbf{x} = \mathbf{b}$$

can readily be interpreted to mean the multiplication of a matrix and a vector that returns a vector. Unfortunately, we have already used $\mathbf{E}$ to mean the vector electric field and $\mathbf{A}$ to mean the magnetic potential, so it is a bit late to enforce the case-sensitive strategy.

Worse, tensors are not restricted to two-dimensional entities, so we do not have a convenient typographical means of distinguishing one-, two-, three- and four-dimensional objects by their typeface. The most common notational convention used in physics explicitly expresses the equations in component form, using superscript indices to denote contravariant components and subscript components to denote covariant components. The number of indices denotes the dimension or rank of the tensor. For example, $R^j{}_{klm}$ denotes a fourth-rank tensor with one contravariant index and three covariant indices. Our matrix equation now reads as follows:

$$\sum_j A_{ij} x_j = b_i.$$

This strategy does not lead to particularly elegant representations. Einstein became so distressed over repeating all of the summation symbols in his work that he invented a new convention: repeated indices are implicitly summed:

$$A^{ij} x_j = b^i.$$

All three of these equations are intended to mean the same thing.

Mathematicians have developed somewhat more elegant forms but are generally much more abstract, making it difficult for non-experts to follow. For this first time through the material, we shall hew to the physics practice, using the component notation but we shall forego the use of the Einstein summation convention. We'll leave it to the students to clean up the formulas as we proceed. Note that a particularly unwieldy result of this notational practice is that by the symbol v^i, we can mean both the ith component of a contravariant vector $\mathbf{v}$ or the vector itself. Alas, one will often have to infer the meaning from context but we shall attempt to clarify as necessary.

Let us briefly define mathematically what we mean by covariant and contravariant. If we have two separate coordinate systems $\mathbf{x} = (x^1, \ldots, x^N)$ and $\mathbf{y} = (y^1, \ldots, y^N)$ then the contravariant vector whose components are v^i will transform as follows:

(2.42)
$$v^i(\mathbf{y}) = \sum_{j=1}^{N} \frac{\partial y^i}{\partial x^j} v^j(\mathbf{x}).$$

Similarly, the covariant vector whose components are denoted by v_i will transform as follows:

(2.43)
$$v_i(\mathbf{y}) = \sum_{j=1}^{N} \frac{\partial x^j}{\partial y^i} v_j(\mathbf{x}).$$

It is a bit difficult to see amidst all the indices but there is an inverse relationship between the coefficients:

$$\frac{\partial y^i}{\partial x^j} = \left[\frac{\partial x^j}{\partial y^i} \right]^{-1}.$$

We won't expand further on this at the moment but it does formally define what we mean by tensors. We should note also that we have used a notation in equations 2.42 and 2.43 where we think of the transformed vector as an explicit function of the new coordinates $\mathbf{y}$, where the original vector was a function of the coordinates $\mathbf{x}$. This strategy is not standard, often the transformed vector is $\mathbf{x}'$, but saves us from further decorating the symbol v with a prime or tilde to denote that a transformation has occurred.

Let us now revisit the Maxwell equations 2.11–2.14 and see what happens if we recast them into tensor form. First, we note that we must work in four-dimensional spacetime and will choose an invariant interval $ds^2 = c^2 dt^2 - dx^2 - dy^2 - dz^2$. This choice makes timelike vectors positive and spacelike vectors negative but students should beware that this choice is by no means unanimous. Other authors will work with the signs reversed. This will lead to differences in intermediate results but same answers eventually. There are also differences as to what to call the terms. Many authors consider $c\,dt = x^0$, the zeroth element of the position vector, which has spatial components x^1, x^2 and x^3. Others will place time into the fourth position $c\,dt = x^4$. As the *Mathematica* software utilizes indices that run from 1 to N, we'll just call $c\,dt$ the first element of the position vector and the x-value will occupy the second.

The Maxwell equations represent first order differential equations for the fields. The obvious choice for a four-dimensional derivative operator is the following:

(2.44)
$$\frac{\partial}{\partial x^i} = \left\{ \frac{\partial}{c\partial t}, \frac{\partial}{\partial x}, \frac{\partial}{\partial y}, \frac{\partial}{\partial z} \right\}.$$

EXERCISE 2.19. A more compact notation defines the differential operator ∂_i as

$$\partial_i \equiv \frac{\partial}{\partial x^i}.$$

Convince yourself that ∂_i is a covariant vector, hence the lower index.[5] Hint: refer to the definitions in equation 2.42 and 2.43.

If we now revisit equations 2.27 and 2.26, we would like to define a four-potential from V and $\mathbf{A}$. From equation 2.27, we know that the dimension of the gradient of the potential V must be $(V{\cdot}m^{-1})$. Furthermore, we can infer that the vector potential $\mathbf{A}$ must have dimension of $(V{\cdot}s{\cdot}m^{-1})$. As a result, we can define a four-potential A that has consistent dimensions as follows:

$$(2.45) \qquad A^i = \left(V/c, A_x, A_y, A_z\right) = \left(V/c, \mathbf{A}\right).$$

The covariant potential will be given by the following:

$$(2.46) \qquad A_i = \left(V/c, -A_x, -A_y, -A_z\right) = \left(V/c, -\mathbf{A}\right).$$

What happens if we now apply the differential operator to the four potential? We find, in fact, the following:

$$(2.47) \qquad \frac{\partial A_k}{\partial x^i} = \begin{bmatrix} \dfrac{\partial V/c}{c\partial t} & -\dfrac{\partial A_x}{c\partial t} & -\dfrac{\partial A_y}{c\partial t} & -\dfrac{\partial A_z}{c\partial t} \\[2mm] \dfrac{\partial V/c}{\partial x} & \dfrac{\partial A_x}{\partial x} & \dfrac{\partial A_y}{\partial x} & \dfrac{\partial A_z}{\partial x} \\[2mm] \dfrac{\partial V/c}{\partial y} & \dfrac{\partial A_x}{\partial y} & \dfrac{\partial A_y}{\partial y} & \dfrac{\partial A_z}{\partial y} \\[2mm] \dfrac{\partial V/c}{\partial z} & \dfrac{\partial A_x}{\partial z} & \dfrac{\partial A_y}{\partial z} & \dfrac{\partial A_z}{\partial z} \end{bmatrix}.$$

Now, let us subtract the transpose of this object:

$$(2.48)$$

$$\frac{\partial A_k}{\partial x^i} - \frac{\partial A_i}{\partial x^k} = \begin{bmatrix} 0 & -\dfrac{\partial V/c}{\partial x} - \dfrac{\partial A_x}{c\partial t} & -\dfrac{\partial V/c}{\partial y} - \dfrac{\partial A_y}{\partial z} & -\dfrac{\partial V/c}{c\partial t} - \dfrac{\partial A_z}{c\partial t} \\[2mm] \dfrac{\partial V/c}{c\partial t} + \dfrac{\partial A_x}{\partial x} & 0 & \dfrac{\partial A_x}{\partial y} - \dfrac{\partial A_y}{\partial x} & \dfrac{\partial A_x}{\partial z} - \dfrac{\partial A_z}{\partial x} \\[2mm] \dfrac{\partial V/c}{c\partial t} + \dfrac{\partial A_y}{\partial y} & \dfrac{\partial A_y}{\partial x} - \dfrac{\partial A_x}{\partial y} & 0 & \dfrac{\partial A_y}{\partial z} - \dfrac{\partial A_z}{\partial y} \\[2mm] \dfrac{\partial V/c}{c\partial t} + \dfrac{\partial A_z}{\partial z} & \dfrac{\partial A_z}{\partial x} - \dfrac{\partial A_x}{\partial z} & \dfrac{\partial A_z}{\partial y} - \dfrac{\partial A_y}{\partial z} & 0 \end{bmatrix}.$$

---

[5]The operator is also known to mathematicians as a 1-form. There is a veritable ocean of mathematical literature on the extension of simple, ordered tuples (vectors) into more complex entities (tensors).

We shall define this last object to be the electromagnetic field tensor F. We can recognize the components of the tensor:

$$(2.49) \qquad F_{ik} \equiv \frac{\partial A_k}{\partial x^i} - \frac{\partial A_i}{\partial x^k} = \begin{bmatrix} 0 & E_x/c & E_y/c & E_z/c \\ -E_x/c & 0 & -B_z & B_y \\ -E_y/c & B_z & 0 & -B_x \\ -E_z/c & -B_y & B_x & 0 \end{bmatrix}.$$

EXERCISE 2.20. Write out the components of equations 2.26 and 2.27 and convince yourself that the identifications made in equation 2.49 are correct.

We haven't yet shown that F is actually a tensor but it is not too difficult to do so. We want now to take the four-divergence of the field tensor but, to do so, we must apply one of the rules associated with tensors. To form the equivalent of a dot product, we utilize the metric tensor g:

$$(2.50) \qquad \sum_{i=1}^{4} \sum_{k=1}^{4} g_{ik} a^i b^k = \sum_{i=1}^{4} a^i b_i = \sum_{k=1}^{4} a_k b^k.$$

Here, we note in the last two terms that the transformation of a vector by the metric tensor converts it from a contravariant to covariant vector, lowering the index. We can also define the conjugate metric tensor g^{ik} that is the inverse of g_{ik}. We can show

$$\sum_{j=1}^{N} g_{ij} g^{kj} = \delta_i^k = \delta_{ik},$$

where here the delta function is one when $i = k$ and zero otherwise.

EXERCISE 2.21. Formally, raising and lowering indices can be accomplished by utilizing the metric tensor g, where here we are using the form:

$$g_{ij} = g^{ij} = \begin{bmatrix} 1 & 0 & 0 & 0 \\ 0 & -1 & 0 & 0 \\ 0 & 0 & -1 & 0 \\ 0 & 0 & 0 & -1 \end{bmatrix}.$$

If we define the contravariant vector $x^i = (ct, x, y, z)$, what is the covariant vector x_i? We defined $A^i = (V/c, A_x, A_y, A_z)$. What is the covariant vector A_i?

The four-divergence of the field tensor is computed as follows:

$$\sum_{i=1}^{4}\sum_{l=1}^{4}g^{il}\frac{\partial}{\partial x^l}F_{ik} = \left[\frac{\partial}{c\partial t}, -\frac{\partial}{\partial x}, -\frac{\partial}{\partial y}, -\frac{\partial}{\partial z}\right]\begin{bmatrix} 0 & E_x/c & E_y/c & E_z/c \\ -E_x/c & 0 & -B_z & B_y \\ -E_y/c & B_z & 0 & -B_x \\ -E_z/c & -B_y & B_x & 0 \end{bmatrix}$$

$$(2.51) \qquad = \begin{bmatrix} \dfrac{\partial E_x}{c\partial x} + \dfrac{\partial E_y}{c\partial y} + \dfrac{\partial E_z}{c\partial z} \\[2mm] \dfrac{\partial E_x}{c^2\partial t} - \dfrac{\partial B_z}{\partial y} + \dfrac{\partial B_y}{\partial z} \\[2mm] \dfrac{\partial E_y}{c^2\partial t} - \dfrac{\partial B_x}{\partial z} + \dfrac{\partial B_z}{\partial x} \\[2mm] \dfrac{\partial E_z}{c^2\partial t} - \dfrac{\partial B_y}{\partial x} + \dfrac{\partial B_x}{\partial y} \end{bmatrix}$$

If we now recall that $\epsilon_o\mu_o = 1/c^2$, we can recognize these results as just the component terms of the Maxwell equations 2.11 and 2.14. We are lacking the source terms, though. If we define the four-current density J as follows:

$$(2.52) \qquad J^i = (c\rho, J_x, J_y, J_z),$$

then we can write Maxwell's equations in the following form:

$$(2.53) \qquad \sum_{i=1}^{4}\frac{\partial}{\partial x_i}F_{ij} = \mu_oJ_j \quad \text{and} \quad \sum_{i=1}^{4}\frac{\partial}{\partial x_i}G_{ij} = 0,$$

where we have made use of the dual electromagnetic tensor G:

$$(2.54) \qquad G_{ik} = \begin{bmatrix} 0 & B_x & B_y & B_z \\ -B_x & 0 & E_z/c & -E_y/c \\ -B_y & -E_z/c & 0 & E_x/c \\ -B_z & E_y/c & -E_x/c & 0 \end{bmatrix}.$$

EXERCISE 2.22. Check the dimensionality of the components of the four-current vector J^i.

EXERCISE 2.23. Expand Maxwell's equations 2.11–2.14 into components and compare to the terms of equation 2.53. Convince yourself that they are just different representations of the same terms.

EXERCISE 2.24. A more common expression of Maxwell's equations is given as follows:

$$\partial_i F^{ik} = \mu_o J^k \quad \text{and} \quad \partial_i G^{ik} = 0,$$

where we have used the Einstein summation convention and the contravariant form of the field tensor and four-current. Write out the components and show that this results in the same equations (up to an overall sign) as those in equation 2.53.

The purpose of this exercise has been to put Maxwell's equations into tensor form because it is in this form that Lorentz transformations have a simple expression. The Lorentz transformations are just a special set of coordinate transformations that consist of the rotation matrices that mix the spatial components of a four-vector or four-tensor and boost matrices that mix the time and space coordinates. In this representation, the Maxwell equations 2.53 are manifestly invariant under Lorentz transformation. The components F_{ij} will be altered but the form of the equation does not change.

EXERCISE 2.25.   Consider a Lorentz boost along the x-axis. This is described by the following matrix:

$$B_{ik} = \begin{bmatrix} \cosh\zeta & \sinh\zeta & 0 & 0 \\ \sinh\zeta & \cosh\zeta & 0 & 0 \\ 0 & 0 & 1 & 0 \\ 0 & 0 & 0 & 1 \end{bmatrix}.$$

What is the Lorentz transform of the four-current density J_i? What is the Lorentz transformation of the field tensor F_{ik}?

EXERCISE 2.26.   There are two scalar invariants that can be constructed from the field tensor F and its dual G. What are the values of the following expressions:

$$\sum_{i=1}^{4}\sum_{k=1}^{4} F^{ik}F_{ik} \quad \text{and} \quad \sum_{i=1}^{4}\sum_{k=1}^{4} F^{ik}G_{ik}?$$

2.5. Solitons

We have seen that Maxwell's equations provide an accurate representation of electromagnetic phenomena, making numerous predictions that have been repeatedly verified experimentally. Moreover, when cast into tensor form, the equations are consistent with Einstein's theory of special relativity. The problem that we face in trying to develop a classical model for the photon is that the properties of a photon are largely unspecified. This is due primarily to the fact that photons classically do not interact with one another. Hence, we cannot scatter a photon from another photon and learn anything about its structure. We can speculate that photons are somehow pulses of electromagnetic energy that have finite spatial extent.

The notion of a finite blob of electromagnetic energy arose initially when the German physicist Max Planck considered the problem of blackbody radiation from a statistical point of view. If one can construct the partition function $\mathcal{Z}$, which is quite literally the sum over all possible states, then one can determine all of the interesting thermodynamic properties of a system. Planck provided the *ansatz* that electromagnetic energy could be dealt with in a similar fashion as was used to enumerate the distribution of particles in a box: that is, the distribution of electromagnetic energy within the box was given by the following formula:

$$(2.55) \qquad \mathcal{Z} = \sum_{n=0}^{\infty} e^{-nh\nu/kT} = \frac{1}{1 - e^{-h\nu/kT}},$$

where $h\nu$ is the energy of the photons, T is the absolute temperature and k is a constant, now known as Boltzmann's constant. From this result, we can obtain the average number of photons $\bar{n}$ by noticing that

$$\frac{\partial \mathcal{Z}}{\partial h\nu} = -\frac{n}{kT}.$$

From this, we obtain Planck's result:

$$(2.56) \qquad \bar{n} = \frac{1}{e^{h\nu/kT} - 1}.$$

Planck's suggestion turns out to solve a conundrum that arose when earlier attempts at trying to explain the behavior of blackbody radiation tried to integrate over the electromagnetic energy in the box. This strategy results in a divergent integral, whereas Planck's summation was finite.

EXERCISE 2.27. Plot the function $f(x) = 1/(\exp(x) - 1)$.

So, the introduction of the photon into physics literature was really a numerical device that succeeded because it (subtly) introduced a frequency cutoff into the partition function. This is not particularly obvious but was later appreciated by Einstein in his *annus mirabilis* paper on the subject, where he introduced the nomenclature Lichtquant to describe the quantized portion of electromagnetic energy.

If asked to describe a photon, the image that undoubtedly springs to mind is a small glowing ball that travels at the speed of light. That is often how photons are depicted but can we be more quantitative about this picture? Can we find solutions of Maxwell's equations that are spatially confined for all time? Some years ago, the physicist James Brittingham conducted a search for just this, subject to the requirements that solutions

(1) satisfy the homogeneous Maxwell's equations,
(2) have a three-dimensional pulse structure,
(3) have no charge,

(4) move at light velocity in straight lines,
(5) are nondispersive and
(6) have finite energy.

If we assume propagation in the z-direction, there are well-known plane wave solutions that are functions of $z + ct$ and $z - ct$ that represent left- and right-propagating solutions, respectively. Brittingham looked for solutions that were, instead, functions of the product $f_1(z+ct)f_2(z-ct)$. He discovered that the following function is a solution of the homogeneous wave equation:

$$(2.57) \qquad \Psi(t,\zeta,z) = \frac{1}{4\pi i} \int_{-\infty}^{\infty} dq\, w(q)\, e^{iq(z+ct)} \frac{e^{-q\zeta^2/\eta}}{\eta},$$

where $\eta = z_0 + i(z - ct)$ and $w(q)$ is an arbitrary weighting function. An example of such a solution is depicted in figure 2.9.

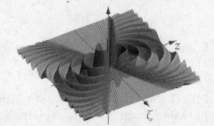

FIGURE 2.9. The simple, rotationally symmetric focus wave mode has a Gaussian profile in the transverse ζ-direction and falls off exponentially in the z-direction.

EXERCISE 2.28.   Define the integrand of the function $\Psi(t,\zeta,z)$ in *Mathematica*. (Ignore the weighting function w.) Use the D function to show that Ψ is a solution to the wave equation in cylindrical coordinates:

$$\left[\frac{\partial^2}{\partial\zeta^2} + \frac{1}{\zeta}\frac{\partial}{\partial\zeta} + \frac{\partial^2}{\partial z^2} - \frac{1}{c^2}\frac{\partial^2}{\partial t^2}\right]\Psi(t,\zeta,z) = 0.$$

EXERCISE 2.29.   Plot the function $\Psi(t,\zeta,z)$ over the domain $-5 \le \zeta \le 5$ and $-5 \le z \le 5$ for $q = 2\pi$ and $z_0 = 0.1$. How does the function change as t increases from 0 to 1 in steps of 0.1?

Brittingham termed the solutions focus wave modes and it has subsequently been demonstrated that these satisfy all of the original conditions except finite energy. The focus wave modes are infinite-energy solutions to Maxwell's equations. So too, are the plane wave solutions that we have discussed previously. We believe that real finite-energy solutions can be constructed from superpositions of plane waves, so it may prove that focus wave modes provide a new basis for studying pulse solutions of Maxwell's equations. That is, with suitable weighting of the focus wave modes or some other, undiscovered solutions, one may yet find a classical description of the photon.

A revisit to figure 1.12, though, calls such a quest starkly into question. Thompson's initial diffraction experiment indicated that single photons produce diffraction patterns that are independent of light intensity. As a result, the photon wave function must have some broader spatial extent than defined by the characteristic wavelength, or else it wouldn't be able to diffract through spatially separated slits. On a larger scale, it is well established that antennas radiate in patterns that can be described as diffraction limited. That is, if the aperture of a radiating body has a characteristic size of d, then the far field radiation will spread with an angle θ that is of the order

$$(2.58) \qquad\qquad \sin\theta \approx \frac{\lambda}{d},$$

where λ is the wavelength. Depending upon the shape of the aperture and other details, there can be a proportionality factor that can range from one to maybe four or so, but not orders of magnitude.

> EXERCISE 2.30. Suppose that a 600 nm photon is emitted from our sun ($d = 7 \times 10^8$ m). What is the diffraction limit angle θ, from equation 2.58 assuming a proportionality factor of 1? At the earth-sun distance ($R_E = 1.5 \times 10^{11}$ m), what is the spatial extent of this photon? At the average Pluto-sun distance ($R_P = 6 \times 10^{12}$ m), what is the spatial extent? The nearest star is approximately 4×10^{16} m distant. When the photon arrives in that solar system (in just over four years), what would be its spatial extent?

Without experimental data to constrain such speculations, physicists have generally not pursued the idea of a classical photon. A number, though, have investigated the general idea of a spatially bounded wavelike entity and have based much of their efforts on the 1834 observation of a localized disturbance in the Union Canal in Scotland. John Scott Russell noted that a barge being pulled through the canal had a mound of water piled in front of it that continued propagating down the canal even after the barge had been stopped. Russell tracked the mound for a couple of miles and reported his findings to the Royal Society and went on to conduct experiments in a wave tank that verified his observations.

Russell's wave of translation, as he called it, can be modelled with a one-dimensional non-linear equation and is the first example of what we now call soliton solutions. Solitons represent spatially bounded solutions of a

non-linear system of equations that serve as models for elementary particles. The equation studied most frequently is the Korteweg-de Vries equation, developed to explain Russell's observations. The equation was initially stated by the French mathematician Joseph Boussinesq in 1871 and
subsequently rediscovered by the Dutch mathematicians Diederik Korteweg and his student Gustav de Vries in 1897.[6]

$$(2.59) \qquad \frac{\partial \phi(t,x)}{\partial t} + \frac{\partial^3 \phi(t,x)}{\partial x^3} - 6\phi(t,x)\frac{\partial \phi(t,x)}{\partial x} = 0$$

is the canonical form of the equation, where $\phi(t,x)$ represents the wave
amplitude. One possible solution is given by the following:

$$(2.60) \qquad \phi(t,x) = -\frac{c}{2}\operatorname{sech}^2\left[\sqrt{c}(x-ct-a)/2\right],$$

where the wave will propagate to the right with a velocity c and a is an
arbitrary constant.

EXERCISE 2.31. Use the Animate function to plot the function $\phi(t,x)$.
Use $c = 1$ and $a = 0$ and plot ϕ over the domain $-20 \le x \le 20$ and for
the times $0 \le t \le 20$.

EXERCISE 2.32. Suppose that you have two solutions to the Korteweg-
de Vries equation,

$$\phi_1(t,x) = -\frac{c_1}{2}\operatorname{sech}^2\left[\sqrt{c_1}(x-c_1 t)/2\right] \quad \text{and}$$

$$\phi_2(t,x) = -\frac{c_2}{2}\operatorname{sech}^2\left[\sqrt{c_2}(x-c_2 t)/2\right].$$

Use the D function to demonstrate that each satisfies the Korteweg-
de Vries equation. Now show that the sum $\phi_1 + \phi_2$ does not.

Equation 2.60 can be found by explicitly trying to construct solutions of
the form $\phi(t,x) = \phi(ct - x)$. A general solution for N propagating solitons, each with a different characteristic velocity, was obtained in 1967 by
Clifford Gardner, John Greene, Martin Kruskal and Robert Miura who utilized a technique now known as the inverse scattering transform.[7] Their
inverse scattering methodology has been systematized and expanded to a
host of other partial differential equations.

---

[6]Boussinesq published "Théorie de lâĂŹintumescence liquide appelée onde solitaire ou de
translation, se propageant dans un canal rectangulaire" in the *Comptes Rendus Hebdomi-
naires des Séances de l'Académie des Sciences* in 1871. Korteweg and de Vries published "On
the change of form of long waves advancing in a rectangular canal and on a new type of long
stationary waves" in the *Philosophical Magazine* in 1895.
[7]Gardner *et al.* published "Method for solving the Korteweg-de Vries equation" in the *Physical Review Letters* in 1967.

We can construct the N-soliton solution as follows:

(1) Define an array P that contains the square roots of the soliton velocities c_i

$$P = \left\{\sqrt{c_1}, \ldots, \sqrt{c_N}\right\}.$$

(2) Define the matrix M where the components are given by the following:

$$M_{ik} = \delta_{ik} + \frac{2\sqrt{P_i P_k}}{P_i + P_k} e^{(P_i + P_k)x - (P_i^3 + P_k^3)t - Q_i - Q_k},$$

where the Q_i are relative phases of the different waves.

(3) The solution can be obtained from the following:

(2.61)
$$\phi(t, x) = 2\frac{\partial^2}{\partial x^2} \ln\left[\det M\right],$$

where $\det M$ is the determinant of the matrix M.

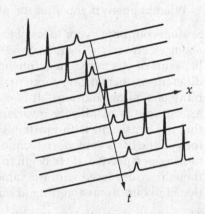

FIGURE 2.10. Two solitons of different velocities coincide at time zero. Close inspection of the results indicates a phase shift arises due to the interaction.

An example of two solitons, with velocities $c = \{1, 4\}$, is illustrated in figure 2.10. The soliton with the larger velocity also has the larger amplitude. One can see from the figure that the two separate solitons pass through one another, largely unaffected by the interaction. In fact, there is a phase shift associated with the interaction.

EXERCISE 2.33. Use equation 2.61 to compute the two-soliton solution for $P = \{1, 2\}$. Use the Animate function to plot the results over the time interval $-10 \le t \le 20$ and for the domain $-50 \le x \le 50$. Now add the single soliton solutions from equation 2.60 to the plot and adjust their phases so that the initial pulses are aligned. What happens after the interaction at $t = 0$?

The use of the Korteweg-de Vries equation to model simple bounded solutions of a wave equation represents a common practice in physics: use a (highly) simplified model to test ideas and behaviors before tackling a complex problem. The Korteweg-de Vries equation has non-trivial analytic solutions. If you want to solve the equation numerically, you can test your solver against known solutions. If you are ultimately interested in solving a more complex set of equations, you should have some idea about the robustness of your solver based on your experience with a simpler system.

In studying the problem of constructing solutions to the Korteweg-de Vries equation, Gardner *et al.* found a general strategy for solving non-linear partial differential equations that, in some sense, generalizes the Fourier transform method. Their efforts have provided a significant new tool for mathematical physicists, with applicability far beyond just the Korteweg-de Vries equation.

EXERCISE 2.34. Construct a three-soliton solution for $P = \{1, 1.5, 2\}$. What happens if you alter the phases from the nominal $Q = \{0, 0, 0\}$?

So, to recount the major ideas of this brief sojourn through what we know about electromagnetism, Maxwell's equations describe the macroscopic behavior of electromagnetic phenomena. They can be written in several different mathematical representations, each of which harbors the possibility of finding solutions. It is possible to write Maxwell's equations in a form that is manifestly invariant to Lorentz transform, meaning that they are consistent with Einstein's special theory of relativity. They have formed the basis of modern technology and are quite useful from an engineering perspective. If we begin to drill down into the microscopic world, though, we are faced with the same problem of granularity that we mentioned in our discussion of sand dunes.

The constitutive relations $\mathbf{D} = \epsilon \mathbf{E}$ and $\mathbf{B} = \mu \mathbf{H}$ are the macroscopic properties that arise from the ensemble average over many individual photons. Solutions to Maxwell's equations are additive: if $\mathbf{E}_1$ and $\mathbf{E}_2$ are solutions, then $\mathbf{E}_1 + \mathbf{E}_2$ is a solution. There are many radio stations broadcasting simultaneously: their signals can be detected and processed independently. As a result, we do not have a classical description of a photon. That is a construct that arises only when we take it in context with charged particles. This is a point to which we shall return subsequently.

III

On the Nature of the Electron

The creation of cathode ray tubes in the late nineteenth century caused a public sensation. Until that time, one obtained light by burning something: wood, oil or gas, typically. In a cathode ray tube, one found light emitted from a glass cell without any form of combustion at all. Today, we are significantly less impressed with electric lights and understand that the cathode ray tubes actually contain a beam of electrons that impinges upon atoms within the tube. These collisions excite the atoms to higher energy states, which subsequently decay back to the ground state with the emission of a photon.

Rutherford's 1911 discovery that the atom has a positively charged nucleus that is five orders of magnitude smaller than the nominal atomic size generated a flurry of theoretical efforts to describe the properties of atoms. The most obvious strategy was to consider the motion of an electron acting in the field of a central force. In this approach, one could replicate the previous work on masses in a gravitational field, thinking of the atom as a miniature planetary system of some form. There were differences, though, that precluded an exact mapping of Newtonian gravity onto the atomic problem. It was known that electrons displayed a wavelike nature and what evolved was a strategy that focused on developing so-called wave functions. The interpretation of these wave functions was that their square magnitude represented the probability of finding an electron somewhere in space. If we consider the nucleus to be very (negligibly) small, we should not be surprised that the early attempts at modelling the atom utilized basis functions that exploited spherical symmetry. The natural set of basis functions in spherical coordinates are the Laguerre polynomials $L_n(r)$ for the radial component of the wave function and the associated Legendre polynomials $Y_{lm}(\theta,\varphi)$ for the azimuthal and polar coordinates.

Basis functions are defined by a set of indices (n,l,m):

$$f_{nlm}(r,\theta,\varphi) = L_n(r)\,Y_{lm}(\theta,\varphi),$$

© Mark A. Cunningham 2018
M.A. Cunningham, *Beyond Classical Physics*,
Undergraduate Lecture Notes in Physics,
https://doi.org/10.1007/978-3-319-63160-8_3

where the indices must satisfy some constraints. For each value of n, a positive integer, l can take on integral values from 0 to $n-1$. For each value of l, m can take on integral values from $-l$ to l. As a result, for $n = 1$ there is only a single function: f_{100}. For $n = 2$, there are four (1+3) functions: f_{200}, f_{21-1}, f_{210} and f_{211} and for $n = 3$, there are nine (1+3+5) functions.

EXERCISE 3.1. Plot the Laguerre polynomials L_n for $n = 1, 2$ and 3 over the range $0 \le r \le 10$. Now plot the product of the Laguerre polynomials with the weighting function e^{-r}.

EXERCISE 3.2. Use the SphericalPlot3D function to study the behavior of the spherical harmonic functions $Y_{lm}(\theta, \phi)$. For each integral value of l, m can take on values of $m = -l, -l+1, \ldots, l-1, l$. Note that the SphericalHarmonicY function is complex; plot the real part.

If we examine the ionization energies of the (neutral) elements, as depicted in figure 3.1, we observe that the energy peaks at specific atomic numbers. For these cases, the electron is more tightly bound to the nucleus than in adjacent elements; this occurs at values of the atomic number of $Z = 2, 10, 18, 36, 54, 80$ and 86. If we conduct a bit of speculative numerology (hoping that the eigenfunctions that describe the electron wavefunctions are very close to the basis functions), then the $n = 1$ level should be able to describe the state of a single electron. Adding the $n = 2$ level functions should enable us to describe an additional four (4=1+3) electrons. We could naïvely suggest that something special might happen at magic numbers of 1 and 5 (=1+4). What we observe from the figure is that the ionization energy peaks at twice the naïve values: at 2 and 10 instead of 1 and 5. We can infer that the data are suggesting that we can somehow place two electrons in each level.

FIGURE 3.1. The energy required to remove one electron from the neutral atom of atomic number Z displays significant structure. Notably, the energy peaks for the noble gas elements and mercury $(Z = 80)$.

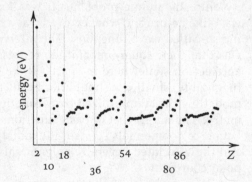

EXERCISE 3.3. Thereafter, the numerology becomes a bit more complex. Adding the $n = 3$ levels can provide us with another nine

functions but we don't see anything special at $Z = 14$ (=1+4+9) or 28 but there is another magic number at 18, which could correspond to just the $l = 0$ and 1 levels from the $n = 3$ functions: 18 = 2[1+4+(1+3)]. The next magic number is 36, which we can obtain from the $n = 1, 2$ and 3 levels plus the $l = 0$ and 1 levels from $n = 4$: 36=2[1+4+9+(1+3)]. Can you obtain a plausible explanation for the magic numbers 54, 80 and 86?

We can recall that Stern and Gerlach conducted an experiment to measure the intrinsic magnetism associated with free electrons and observed that the electrons possessed an intrinsic magnetic dipole moment and that it was quantized.[1] In passing through the (non-uniform) magnetic field, the electron beam separated into two components. This property of electrons is now known as **spin**, which is a rather unfortunate name. From figure 1.12, we know that the electron is a wave, not a small ball of some miniscule dimension. As a result, the intrinsic magnetic field does not result from the rotation of a small charge density, as aesthetically pleasing as that idea may be. In large measure, we do not understand why the electron has spin but we know that it does.

3.1. Dirac Equation

When the British physicist Paul Dirac sat down to develop a new theory of the electron, he mentioned this *duplexity* problem. Apparently, if we are to describe the states of an electron in a central force, it is not sufficient to use just the indices (n, l, m). We must augment them with an additional index s to define the spin: (n, l, m, s). While groups headed by Erwin Schrödinger and Werner Heisenberg were making progress in their development of a quantum theory of atoms, what Dirac observed was that they were not constructing models that were consistent with Einstein's theory of relativity. As a result, the Schrödinger and Heisenberg models represented only a low-velocity, limiting case of a more general theory. So, Dirac set out to make a relativistic theory of the electron.

In order to have a theory that is compatible with Einstein's Special Theory of Relativity, it must be invariant under Lorentz transformations. If we utilize the tensor notation, we can attempt to write this new theory in a form that is manifestly invariant. That is, we cannot ask the question: How does the state of the system evolve in time? Time is simply the first component in the four-vector that describes spacetime. Moreover, time does not have a global definition, independent of the frame of reference: only local values of time are meaningful. What we must ask instead is the

---

[1]Otto Stern and Walther Gerlach published "Das magnetische Moment des Silberatoms" in the *Zeitschrift für Physik* in 1922.

question: How does the system evolve along a spacetime path? Mathematically, we must make the following replacement:

$$\frac{\partial}{\partial t} \longrightarrow \frac{\partial}{\partial x^i}.$$

This is precisely what we did when we formulated the relativistic form of Maxwell's equations.

A reasonably obvious choice for a relativistic quantum theory was explored by Oskar Klein and Walter Gordon.[2] They made use of the fact that the wave equation is Lorentz invariant and, hence, postulated that the following equation could describe the electron:

$$(3.1) \qquad \left[\sum_{i=1}^{4} \sum_{k=1}^{4} g^{ik} \frac{\partial}{\partial x^i} \frac{\partial}{\partial x^k} + \frac{m^2 c^2}{\hbar^2} \right] \psi(\mathbf{x}) = 0,$$

where the wave function ψ is a function of the four-vector $\mathbf{x}$, m is the mass of the electron and $\hbar$ is defined as Planck's constant h divided by 2π.[3] Because all of the indices are summed away, the terms within the brackets represent a scalar quantity and, thus, is manifestly Lorentz invariant.

EXERCISE 3.4. Rewrite the Klein-Gordon equation 3.1 using shorthand notation: Use the Einstein summation convention and the ∂_i derivative operator. An additional simplification, common in particle theory, is to also use a unit system in which $\hbar = 1$ and $c = 1$. What is this simplest form of the Klein-Gordon equation?

EXERCISE 3.5. Expand the Klein-Gordon equation into its explicit components and show that you recover the wave equation, with an additional mass term.

To derive a probabilistic interpretation of the wave function ψ, which is assumed to be a complex function, let us multiply the Klein-Gordon equation 3.1 by the complex conjugate and then subtract the conjugate of the

---

[2] Klein and Gordon published separate papers in the *Zeitschrift für Physik* in 1926, as did Vladimir Fock, and several others, including Louis de Broglie.
[3] The quantity $h/2\pi$ occurs frequently, so the shorthand notation $\hbar$ is commonplace.

whole expression:

$$\psi^*(\mathbf{x}) \left[\sum_{i=1}^{4} \sum_{k=1}^{4} g^{ik} \frac{\partial}{\partial x^i} \frac{\partial}{\partial x^k} + \frac{m^2 c^2}{\hbar^2} \right] \psi(\mathbf{x})$$

$$- \psi(\mathbf{x}) \left[\sum_{i=1}^{4} \sum_{k=1}^{4} g^{ik} \frac{\partial}{\partial x^i} \frac{\partial}{\partial x^k} + \frac{m^2 c^2}{\hbar^2} \right] \psi^*(\mathbf{x})$$

$$(3.2) \qquad = \sum_{i=1}^{4} \sum_{k=1}^{4} g^{ik} \frac{\partial}{\partial x^i} \left[\psi^*(\mathbf{x}) \frac{\partial}{\partial x^k} \psi(\mathbf{x}) - \psi(\mathbf{x}) \frac{\partial}{\partial x^k} \psi^*(\mathbf{x}) \right].$$

The result here is analogous to the charge conservation equation we know from electromagnetics:

$$(3.3) \qquad \frac{\partial \rho}{\partial t} - \nabla \cdot \mathbf{J} = 0,$$

where the time rate of change of charge density within some volume must equal the current flux through the surface of that volume. The time component of equation 3.2 is just the following:

$$(3.4) \qquad \frac{1}{c^2} \frac{\partial}{\partial t} \left[\psi^*(\mathbf{x}) \frac{\partial}{\partial t} \psi(\mathbf{x}) - \psi(\mathbf{x}) \frac{\partial}{\partial t} \psi^*(\mathbf{x}) \right].$$

We would like to interpret the quantity within the brackets as the probability density for the electron. Unlike the charge density ρ, that can be negative, we require the probability density to be positive-definite. Unfortunately, the expression in equation 3.4 is not always positive. Thus, we would lose the probabilistic interpretation of the wave functions defined by the Klein-Gordon equation.

EXERCISE 3.6. Use *Mathematica* to show that equation 3.2 is correct. Hint: compute terms corresponding to the left-hand and right-hand sides of the equation. Subtract and show that you obtain zero.

EXERCISE 3.7. Write out the components of equation 3.2 in terms of the explicit time derivatives and the gradient ∇ operator. Compare the terms to those in equation 3.3.

There are other problems with the Klein-Gordon model. If we push a bit further, we could also calculate some of the predictions for energy levels in the hydrogen atom made by the Klein-Gordon equation. These do not agree with the experimental data. So, the Klein-Gordon equation does not represent the correct relativistic description of the electron.

In 1928, Dirac concluded that part of the difficulty with the Klein-Gordon equation arose from the fact that the equation was second-order in time, whereas the Schrödinger equation contained only a first-order derivative with respect to time. As the Schrödinger approach gave better agreement with experimental data, Dirac attempted to derive a relativistically correct equation that was a first-order equation for the wave function.[4]

As we have seen, if we want to develop a theory that is consistent with Einstein's theory of Special Relativity, we should use a tensor formulation. Rather than consider the time evolution of the system, we must consider the spacetime evolution. In this sense, the appropriate derivative operator is the covariant vector $\partial/\partial x^i$ but just the four-gradient applied to the wave function does not lead to a reasonable description of the electron.

Dirac had the imaginative idea to construct a Lorentz scalar from the four-gradient and another four-vector. He recognized that this new vector could not simply be a constant vector, as the result of forming the inner product of a constant vector with the four-gradient would not produce a Lorentz scalar. Instead, Dirac suggested that the new four-vector must be a vector of matrices. There is some ambiguity in the particular choices of matrices: the size must be even but 2x2 matrices don't work. Ultimately, Dirac decided that 4x4 matrices would be adequate but it is possible for the dimension to be larger. Dirac's matrices can be written as follows[5]:

$$\gamma^1 = \begin{bmatrix} 1 & 0 & 0 & 0 \\ 0 & 1 & 0 & 0 \\ 0 & 0 & -1 & 0 \\ 0 & 0 & 0 & -1 \end{bmatrix}, \qquad \gamma^3 = \begin{bmatrix} 0 & 0 & 0 & -i \\ 0 & 0 & i & 0 \\ 0 & i & 0 & 0 \\ -i & 0 & 0 & 0 \end{bmatrix},$$

$$(3.5) \qquad \gamma^2 = \begin{bmatrix} 0 & 0 & 0 & 1 \\ 0 & 0 & 1 & 0 \\ 0 & -1 & 0 & 0 \\ -1 & 0 & 0 & 0 \end{bmatrix}, \qquad \gamma^4 = \begin{bmatrix} 0 & 0 & 1 & 0 \\ 0 & 0 & 0 & -1 \\ -1 & 0 & 0 & 0 \\ 0 & 1 & 0 & 0 \end{bmatrix}.$$

Dirac then proposed that the equation of a free electron would take the following form:

$$(3.6) \qquad \left[i\hbar \sum_{i=1}^{4} \sum_{s=1}^{4} (\gamma^i)_{rs} \frac{\partial}{\partial x^i} - mc \right] \psi_s(\mathbf{x}) = 0,$$

where now ψ has four components:

$$\psi(\mathbf{x}) = \begin{bmatrix} \psi_1(\mathbf{x}) \\ \psi_2(\mathbf{x}) \\ \psi_3(\mathbf{x}) \\ \psi_4(\mathbf{x}) \end{bmatrix}.$$

----

[4]Erwin Schrödinger and Paul A. M. Dirac shared the Nobel Prize in Physics in 1933 "for the discovery of new productive forms of atomic theory."
[5]Recall that our indices run from 1 to 4 with time as the first index. Other authors use different notation.

Here, we have run headlong into another notational difficulty. We have been reasonably consistent of late to use subscript and superscript indices to mean contravariant and covariant. Each of the gamma matrices is a constant matrix that does not transform under a Lorentz transformation; that is, they do not depend on the choice of coordinate system. The four gamma matrices together do transform as a Lorentz vector. Technically, the gamma matrices, with the identity matrix and $\gamma^5 = i\gamma^1\gamma^2\gamma^3\gamma^4$, form what is known to mathematicians as a Clifford algebra and the wave function is called a **spinor**. To distinguish these non-tensor indices, we have used letters from elsewhere in the Latin alphabet. So, $(\gamma^i)_{rs}$ means the element of the rth row and sth column of the ith component of the contravariant four-vector γ. The sum over s in equation 3.6 is a matrix-spinor multiplication, which results in a spinor.

> EXERCISE 3.8.   Rewrite the Dirac equation 3.6 using the Einstein summation convention and the ∂_i operator. The American physicist Richard Feynman grew weary of even this compact notation, and invented his own additional shorthand:
>
> $$\gamma^i \partial_i \equiv \partial\!\!\!/.$$
>
> Rewrite the Dirac equation, using $\hbar = c = 1$ and the Feynman slash notation.

From solutions $\psi(\mathbf{x})$ of the Dirac equation, one computes the probability density by multiplying ψ by its Hermitian conjugate, the transpose of the complex conjugate:

$$\psi^\dagger(\mathbf{x}) = \begin{bmatrix} \psi_1^*(\mathbf{x}) & \psi_2^*(\mathbf{x}) & \psi_3^*(\mathbf{x}) & \psi_4^*(\mathbf{x}) \end{bmatrix}.$$

Dirac arrived at this particular choice for the gamma matrices by demanding that the square of his equation recover the Klein-Gordon equation. This ensured that his solutions would satisfy the Einstein mass energy formula: $E^2 = p^2c^2 + m^2c^4$. The gamma matrices are required to have specific properties but are not uniquely defined.

> EXERCISE 3.9.   Define the gamma matrices in *Mathematica*. What are the matrix products $(\gamma^i)^2$? The gamma matrices anticommute. Show that if $i \neq k$ that $\gamma^i\gamma^k = -\gamma^k\gamma^i$.

> EXERCISE 3.10.   Multiply the Dirac equation 3.6 on the left by the following: $(i\hbar\gamma^i\partial_i + mc)$. Use the properties of the gamma matrices to demonstrate that the spinor $\psi(\mathbf{x})$ and, hence, each of its components independently, satisfies the Klein-Gordon equation.

We can recall that Dirac was concerned about the duplexity problem, wherein there was an additional quantum number that described electrons. He wanted to obtain a solution that incorporated the spin quantum number in a natural fashion. To make the wavefunctions consistent with Lorentz invariance, Dirac required that the wavefunction be a four-component spinor. This is two components too many.

To see how Dirac chose to accommodate this issue, we can look at the plane wave solutions of the Dirac equation. The four-dimensional Fourier transform of $\psi(\mathbf{x})$ is just

$$(3.7) \qquad \psi(\mathbf{x}) = \frac{1}{(2\pi)^2} \int d^4p \, \exp\left[\frac{i}{\hbar} \sum_{i=1}^{4} \sum_{k=1}^{4} g_{ik} p^i x^k\right] \widetilde{\psi}(p),$$

where $\widetilde{\psi}(p)$ is a function only of the transform variable p.[6] The sum in the exponential is actually just $Et - p_x x - p_y y - p_z z$, so our notation is certainly not compact. Consider the first component of the derivative operator:

$$(3.8) \qquad \gamma^1 \frac{\partial}{c\partial t} = \mathrm{diag}\left[\frac{\partial}{c\partial t}, \frac{\partial}{c\partial t}, -\frac{\partial}{c\partial t}, -\frac{\partial}{c\partial t}\right],$$

where diag means a 4×4 matrix with only nonzero elements on the diagonal. When applied to the transform of ψ from equation 3.7, we find
(3.9)

$$i\hbar\gamma^1 \frac{\partial}{c\partial t} e^{i(Et - p_x x - p_y y - p_z z)/\hbar} = -\mathrm{diag}\left[\frac{E}{c}, \frac{E}{c}, -\frac{E}{c}, -\frac{E}{c}\right] e^{i(Et - p_x x - p_y y - p_z z)/\hbar}.$$

The first two components have positive energy and the second two have a negative energy.

EXERCISE 3.11. Define the function

$$F(p) = i\hbar e^{i(Et - p_x x - p_y y - p_z z)/\hbar}$$

in *Mathematica* and apply the derivative operator $\partial/\partial x^i$ to produce a four vector. (Divide by $F(p)$ to produce a matrix that contains only the momenta, as in equation 3.9.) Now compute the 4×4 matrix $\mathbf{M}$ that is defined in Equation 3.6.

Consider the four spinors:

$$\widetilde{\psi}_1 = \begin{bmatrix} 1 \\ 0 \\ \frac{p_z}{-E/c+mc} \\ \frac{p_x + i p_y}{-E/c+mc} \end{bmatrix}, \quad \widetilde{\psi}_2 = \begin{bmatrix} 0 \\ 1 \\ \frac{p_x - i p_y}{-E/c+mc} \\ \frac{-p_z}{-E/c+mc} \end{bmatrix}, \quad \widetilde{\psi}_3 = \begin{bmatrix} \frac{-p_z}{E/c+mc} \\ \frac{-p_x - i p_y}{E/c+mc} \\ 1 \\ 0 \end{bmatrix}, \quad \widetilde{\psi}_4 = \begin{bmatrix} \frac{-p_x + i p_y}{E/c+mc} \\ \frac{p_z}{E/c+mc} \\ 0 \\ 1 \end{bmatrix}.$$

---

[6]Here, we have used the momentum p as the transform variable instead of the more usual wavenumber k, where $\mathbf{p} = \hbar\mathbf{k}$.

Show that the matrix products of the momentum operator matrix $\mathbf{M}$ with the spinors all vanish, provided that $E^2 = m^2 c^4 + p^2 c^2$.

At first, Dirac proposed that we simply ignore the negative-energy solutions as being unphysical. There is certainly precedent for this idea. We have, in elementary kinematics, had occasion to discount the existence of negative-time solutions that arise when computing the time it takes for a ball to fall from some height. Perhaps Dirac's negative energy solutions fall into that same category of extraneous solutions that can or should be ignored.

This initial interpretation of the Dirac equation 3.6 was called into question by the discovery of the positron by Carl Anderson in 1932.[7] After completing his doctoral thesis in 1931, Anderson constructed a cloud chamber and began an investigation of so-called cosmic rays. He and his mentor Robert Millikan were able to distinguish the thin tracks of the negatively charged electrons from the thicker tracks of positively charged protons from the curvature of the tracks in a strong magnetic field.

What puzzled Anderson and Millikan was the existence of thin tracks of positively charged particles that appear to have the mass of an electron. Today, we call these particles **antimatter** and a host of antiparticles, including antiprotons, have been observed. A potential reinterpretation of the Dirac equation assigns the positron solutions to the negative-energy solutions, thus accounting for all four of the spinor terms. While Dirac did not predict the existence of antimatter, it appears that his equation might be able to accommodate it without contrivance. We will revisit this idea subsequently.

3.2. Gauge Theory

Equation 3.6 defines the spacetime evolution of the spinor wavefunction of a free electron. What we would now like to do is consider the motion of an electron in an external electromagnetic field, with an eye towards producing an initial model for the hydrogen atom. Here, we are specifically thinking of treating the interaction of the electron with the electromagnetic field of the hydrogen nucleus (proton) but possibly also macroscopic fields produced by macroscopic sources. This turns out to be relatively simple and provides a framework for additional interactions.

---

[7]Anderson published "The apparent existence of easily deflectable positives" in *Science* in 1932 followed by "The positive electron" in the *Physical Review* in 1933. He was awarded the Nobel Prize in Physics in 1936 "for his discovery of the positron." Anderson shared the prize with Victor Franz Hess, who was cited "for his discovery of cosmic radiation."

We first remark that it was recognized very early in the development of quantum theory that the wavefunctions are not uniquely defined. This is due to the fact that all observable quantities, like the probability density, are constructed from the square of the wavefunction:

$$(3.10) \qquad P(\mathbf{x}) = |\psi(\mathbf{x})|^2 = \psi^\dagger(\mathbf{x})\psi(\mathbf{x}),$$

where the dagger here again means the Hermitian conjugate for the spinor wavefunctions of the Dirac equation and just the complex conjugate otherwise. Consequently, if we replace the initial wavefunction with one that is multiplied by an arbitrary phase, $\psi(\mathbf{x}) \to \psi(\mathbf{x})e^{i\theta}$, then the probability density defined in equation 3.10 is unaffected.

In 1918, the German mathematician Hermann Weyl attempted to unify Einstein's new geometrical theory of gravity with electromagnetism, motivated by similarities of the mathematical structures of the two theories.[8] What Weyl considered is what is now known as a local symmetry, where the phase factor θ is not simply a global constant, but is actually a function of space $\theta = \theta(\mathbf{x})$. Again, even if θ is a function of space, the probability density will not depend upon θ.

There is, though, the question of whether or not the transformed wavefunction remains a solution of the Dirac equation. This is obviously satisfied when θ is a constant, but when θ is a function of $\mathbf{x}$, we find an additional term in the Dirac equation due to the derivatives of θ:

$$(3.11) \quad \left[i\hbar \sum_{i=1}^{4} \sum_{s=1}^{4} \left(\gamma^i \right)_{rs} \frac{\partial}{\partial x^i} - mc \right] \psi_s(\mathbf{x}) e^{i\theta(\mathbf{x})}$$

$$= e^{i\theta(\mathbf{x})} \left\{ i\hbar \sum_{i=1}^{4} \sum_{s=1}^{4} \left(\gamma^i \right)_{rs} \left[\frac{\partial}{\partial x^i} + i \frac{\partial \theta(\mathbf{x})}{\partial x^i} \right] - mc \right\} \psi_s(\mathbf{x}).$$

At first glance, one might conclude that the additional terms spoil everything. The wavefunction ψ now solves an equation that is not the Dirac equation but there is a bit of magic available.

If we use what physicists call minimal coupling, we can add the interaction to the electromagnetic field by a modification to the derivative operator:

$$(3.12) \qquad \frac{\partial}{\partial x^i} \to \frac{\partial}{\partial x^i} + i \frac{e}{c} A_i,$$

---

[8]Weyl published his theory combining electromagnetism and gravity in 1918 in the *Sitzungsberichte der Königlich Preussischen Akademie der Wissenschaften zu Berlin*. Weyl investigated the consequences of rescaling the metric tensor $g_{ik} \to \lambda g_{ik}$ by a continous function λ; this has been rendered into English as a gauge transformation.

where e is the fundamental (electron) charge and A_i is the electromagnetic potential. We shall provide further details of why this is possible subsequently but, for the moment, we shall just state the result as fact.

If we now include this in our description of the electron, we can write the following:

$$(3.13) \quad \left\{ i\hbar \sum_{i=1}^{4} \sum_{s=1}^{4} \left(\gamma^i \right)_{rs} \left[\frac{\partial}{\partial x^i} + i\frac{e}{c}A_i \right] - mc \right\} \psi_s(\mathbf{x}) e^{i\theta(\mathbf{x})}$$

$$= e^{i\theta(\mathbf{x})} \left\{ i\hbar \sum_{i=1}^{4} \sum_{s=1}^{4} \left(\gamma^i \right)_{rs} \left[\frac{\partial}{\partial x^i} + i\frac{e}{c}A_i + i\frac{\partial \theta(\mathbf{x})}{\partial x^i} \right] - mc \right\} \psi_s(\mathbf{x}).$$

Recall now that the electromagnetic fields obtained from the potentials also had an ambiguity (see equation 2.28): replacing the potentials A_i with $A_i - \partial \phi / \partial x^i$ produces the same fields as the original potentials. If we define $\theta = -e\phi/c$, then the gauge transformation of the wavefunction also produces a gauge transformation in the potentials. Consequently, the spinor solutions are solutions of the Dirac equation, provided that we utilize the appropriate, covariant derivative.

We should probably place some sort of gaudy marker in the margin of the text at this point but we shall allow readers to perform that task in whatever sort of fashion suits their personal taste. We have arrived at a truly significant juncture. As we shall see, essentially all theories of elementary particles are based on the ideas that we have just covered:

(1) the equations of motion are invariant under Lorentz transformations,
(2) particles are described by wavefunctions that are invariant to gauge transformations,
(3) particles interact with fields that are also invariant to gauge transformations and
(4) these gauge invariances give rise to Noether's conserved currents.

For Maxwell's equations, we know that electric charge is conserved. This is, of course, an experimentally observed property of matter, that charge is neither created nor destroyed but, within the theory that describes electromagnetic phenomena, this property arises from the invariance of the equations to the gauge transformations.

We are now in possession of a mathematical mechanism to provide for other, experimentally observed conserved quantities. As the high-energy particle experimental community continues to conduct experiments in which various particles are forced to interact and the results of those interactions are observed, we shall find that there are other such conservation laws observed. We can build a theory that contains those properties simply by introducing a gauge transformation, via the covariant derivative. We shall encounter the covariant derivative again later in the text when we discuss Einstein's general theory of relativity, as it arises when we attempt to define derivatives that do not depend on the coordinate system in curved spaces. Einstein struggled for a long time to understand the nuances of these subtle, yet complex, ideas, so we shall postpone that discussion for a time. Nevertheless, it was Einstein who brought tensor calculus to physics and Weyl who recognized that the so-called minimal coupling of the electromagnetic field was precisely the covariant derivative that Einstein used in his general theory of relativity.

At this point, we can utilize some of the mathematical advances developed during the nineteenth century that will aid our progress. We have noted previously that mechanical systems were characterized by an energy $\mathcal{E}$ that is conserved and is generally made up of a kinetic T part that depends on velocities v_j and a potential $\mathcal{U}$ part that depends on positions x_j. The French mathematician Simon Pierre Lagrange found that one could derive the equations of motion of a system by defining a new function (now known as the Lagrangian) $\mathcal{L} = T - \mathcal{U}$. The equations of motion can now be obtained from derivatives of the Lagrangian:

$$(3.14) \qquad \frac{\partial \mathcal{L}}{\partial x_j} - \frac{d}{dt}\frac{\partial \mathcal{L}}{\partial v_j} = 0.$$

EXERCISE 3.12. As a simple example, consider the one dimensional harmonic oscillator. The kinetic energy of a mass m is just $T = \frac{1}{2}mv^2$ and the potential energy will be $\mathcal{U} = \frac{1}{2}k(x - x_0)^2$, where k is the spring constant and x_0 is the equilibrium point. Use equation 3.14 to show that the following equation describes the motion of the mass:

$$m\frac{d^2x}{dt^2} + k(x - x_0) = 0.$$

Moreover, there is an even more abstract principle at work, which was identified by the Irish physicist William Rowan Hamilton. If we define the **action** as follows:

$$(3.15) \qquad \mathcal{S} = \int_a^b dt\,\mathcal{L},$$

then the equations of motion can be derived from minimizing the action with respect to the pathway from a to b:

$$(3.16) \qquad \frac{\delta \mathcal{S}}{\delta x(t)} = 0.$$

We have used a notation in equation 3.16 where by δ we mean a variation in the action. This idea is, mathematically, essentially the same as the notion of a derivative but with respect to a function of the coordinates, not the coordinates themselves. What Hamilton proposed is that the observed trajectory $x(t)$ of a system, as it evolves from the time a to the time b, is the one that minimizes the action S.

It is also possible to make all of this relativistically correct. We recognize that time is not a relativistically invariant quantity, so we need to replace the integral over time in equation 3.15 with an integral over $d^4\mathbf{x}$ and the Lagrangian becomes a Lagrangian density. As a Lorentz scalar, the Lagrangian density will be invariant under Lorentz transformations, as will be the integral over spacetime.

The action that represents an electron in an electromagnetic field can be written as follows:

(3.17)
$$S = \int d^4\mathbf{x}\, i\hbar \sum_{i,r,s} \psi_r^\dagger(\mathbf{x}) \left\{ \left(\gamma^i\right)_{rs} \left[\frac{\partial}{\partial x^i} + i\frac{e}{c}A_i \right] - mc \right\} \psi_s(\mathbf{x}) - \tfrac{1}{4} \sum_{i,k} F_{ik}F^{ik}.$$

Here, the summations run, as before, over indices from one to four. The Dirac equation can be recovered by considering the variation of S with respect to $\psi^\dagger$ and Maxwell's equations can be recovered by considering the variation of the action with respect to the potentials A_i, which will then include the electron as a source term.

Equation 3.17 represents the action for quantum electrodynamics, one of the most successful physical theories developed to date.[9] The American physicist Richard Feynman developed a new approach to the formulation of quantum mechanics, in what has subsequently been called the path integral method. Feynman utilized an observation made previously by Dirac, that the exponential of the Lagrangian function was "analogous" to the propagator, or kernel function. That is, if one asks what is the electron wavefunction at a time infinitesimally (ϵ) later than some initial time t, then we could write

$$\psi(t+\epsilon, x) = \int dx\, e^{i\epsilon L/\hbar} \psi(t,x).$$

Feynman recognized that, by taking a series of infinitesimal steps, the wavefunction at some finite later time is just the integral of the Lagrangian,

---

[9]Sin-Itiro Tomonaga, Julian Schwinger and Richard P. Feynman were awarded the Nobel Prize in 1965 "for their fundamental work in quantum electrodynamics, with deepploughing consequences for the physics of elementary particles."

which is, of course, the action:

$$(3.18) \qquad \psi(t_b, x) = \int dx\, e^{i\mathcal{S}/\hbar} \psi(t_a, x).$$

The exponential of the action was not simply analogous to the propagator; it was (proportional to) the propagator. This representation of quantum mechanics, based on the action, had enormous aesthetic appeal to Feynman, because it also demonstrated that one could construct a valid theory without resort to computation of the electromagnetic fields. In this particular view, the fields are computational artifices; the only measurable quantities involve particle paths and not any fields that might be present.

One might wonder how valuable an expression like that in equation 3.18 might be in practice, or even how to interpret the exponential of a complicated function. What Feynman found was a natural definition in terms of the series:

$$(3.19) \qquad e^{i\mathcal{S}/\hbar} = 1 + \frac{i}{\hbar}\mathcal{S} + \frac{1}{2}\left(\frac{i}{\hbar}\right)^2 \mathcal{S}\mathcal{S} + \cdots.$$

This is a perturbation expansion, where one might implicitly assume that the series would converge reasonably rapidly. To expedite his calculations, Feynman developed a pictorial representation of the series terms that enabled him to perform the associated bookkeeping. Some examples are illustrated in figure 3.2.

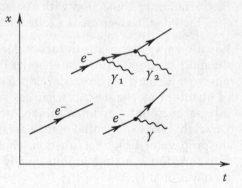

FIGURE 3.2. The pictorial representation of the terms in the series expansion of the exponential operator take the form of solid lines, representing electrons, interacting at vertices with the electromagnetic field, represented by wavy lines.

In Feynman's approach, one defined the electron to be in some initial state ψ_i and, after interaction with the electromagnetic field, was measured to be in a final state. The operator that transformed the system from initial to final was the exponential of the action. This operator is a unitary transformation, meaning that the magnitude of the wave function was unchanged. This is an essential attribute if we are to interpret the wavefunction probabilistically. Feynman's approach here was similar to that of previous workers, like the German theorist Werner Heisenberg, who pioneered a matrix approach to quantum mechanics.[10] Feynman used his pictorial representations not only as a method of bookkeeping but also to help him identify what he believed to be the physical picture represented by the mathematics.

His work, though, relied strongly on a probabilistic interpretation of the wavefunction, where other theorists, notably Schwinger and Tomonaga, had adopted a more formalistic approach to developing a dynamical theory of the electron. Schwinger, in particular, had developed a fully covariant theory using more formal mathematical tools. Initially, there was some confusion over whose approach was better but, in 1949, the American physicist Freeman Dyson published two papers that reconciled affairs: proving that the two approaches were mathematically equivalent.

3.3. Gyromagnetic Ratio

To this point, we have discussed some of the ingredients of the modern theory of the electron but have not discussed why this complicated theory is considered valid. We shall now rectify that situation. In 1922, the German physicists Otto Stern and Walther Gerlach conducted an experiment in which they found that a beam of neutral silver atoms passing through an inhomogeneous magnetic field was separated into two components. If we attribute any intrinsic magnetic field of a silver atom to that generated by an unpaired electron, then Stern and Gerlach's work can be interpreted as measuring the magnetic dipole moment of an electron with the mass of a silver atom instead of the mass of an electron. Subsequent experiments in hydrogen replicated Stern and Gerlach's original findings, without the complications of 46 other electrons. The electron does appear to possess an intrinsic magnetic dipole moment.

In Niels Bohr's initial planetary model of the atom, electrons orbited the nucleus with angular momentum $\mathbf{L}$ quantized to integral multiples of $\hbar$. The magnetic moment $\boldsymbol{\mu}$ of such an electron is given by

$$(3.20) \qquad \boldsymbol{\mu} = g \frac{-e}{2m_e} \mathbf{L} = -g\mu_B \frac{\mathbf{L}}{\hbar},$$

where $\mu_B = e\hbar/2m_e$ is called the Bohr magneton. The factor g, unimaginatively called the g-factor, is present to account for the fact that the experimental observations do not agree with Bohr's simple model. In fact, the experimentally measured moment was about twice the predicted value.

---

[10]Heisenberg was awarded the Nobel Prize in Physics in 1932 "for the creation of quantum mechanics, the application of which has, *inter alia*, led to the discovery of the allotropic forms of hydrogen."

An initial success of the Dirac equation was to produce a value of $g = 2$ when calculating the first-order term in the perturbation expansion. Subsequently, Schwinger provided the first calculation of the second-order correction,

$$(3.21) \qquad\qquad g = 2 + \frac{\alpha}{\pi},$$

where $\alpha = e^2/(4\pi\epsilon_0 \hbar c)$ is called the fine-structure constant.

> EXERCISE 3.13.   Consider a small ring of radius R rotating with a tangential velocity v that contains a uniformly distributed charge Q and mass M. What is the angular momentum L of the ring? The magnetic moment μ for a ring is simply the product of the current I and area A of the ring: $\mu = IA$. What is the gyromagnetic ratio μ/L for this classical ring? What would be the gyromagnetic ratio of a sphere? (Hint: do NOT perform any integrals.)

Recent experiments by the American physicist Gerald Gabrielse and his group members have established the value of g to extraordinary precision: better than a part per trillion. The experiments involve trapping a single electron in external electromagnetic fields, thereby eliminating any contributions from an atomic nucleus. Nominally, the experiment utilizes a Penning trap, named after the Dutch physicist Frans Michel Penning by the German-born physicist Hans Georg Dehmelt.[11] Dehmelt modified Penning's vacuum gauge to serve as an ion trap, using a uniform magnetic field along the z-direction and a quadrupole electrostatic field provided by a cylindrically symmetric cavity. The electrostatic potential is illustrated in figure 3.3.

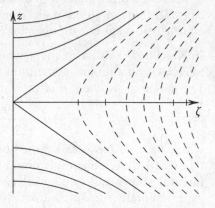

FIGURE 3.3. The quadrupole electrostatic potential in a Penning trap is symmetric around the z-axis. The isopotential surfaces form hyperbolas and the electrodes are biased negative (solid) or positive (dashed).

---

[11]Dehmelt shared one-half of the Nobel Prize in Physics in 1989 with Wolfgang Paul "For the development of the ion trap method." The other half was awarded to Norman F. Ramsey "for the invention of the separated oscillatory fields method and its use in the hydrogen maser and other atomic clocks."

EXERCISE 3.14.   For a classical charge in an electromagnetic field, the Lorentz force law governs the motion. The electric field for the Penning trap is obtained from the negative gradient of the potential $V = V_0(z^2 - (x^2 + y^2)/2)/2d^2$ and the magnetic field can be taken to be constant in the z-direction: $\mathbf{B} = \hat{z}B$. In *Mathematica* we can define the equations of motion as follows:

```
eqs={ x'[t]==vx[t],y'[t]==vy[t],z'[t]==vz[t],
vx'[t]==qm(x[t] Vod/2 + vy[t] B),
vy'[t]==qm(y[t] Vod/2 - vx[t] B),
vz'[t]==-qm(z[t] Vod}

ics={ x[0]==1,y[0]==0,z[0]==0,
vx[0]==0,vy[0]==1,vz[0]==0.05}
```

Here, we have lumped the charge to mass ratio e/m into a parameter qm, the potential strength V_0 and size d into the parameter Vod = V_0/d^2. We can solve these numerically, as follows:

```
soln=NDSolve[Join[eqs,ics]/.{qm->1,B->2,Vod->1},
{x,y,z,vx,vy,vz},{t,0,20}]
```

and plot them with

```
ParametricPlot3D[Evaluate[{x[t],y[t],z[t]}]/.soln,{t,0,20}]
```

Describe the resulting trajectory. How is the trajectory altered as the parameters change?

One still requires four quantum numbers to define the electron states within the potential but these are no longer the same as for a central force. The classical motion of a charged particle in the trap field can be described as harmonic motion in the z-direction and an epicycloidic motion in the x-y plane. This epicycloidic motion can be thought of as being composed of a large-radius circular motion, called the magnetron motion, and a higher-frequency, smaller radius motion called the cyclotron motion. We can assign quantum numbers (c, z, m, s) to the electron states within the cavity field, corresponding to the three motions and the electron spin.

Introducing microwave or radio-frequency fields into the cavity can drive transitions between the various energy levels that the electron can occupy. For the cavities devised by Gabrielse and his team, the largest differences between energy levels were determined by the cyclotron motion, parameterized by c, followed by the spin state s, as depicted in figure 3.4. An additional radio-frequency potential applied to the electrodes provided a mechanism for measuring the state of the electron.

It is remarkable that a signal from a single electron can be detected directly but this has, in fact, been accomplished. The trap itself is cooled to

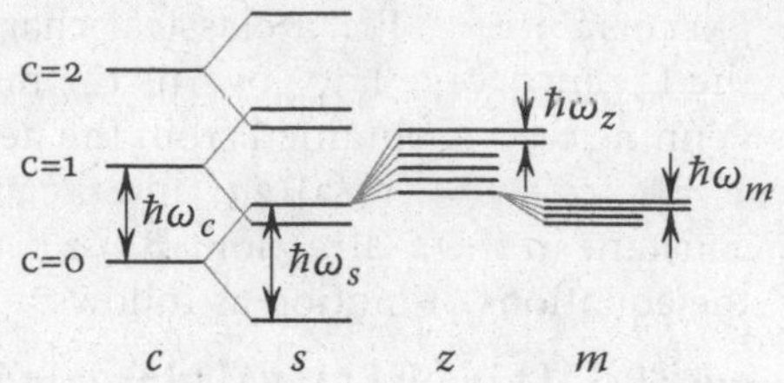

FIGURE 3.4. The level diagram for the Penning trap is defined by quantum numbers (c, z, m, s). For each level with a given c and s, there are manifolds of states of different z and m.

4 K by immersion in liquid helium and the electrons are further cooled within the trap to the order of 100 mK. Electrons within the trap can be further manipulated to eject one at a time until only a single electron remains. From their studies of the levels in what Gabrielse and team call *geonium*, the best value of the g-factor is currently[12]:

$$(3.22) \qquad g/2 = 1.001\,159\,652\,180\,73\,(28),$$

where the numbers in parentheses represent the experimental uncertainty (1 standard deviation). This is a precision of 3 parts in 10^{13}, arguably the most precise experimental number ever.

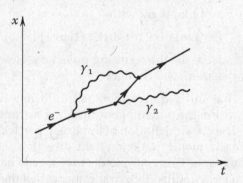

FIGURE 3.5. The anomalous g-factor arises from radiative corrections to the magnetic moment like that illustrated at right. In addition to the coupling to the external field (γ_2) there is an internal loop (γ_1).

The theoretical calculations of the g-factor produce a series in powers of α/π, one term is illustrated in figure 3.5. Schwinger, Tomonaga and Feynman each calculated the first order term and the most recent calculation by the Japanese physicists Tatsumi Aoyama, Masashi Hayakawa, Toichiro Kinoshita and Makiko Nio are complete to tenth order $(\alpha/\pi)^5$. This last result required the evaluation of 12, 672 Feynman diagrams, along with an automated procedure for identifying and computing the integrals.[13] Their results lead to a theoretical prediction of the g-factor as follows:

$$(3.23) \qquad g/2 = 1.001\,159\,652\,181\,78\,(77).$$

[12]This result was reported in *Physical Review Letters* in 2008.
[13]Aoyama *et al.* published their results in *Physical Review Letters* in 2012

Stunningly, the theoretical prediction in equation 3.23 only differs from the experimental measurement in equation 3.22 in the twelfth decimal place. The difference is only about one standard deviation, so to the current level of precision the experimental and theoretical results are in complete accord.

The radiative corrections that give rise to the anomalous magnetic moment of the electron also manifest themselves in atomic spectra. In 1947, the American physicist Willis Lamb utilized microwave techniques to measure the energy levels in hydrogen to high precision.[14] Lamb's experiment demonstrated an extraordinarily innovative strategy to measure the frequency shift in hydrogen. The low-lying energy levels in hydrogen are depicted in figure 3.6. The spectroscopic notation labelling the levels is derived from early measurements in alkali metals in which the spectral lines had particular characteristics. These were defined to be sharp, principal, diffuse and fundamental. Subsequently, the characteristics were identified to be associated with the orbital angular momentum L but the spectrographic identifications have been retained.

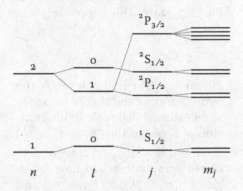

FIGURE 3.6. In the hydrogen atom, the eigenstates are labelled by the total angular momentum J, which is the sum of orbital L and spin S. The ground state has two magnetic substates $m_j = \pm\frac{1}{2}$.

In hydrogen, it was anticipated that the total angular momentum (orbital plus spin) would be the conserved quantity. In this regard, the Dirac equation predicts that the $^2S_{1/2}$ and $^2P_{1/2}$ levels would have the same energy, as both have $J = \frac{1}{2}$. In Lamb and Retherford's experiment, molecular hydrogen was dissociated in a tungsten oven and a jet of atomic hydrogen emerged from a small orifice. Shortly after emerging from the slit, the hydrogen beam was impacted by an electron beam at approximately right angles. The electron beam served to collisionally excite

[14]Lamb and his student Robert Retherford published their findings in the *Physical Review*. Lamb later shared the 1955 Nobel Prize in Physics "for his discoveries concerning the fine structure of the hydrogen spectrum." Lamb shared the prize with Polykarp Kusch, who won "for his precision determination of the magnetic moment of the electron."

higher-energy states in the atomic beam. These higher energy states decay rapidly through electric dipole radiation ($\Delta L = 1$) back to the ground state, with the exception of the $^2S_{1/2}$ state. This metastable state cannot decay through electric dipole processes and, so, persists for some time before decaying through weaker magnetic dipole or electric quadrupole processes.

As a result, hydrogen atoms in the beam that were excited into the $^2S_{1/2}$ state could survive to impact a metal foil beamstop, along with hydrogen atoms in the $^1S_{1/2}$ ground state. The metal foil was biased so that excited state atoms ejected an electron upon impact, whereas ground state atoms did not. Measuring the ejected electron current via a sensitive galvanometer allowed Lamb and Retherford to measure the content of $^2S_{1/2}$ excited state atoms within the beam.

If the hydrogen beam is illuminated with microwave radiation, it is possible to drive transitions between the metastable $^2S_{1/2}$ state and either of the ^2P states. These states then decay rapidly, with the result that one should see a decrease in the ejected electron current to the depopulation of the $^2S_{1/2}$ state. This is precisely what Lamb and Retherford observed.

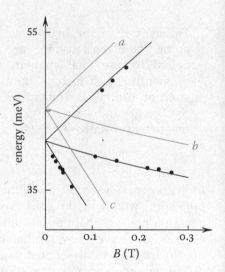

FIGURE 3.7. The energy levels of the hydrogen atom shift in the presence of an external magnetic field. Transitions between the $^2S_{1/2}(m_j = 1/2)$ level and the $^2P_{3/2}$ level were predicted (gray curves) to have energies $a : (m_j = 3/2)$, $b : (m_j = 1/2)$ and $c : (m_j = -1/2)$ substates. Shifting the theoretical curves by 4.135 meV (black) reproduces the experimental measurements (dots).

Applying an external magnetic field shifts the energy levels in hydrogen.[15] Using the Dirac equation, we can predict the energy differences for the $n = 2$ levels in hydrogen as a function of applied magnetic field B. These are the gray curves in figure 3.7. Lamb and Retherford observed

---

[15]The Dutch physicists Pieter Zeeman and Hendrik Antoon Lorentz were awarded the Nobel Prize in Physics in 1902 "in recognition of the extraordinary service they rendered by their researches into the influence of magnetism upon radiation phenomena."

quenching of the ejected electron current at particular microwave frequencies around 10 GHz ($\approx$ 41.35 meV) that changed depending upon the applied magnetic field but the experimental data did not agree with the theoretical predictions. If they shifted the predicted energy levels by about 1 GHz (4.135 meV), then the experimental results aligned with the shifted theoretical curves. This effect is known as the Lamb shift and is a consequence of the radiative corrections in quantum electrodynamics that cause the $^2S_{1/2}$ level to be higher than the $^2P_{1/2}$ level by about 4 meV, as depicted in figure 3.6.

EXERCISE 3.15.   An electric dipole antenna can be constructed from a thin, center-fed wire of total length a. The energy propagating to infinity is represented by the Poynting vector $\mathbf{E} \times \mathbf{H}$. The total power radiated can be calculated by integrating the Poynting vector over a sphere at large distance from the origin and shown to be given by the following:

$$P_e = \frac{(qa)^2}{12\pi\epsilon_0} \frac{\omega^4}{c^3},$$

where q here is representative of the maximal charge on the antenna and ω is the angular frequency. For a small current loop of radius a, the radiated field is a magnetic dipole field, with a power given by the following:

$$P_m = \frac{(q\omega\pi a^2)^2}{12\pi\epsilon_0} \frac{\omega^4}{c^5}.$$

What is the ratio of radiated powers P_m/P_e? What is the value of this ratio at a frequency $\hbar\omega = 10$ eV, if we take a to be the Bohr radius $a = (4\pi\epsilon_0\hbar^2)/(m_e e^2)$?

3.4. Mathematical Difficulties

An essential skill for practicing physicists is the ability to understand the physical meaning of mathematical equations. That is, what do the various entities represent? This ability is not so difficult to develop when dealing with elementary mechanics; the variables represent the position (and orientation) of some object as a function of time in a particular coordinate system. In these cases, one can readily visualize the physical system. For systems involving fields, there is no readily available means of visualizing the fields, so developing a physical understanding of the system is more challenging. For quantum systems, which are probabilistic at their heart, ascribing a physical meaning to the mathematics can prove to be an arduous task.

In truth, the Dirac equation has undergone several reinterpretations from Dirac's initial proposition that the equation represents the equation of

motion for an isolated electron in an electromagnetic field. In developing
the equation, Dirac needed to utilize a four-component spinor to represent
the electron in its two spin states. The additional two components were
seen to have negative energies, so initially they were simply ignored. Over
time, it became difficult to reconcile that these available states would not
be utilized in nature, so a new interpretation arose: the negative energy
states were populated, giving rise to a vacuum state in which the electron
under consideration was actually propagating in the presence of (unseen)
negative-energy electrons. This interpretation conjures up the image of
the vacuum as a seething cauldron of (virtual) particles through which an
electron must propagate.

The subsequent discovery of the positron provided another possible interpretation: the negative energy states represented positrons. As positrons
are not electrons, one could argue that the negative energy states were
simply not available to electrons, only to their antiparticles. One occasionally reads that the Dirac equation predicts the existence of antimatter
but this idea stretches the historical record significantly. It was only in
retrospect that physicists tried to fit positrons into the negative energy
states. An early idea, attributed to the American physicist John Wheeler,
was that positrons could be considered to be electrons travelling back-
wards in time. While this idea was not taken particularly seriously, Feynman adopted the notation in his pictorial representations: antiparticles
are drawn with the propagation direction inverted from those of particles.

In his Nobel Prize lecture, Feynman recounts some of the history of his
path to the theory of quantum electrodynamics. Initially, he was interested in resolving the problem of infinities that arise in classical electrodynamics, before eventually moving into the study of a quantum-mechanical
description of the electron. We have introduced the results of these studies: the creation of a systematic approach to calculating so-called matrix
elements based on a path-integral formulation involving the action. Feynman's approach was focussed on evolution of the wavefunction and, so,
was not in concert with the latest developments in quantum field theory produced by Tomonaga and Schwinger. In their manifestly covariant approach, the wavefunction was replaced with an operator-valued
field. This second-quantization approach was tied to powerful new ideas
in mathematics and, thus, offered the possibility of understanding more
deeply the connections between the physical theory and the underlying
mathematical structure. As a result, Feynman's more intuitive approach
was not immediately embraced by workers in the field.

The discrepancies between Feynman's approach and the operator method-ology were quickly reconciled by Freeman Dyson.[16] Despite their differ-ent mathematical formulations, Dyson demonstrated that the two theories were simply different mathematical representations of the same underly-ing theory. Moreover, Dyson reconciled the problems with infinities that arose in both theories.

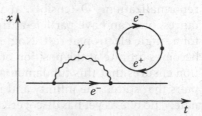

FIGURE 3.8. Possible higher-order graphs include the emission and re-absorption of a (virtual) photon by a propagating electron and sponta-neous pair (e^+e^-) production.

The perturbation series that arises from equation 3.19 can be organized according to the number of vertices that appear in the diagrams. The diagram in figure 3.8 contains four vertices and leads to integrals that are divergent. What Dyson recognized was that one could reorganize the different graphs in another fashion. For any graph, one can categorize the graph by its connectivity. If, like in figure 3.8, there is a disconnected portion (the e^+e^- pair), then the disconnected portion can be eliminated from contributing to the matrix element that defines the propagation of the electron.

Dyson also realized that infinities that arise from terms like that repre-sented by the emission and reabsorption of a photon by a propagating electron, illustrated in the lower half of figure 3.8, serve to modify the basic Green's function, or propagator. If one reorganizes the calculations, one obtains a fundamental graph, like the single-vertex graph illustrated in figure 3.2, that contains an electron line entering and leaving and a single photon. The electron lines should be interpreted not as the simple propagator that arises from the first term in the perturbation series but the sum of all of the contributions from various orders that modify the fundamental. Reorganizing the sum of graphs not by the number of ver-tices within the graph but by the number of lines entering and leaving the graph, Dyson was able to demonstrate that observable quantities derived from the theory were finite.

Dyson's machinations have been given the name **renormalization**. Feyn-man, among others, was skeptical about the validity of shuffling terms

---

[16]Dyson published two papers in 1949 in the *Physical Review*: "The radiation theories of Tomonaga, Schwinger and Feynman," and "The S matrix in quantum electrodynamics."

back and forth, believing that there might be some mathematical sleight-of-hand at work. Nevertheless, renormalization is now an accepted practice within theoretical physics and serves as a prerequisite for putative theories of new phenomena. That is, if a theory cannot be renormalized, then it is unlikely to have any validity as a physical theory.

There have been different attempts to understand the physical need for renormalization. Ostensibly, at very high momenta, or very short distances, one can have particle-antiparticle pair creation. This is true even for a single electron at rest. Nominally, the electric field of a point charge becomes infinite at the position of the charge; the process of renormalization ensures that this fundamental infinity is shielded by a sea of virtual pairs that soften the infinity and result in finite physical fields. Unfortunately, no one as yet has discovered any clear pathway to eliminating point charges from the theory. Feynman proposed replacing the delta functions that define the particle position with some other sort of function, gaussians say, that are finite at the origin and suitably narrow but this complicates the calculations and the functions are not uniquely defined.

As a result, most physicists have simply moved on to other topics, considering the problem of quantum electrodynamics to be essentially solved. There is a theory. It provides extraordinary agreement with experimental measurements. This is enough.

On Atoms

The experiments conducted by Ernest Rutherford's students Geiger and Marsden in 1911, in which they measured the scattering amplitude of α particles from gold foils produced the surprising result that α particles could be deflected to large angles from their original trajectories. Rutherford interpreted these results to mean that atoms possessed a nuclear structure, where the majority of the mass of an atom was concentrated in a small, positively charged core that was surrounded by a cloud of negatively charged electrons.[1] This discovery unleashed a firestorm of theoretical activity aimed at reconciling the obvious planetary models of the atom with the large collection of spectroscopic data that continued to accumulate.

Earlier, in 1885, the Swiss mathematician Johann Balmer discovered a pattern to four lines in the visible spectrum of hydrogen. He found that the wavelengths could be determined from the following formula[2]:

$$(4.1) \qquad \lambda = \frac{hn^2}{n^2 - 4},$$

where n is an integer from the series $n = 3, 4, 5, \ldots$ and h is a constant that Balmer established as $h = 364.56$ nm. From this formula, Balmer predicted that there should be a line for $n = 7$ at 397 nm, which, indeed, had just been observed by the Swedish spectroscopist Anders Ångström.

A more general formula was obtained by the Swedish physicist Johannes Rydberg, whose "Recherches sur la constitution des spectres d'émission desélements chimique" was presented to the Royal Swedish Academy of Science in November, 1889 and subsequently published by the Academy

---

[1]Rutherford published "The scattering of alpha and beta particles by matter and the structure of the atom," in the *Philosophical Magazine* in 1911. Rutherford was awarded the Nobel Prize in Chemistry in 1908 "for his investigations into the disintegration of the elements, and the chemistry of radioactive substances."

[2]Balmer published his "Notiz über die Spectrallinien des Wasserstoffs" in the *Annalen der Physik und Chimie*. He was 60 years old at the time.

© Mark A. Cunningham 2018
M.A. Cunningham, *Beyond Classical Physics*,
Undergraduate Lecture Notes in Physics,
https://doi.org/10.1007/978-3-319-63160-8_4

the following year.[3] In modern notation, the Rydberg formula can be written as follows:

$$(4.2) \qquad \frac{1}{\lambda} = R\left[\frac{1}{n_1^2} - \frac{1}{n_2^2}\right],$$

where R is now known as the Rydberg constant and $n_1 > n_2$ are principal quantum numbers. Balmer's formula could be seen to be a special case of the Rydberg formula, where $n_2 = 2$. At the time, Rydberg's formula was considered something of a novelty. It provided some hints as to the distribution of spectral lines but no real explanation of their reason for existence.

As a result, Rydberg did not benefit significantly from his discovery. He was, at the time of publication, an assistant lecturer (docent) at Lund University but had undertaken all of the responsibilities of the professor of physics position vacated by the retirement of Karl Albert Viktor Holmgren in 1897. The chair position was a royal appointment that was ultimately bestowed on the mathematician Albert Victor Bäcklund in 1900 by Bäcklund's good friend, Oscar II. This appointment was a blow to Rydberg, as Bäcklund's application had not been recommended for review by the referees and, ultimately, the University administration had supported Rydberg's candidacy. Rydberg took a position as an accountant at a local bank to help support his family but was named an extraordinary professor in 1901. His promotion to full professor came, at last, in 1909 but Rydberg received few accolades during his lifetime. He was never named to the Swedish Academy of Sciences and never awarded a Nobel Prize; he was, reportedly, nominated in 1917 but no prize was awarded that year. Rydberg was elected a Fellow of the Royal Society of London shortly before his death in 1919.

4.1. Hydrogen

The first workable model that reconciled Rydberg's spectroscopic formula was provided by the Danish physicist Niels Bohr who envisioned a planetary model for electrons, where the orbits were required to be quantized.[4] Bohr's model provided a definition of the Rydberg constant in

---

[3]Rydberg had, in fact, discovered his formula in 1887, evidenced by an appendix he attached to a request for financial support from the Royal Academy. His "Recherches..." was published in *Kungliga Svenska Vetenskapsakademien, Handlignar*.

[4]Bohr's "On the constitution of atoms and molecules" was published in three parts in the *Philosophical Magazine* in 1913. Bohr was awarded the Nóbel Prize in Physics in 1922 "for his services in the investigation of the structure of atoms and the radiation emanating from them."

equation 4.2 in terms of other fundamental constants:

$$(4.3) \qquad R = \frac{m_e Z^2 e^4}{4\pi \hbar^3},$$

where Z is the nuclear charge.

Bohr's model worked quite well for hydrogen but less well for heavier atoms. So, work to define a comprehensive theory of the atom continued. Most efforts concentrated around adapting classical mechanics, a well-studied mathematical theory. It should be no surprise that, when confronted with new problems, physicists will initially attempt to understand them within the framework of existing tools. This has proven to be a successful strategy in many areas of physics. Sometimes, though, it is necessary to recast things in a different light.

In our current discussion, this recasting was performed by the German physicist Werner Heisenberg in a pivotal paper in 1925.[5] In this remarkable work, Heisenberg set forth the defining ideas underlying quantum theory.

Heisenberg focussed on **observables**. A quantum system may exist in a particular state but the state itself is not something that can be directly observed. What Heisenberg recognized is that it is the transitions between states that give rise to measurable quantities. So, he dispensed entirely with notions like the position of the electron around the nucleus and, instead, focussed on defining the means for computing the transition amplitudes.

Heisenberg began with the statement that the energy of an observed photon, like the ones described by Balmer and Rydberg, represented the difference in energies between two discrete states in the system:

$$(4.4) \qquad \hbar\omega(n, n-\alpha) = \mathcal{E}(n) - \mathcal{E}(n-\alpha),$$

where the state labelled by the quantum number n decays to a lower state $n - \alpha$. Note that the frequency $\omega(n, n-\alpha)$ depends upon both the initial and final states, not just on one or the other. Classically, if one wants to ask about the position of an electron in the state n, then one would utilize the Fourier transform:

$$(4.5) \qquad x(n, t) = \sum_{\alpha=-\infty}^{\infty} X_\alpha(n) e^{i\omega(n)\alpha t}.$$

---

[5]Heisenberg's "Über quantentheoretische Umdeutung kinematischer and mechanischer Beziehungeen," was published in the *Zeitschrift für Physik* in 1925. Heisenberg was awarded the Nobel Prize in Physics in 1932 "for the creation of quantum mechanics, the application of which has, *inter alia*, led to the discovery of the allotropic forms of hydrogen."

Heisenberg, instead, suggests that such things could not ever be directly observed but what could be observed would be the transition amplitudes $X(n, n-\alpha)$ that depend on both the initial and final state. The observable quantity $|x(t)|^2$ would then be proportional to the following:

$$|x(t)|^2 \equiv \sum_\beta Y_\beta(n, n-\beta)e^{i\omega(n,n-\beta)t}$$

$$(4.6) \qquad = \sum_\alpha \sum_\beta X(n, n-\alpha)e^{i\omega(n,n-\alpha)t} X(n-\alpha, n-\beta)e^{i\omega(n-\alpha,n-\beta)t}.$$

Rydberg and the Swiss physicist Walther Ritz had developed a principle based on their observation of spectral lines that the lines in each element contain frequencies that are the sum or difference of other lines. Viewed in this context, we see that the Rydberg-Ritz combination principle implies the following:

$$(4.7) \qquad \omega(n, n-\beta) = \omega(n, n-\alpha) + \omega(n-\alpha, n-\beta).$$

As a result, the exponentials in equation 4.6 can be eliminated and we are left with Heisenberg's result for the combination of transition amplitudes:

$$(4.8) \qquad Y(n, n-\beta) = \sum_\alpha X(n, n-\alpha)X(n-\alpha, n-\beta).$$

Heisenberg did not really derive equation 4.8, instead he simply stated that it was "almost necessary."

When he finished the paper, Heisenberg gave it to the German physicist Max Born, who had supervised Heisenberg's habilitation research and was the chair in physics at Göttingen, for submission to the *Zeitschrift für Physik*. Born very quickly recognized equation 4.8 as the definition of matrix multiplication. With his former student Pascual Jordan, who had also studied under the mathematician Richard Courant, Born began to flesh out Heisenberg's idea. They published "Zür Quantenmechanik" just two months after Born first received Heisenberg's paper. Shortly thereafter, Heisenberg returned to Göttingen from visiting Niels Bohr in Copenhagen, whereupon he joined Born and Jordan in publishing "Zur Quantenmechanik II" by the end of the year.[6] The identification that the rules of quantum mechanics could be cast into a matrix form meant that

---

[6]Born was ultimately awarded the Nobel Prize in Physics in 1954 "for his fundamental research in quantum mechanics, especially for his statistical interpretation of the wavefunction. Born shared the prize with the German physicist Walther Bothe who was cited "for the coincidence method and his discoveries made therewith."

physicists could apply all of the mathematical tools that had been developed in the course of studying matrix algebras. Indeed, as Born and Jordan demonstrated, and shortly thereafter Dirac,[7] the matrix formulation provided a systematic framework for computing the transition elements, now known as matrix elements. Computing energy levels could be cast as a matrix eigenvalue problem, for which the mathematical groundwork had already been established. Rapid progress ensued.

> EXERCISE 4.1. Construct 3×3 matrices $A = \{A_{ik}\}$ and $B = \{B_{ik}\}$, where $i, k = 1, 2, 3$. Compute the products AB and BA. What are the values of $(AB)_{ik}$ and $(BA)_{ik}$?

Real, symmetric matrices and Hermitian matrices share the property that they can be converted by a similarity transformation to be diagonal, with real-valued elements.[8] (Hermitian matrices are those matrices that are equal to the transpose of their complex conjugates: $A_{ik} = A_{ki}^*$.) That is, for a given matrix A, we can find a unitary matrix V where the following is true:

$$(4.9) \qquad \Lambda = V^\dagger A V.$$

Here, the matrix Λ has the form $\Lambda = \text{diag}(\lambda_1, \lambda_2, \ldots, \lambda_n)$. The values λ_i are known as the eigenvalues of the matrix and the columns of V are the eigenvectors of A.

> EXERCISE 4.2. Consider the following matrix:
>
> $$A = \begin{bmatrix} 3 & 1+i & 0 \\ 1-i & 2 & 0.4 \\ 0 & 0.4 & 1 \end{bmatrix}.$$
>
> Use the Eigenvectors function to compute the eigenvectors of A. (Note that the *Mathematica* function returns the eigenvectors as a list. The matrix V is the transpose of the returned list.) Verify that the eigenvalues are real by computing equation 4.9. Demonstrate that V is unitary: $V^\dagger V = I$. Is $VV^\dagger$ also equal to the identity matrix? (Note: use the Chop function to suppress numerical noise in the results.) Show that, for each eigenvector v_i, we have $Av_i = \lambda_i v_i$.

Somewhat surprisingly, these results are also true for infinite-dimensional matrices, although as a practical matter one cannot solve an infinite-dimensional system in finite time. Hence, we shall most often truncate the

---

[7]Dirac's "The fundamental equations of quantum mechanics" was published in the *Proceedings of the Royal Society* in 1925 and "On the theory of quantum mechanics" was published in the same journal the subsequent year.

[8]Technically, a similarity transformation takes the form $A' = B^{-1}AB$, where the matrices A and A' are called similar. For unitary matrices, $U^\dagger \equiv U^{-1}$.

matrices to some finite size. This truncation has consequences but we shall defer discussion on this point for the time being.

The matrix formulation seems to be adequate to describe bound states of the hydrogen atom, which can be characterized as having negative energy with respect to the energy of a proton and an electron separated to large (infinite) distance. As depicted in figure 3.6, states identified by lines in the hydrogen spectrum lie at distinct, discreet energies. There are, though, states with positive energy in which the proton and electron are not bound; for these states the quantum numbers are not integers but can be real numbers. The Hungarian-American mathematician John von Neumann, a student of Hilbert, ultimately provided a solid mathematical foundation for quantum mechanics that encompassed both the discrete bound and continuous unbound states.[9] Moreover, von Neumann recognized that quantum states could be understood as vectors in an infinite-dimensional Hilbert space. The key characteristic of a Hilbert space, that extends the notion of a Euclidean space into functional spaces, is the existence of an inner product, i.e., one can compute a distance between two points within the space.

EXERCISE 4.3. The inner product of two vectors is a scalar computed by summing the product of the components:

$$\mathbf{x} \cdot \mathbf{y} = \sum_i x_i y_i.$$

Hilbert spaces provide for more elaborate definitions of the inner product. For example, functions that vanish on the boundary of the domain $0 \le x \le 1$ can expanded in a Fourier sine series:

$$f(x) = \sum_{n=1}^{\infty} a_n \sin(\pi n x).$$

The coefficients a_n can be obtained by computing the inner product (an integral) with the basis functions:

$$a_n = 2 \int_0^1 dx\, f(x) \sin(\pi n x).$$

Consider the function $f(x) = 1 - 16(x - 1/2)^4$. Compute the coefficients a_n. Compute approximations to $f(x)$ by summing 1, 5 and 11 terms. Plot the approximations and $f(x)$. Plot the differences between the approximations and the function.

---

[9]Von Neumann's *Mathematische Grundlagen der Quantenmechanik* was published in 1932.

EXERCISE 4.4. Functions that possess vanishing derivatives on the boundary of the domain $0 \le x \le 1$ can be expanded in a Fourier cosine series:

$$f(x) = a_0 + \sum_{n=1}^{\infty} a_n \cos(\pi n x).$$

The coefficients a_n can be obtained from the orthogonality properties of the basis functions:

$$a_0 = \int_0^1 dx\, f(x) \quad \text{and} \quad a_n = 2 \int_0^1 dx\, \cos(\pi n x) f(x).$$

Consider the function $f(x) = 1 + 4(5x^3 - 6x^4 - 3x^5 + 4x^6)$. Compute the coefficients a_n. Compute approximations to $f(x)$ by summing 3, 7 and 11 terms. Plot the approximations and $f(x)$. Plot the differences between the approximations and the function.

EXERCISE 4.5. The Laguerre polynomials $L_n(x)$ form a basis for the domain $0 \le x < \infty$ with a weighting function e^{-x}. That is, functions can be represented as a sum of Laguerre polynomials:

$$f(x) = \sum_{n=1}^{\infty} a_n L_n(x).$$

The coefficients a_n can be obtained through the orthogonality of the L_n:

$$a_n = \int_0^{\infty} dx\, e^{-x} L_n(x) f(x).$$

Show that $L_2(x)$ and $L_3(x)$ are orthogonal. Consider now the function $f(x) = \sin 3x$. Determine the coefficients a_n. Plot $f(x)$ and the approximations to $f(x)$ obtained by summing 10, 20 and 40 terms. Consider the domain $0 \le x \le 10$.

It is likely that many students will be dismayed by the notion we have expanded the definition of inner product to include integration, which is, undeniably, much more complicated than multiplication. As Hilbert and von Neumann were able to demonstrate, algebraically these operations have the same mathematical structure and permit the expansion of the concept from the field of real numbers to functions. Indeed, the whole of quantum mechanics can be described as Hermitian operators acting on states in a Hilbert space. The Hermitian bit is required if we want to obtain real eigenvalues, i.e., observable quantities. While the mathematical foundations are more complex, there are a host of new mathematical tools at one's disposal to help understand the structure of the theory without forever being bogged down in calculational details.[10]

---

[10]Part of the somewhat tepid response to Feynman's original articles on quantum electrodynamics arose because he did not frame his theory in terms of operators on a functional

These rather remarkable results demonstrate that the ideas of a coordinate basis can be extended into functional spaces. This provides a systematic strategy for solving partial differential equations, for example. The orthogonal function decompositions that we encountered in the previous exercises are only possible because the sine, cosine and Laguerre polynomial functions provide bases on particular domains. That means that *any* (smooth) function on the domain can be expanded in terms of a unique set of coefficients. This will, of course, be most useful if the number of terms required for a particular accuracy is small.

In 1939, Dirac created a new notation, known as the bra-ket notation that concisely denotes the concepts we are discussing. Vectors in the Hilbert space are defined as kets: $|\alpha\rangle$, labelled here by a single character but will generally have many parameters that define the state. Dual vectors are defined as bras: $\langle\beta|$. The inner product is concisely defined as $\langle\beta|\alpha\rangle$. Students will have to be wary that this notation implicitly involves integrations over functional spaces. Nevertheless, in this notation, the matrix elements of an (Hermitian) operator H will be defined as follows:

$$(4.10) \qquad H_{\beta\alpha} = \langle\beta|H|\alpha\rangle,$$

where we note that $H_{\beta\alpha} = H_{\alpha\beta}^*$.[11] The connection to the work of Erwin Schrödinger, whose initial development of the wavefunction led to early successes in quantum theory is, in Dirac's notation, simply given by the following:

$$(4.11) \qquad \psi_\alpha(\mathbf{r}) = \langle\mathbf{r}|\alpha\rangle.$$

That is, the wavefunction (labelled by the quantum numbers α) is simply the representation of the state $|\alpha\rangle$ on the coordinate space. This relatively simple connection explains why, despite the vastly different formulations of quantum mechanics by Heisenberg and Schrödinger, one obtains precisely the same results from both approaches. The Heisenberg and Schrödinger methods are simply different mathematical representations of the same system of equations.

We haven't as yet discussed Schrödinger's contributions to quantum theory in any detail but his method provided a means for studying the time evolution of the wave function $\psi(t, \mathbf{r})$, using the Hamiltonian of the system as the basis for his approach.[12] Recall that the Lagrangian $\mathcal{L}$ is defined as

---

space, as did Schwinger and Tomonaga. Dyson, nonetheless, recognized the connection and provided impetus to using Feynman's techniques as computational aids.

[11] We use a notation here that emphasizes the matrix element mathematical structure but α and β are not, in general, integer indices.

[12] Schrödinger published "Quantisierung als Eigenwertproblem" in *Annalen der Physik* in four parts in 1926. Schrödinger and Dirac were awarded the Nobel Prize in Physics in 1933 "for the discovery of new productive forms of atomic theory."

the difference between the kinetic and potential energies of the system: $\mathcal{L} = \mathcal{T} - \mathcal{U}$. The Hamiltonian $\mathcal{H}$ is the sum: $\mathcal{H} = \mathcal{T} + \mathcal{U}$. Application of the Hamiltonian operator in quantum mechanics obtains the energy of the quantum state:

$$(4.12) \qquad\qquad \langle \alpha | \mathcal{H} | \alpha \rangle = \mathcal{E}_\alpha.$$

Schrödinger further proposed that the Hamiltonian defines time evolution through the following equation:

$$(4.13) \qquad\qquad i\hbar \frac{\partial}{\partial t} \psi(t, \mathbf{r}) = \mathcal{H}\psi(t, \mathbf{r}).$$

This equation has a formal solution that can be written immediately:

$$(4.14) \qquad\qquad \psi(t, \mathbf{r}) = e^{-i\mathcal{H}t/\hbar} \psi(0, \mathbf{r}).$$

This result leads to the notion that the Hamiltonian is the (time) propagator for the wavefunction. We can also see that, if we are to interpret the wavefunction in a probabilistic sense, that the operation of the Hamiltonian operator will not affect the magnitude of the wavefunction: it is clearly a unitary operator.

Schrödinger was able to compute the spectrum for the hydrogen atom using his methodology by representing the electromagnetic field of the proton as simply the Coulomb field of a point charge. He found that the states would have the following energies:

$$(4.15) \qquad\qquad \mathcal{E}_n = -\frac{m_e Z^2 e^4}{2\hbar^2 n^2}.$$

Taking differences between states with different values of the principal quantum number n yields the experimentally supported Rydberg formula in equation 4.2.

As Heisenberg had noted in his initial paper, and subsequently refined with Born and Pascual, the use of matrix algebra meant that multiplication of transition amplitudes did not commute. If A and B are matrices, then, in general, $AB \neq BA$. In particular, for the position x_i and momentum p_k, we find the following:

$$(4.16) \qquad\qquad x_i p_k - p_k x_i = i\hbar \delta_{ik}.$$

This remarkable formula encapsulates Heisenberg's concept of observable elements. This particular example indicates that, at the microscopic level, quantum states cannot be, simultaneously, eigenstates of both position and momentum. From a physical viewpoint, we can interpret this to mean that, if we want to measure something about a quantum state, we must interact with it via some external means: bounce a photon from it, for example. The process of interaction, though, changes the internal quantum state, so there are limits to our ability to extract information about the precise state of the system. This limitation has been deemed the Heisenberg *uncertainty principle*.

If we look a bit more closely at equation 4.16, we can notice that $\hbar$ is a small number. So, in the limit of macroscopic positions and momenta, we see that the position and momentum (almost) do commute; one can simultaneously determine the position and momentum of a macroscopic object. Niels Bohr called this property the **correspondence principle**. In the limit of macroscopic objects, the quantum theory should reproduce Newtonian mechanics. Heisenberg's matrix formulation of quantum mechanics includes this principle implicitly.

4.2. Many-body Problems

The concept that quantum states can be described as vectors within a Hilbert space and that observables are obtained through the action of Hermitian operators has proven to be a very powerful approach to representing quantum phenomena. We have seen that one can obtain the energies of the hydrogen atom but it is also possible to study transitions between the states. These arise primarily from electric dipole radiation but can also be obtained through electric quadrupole or magnetic dipole radiation, although at reduced intensities. Indeed, the relative intensities of lines in the hydrogen spectrum are largely explained by computing the matrix elements associated with these operators.

Moreover, as spectroscopists developed more precise instruments, it became clear that spectral lines actually possessed structure, known today as fine structure and hyperfine structure, that reflects transitions between the the sublevels associated with the hydrogen energy levels depicted in figure 3.6. The energies associated with these transitions depend upon applied external fields. This is not terribly surprising. If we go back to the Dirac equation, there is a direct coupling to the external potential A_i and, for large enough external fields, it is a reasonable conclusion that the electron wavefunction will be affected.

Indeed, application of external fields provides us with a means for systematically exploring potential terms in the Hamiltonian. We've seen that the central force provided by the proton yields an energy that depends solely upon the principal (radial) quantum number n. Adding an additional electric or magnetic field will break the rotational symmetry and give rise to terms that will be dependent upon the other quantum numbers. Quantum mechanics was generally quite successful in explaining these effects in hydrogen.

The list of elements extends significantly beyond hydrogen, though. Today, we entertain the idea that an atom is composed of a nucleus of charge Ze and Z electrons. As Newton and others discovered in their studies of the gravitational three-body problem, there are no analytic solutions available to the equations of motion when the number of interacting bodies exceeds two. Thus, even for helium, with just two electrons, one cannot simply hope to find a quantum analog of a well-known, classical solution.

EXERCISE 4.6. Singly ionized helium has only a single electron. The $n = 2, 3, 4$ levels have energies of 40.81, 48.37 and 51.02 eV above the ground state. Is this consistent with the Rydberg formula?

1 H																	2 He
3 Li	4 Be											5 B	6 C	7 N	8 O	9 F	10 Ne
11 Na	12 Mg											13 Al	14 Si	15 P	16 S	17 Cl	18 Ar
19 K	20 Ca	21 Sc	22 Ti	23 V	24 Cr	25 Mn	26 Fe	27 Co	28 Ni	29 Cu	30 Zn	31 Ga	32 Ge	33 As	34 Se	35 Br	36 Kr
37 Rb	38 Sr	39 Y	40 Zr	41 Nb	42 Mo	43 Tc	44 Ru	45 Rh	46 Pd	47 Ag	48 Cd	49 In	50 Sn	51 Sb	52 Te	53 I	54 Xe
55 Cs	56 Ba		72 Hf	73 Ta	74 W	75 Re	76 Os	77 Ir	78 Pt	79 Au	80 Hg	81 Tl	82 Pb	83 Bi	84 Po	85 At	86 Rn
87 Fr	88 Ra		104 Rf	105 Db	106 Sg	107 Bh	108 Hs	109 Mt	110 Ds	111 Rg	112 Cn	113 Nh	114 Fl	115 Mc	116 Lv	117 Ts	118 Og

57 La	58 Ce	59 Pr	60 Nd	61 Pm	62 Sm	63 Eu	64 Gd	65 Tb	66 Dy	67 Ho	68 Er	69 Tm	70 Yb	71 Lu
89 Ac	90 Th	91 Pa	92 U	93 Np	94 Pu	95 Am	96 Cm	97 Bk	98 Cf	99 Es	100 Fm	101 Md	102 No	103 Lr

FIGURE 4.1. The elements are aligned vertically into columns with similar chemical properties. Hydrogen-like elements are on the left and noble gases are on the right. Traditionally, the Lanthanide and Actinide series are excised from the table to avoid having the width extend too far to be printable.

A significant clue as to the structure of the wavefunctions of multi-electron atoms comes from the periodic table illustrated in figure 4.1. Chemists discovered that elements tended to combine in a manner that prescribed integer relationships: CH_4, NH_3 and OH_2, for example. Elements that reacted with similar integer relationships could be ordered into columns.

All elements in the first column form molecules (hydrides) of the form XH, where X is any of the elements.

The observed chemical relationships means that, in multi-electron atoms, not all electrons can have the same quantum number as the hydrogen ground state. Indeed, as we have mentioned earlier, a model that explains the behavior is one in which electrons are distributed individually into the quantum states defined previously, with the additional caveat that there is a duplexity arising from the electron spin.

The most obvious way to describe multi-electron states is to use define them as tensor products of single-electron states. Mathematically, such a space is known as a Fock space, after the Russian physicist Vladimir Aleksandrovich Fock.[13] Given states $|\alpha\rangle$ in a Hilbert space for a single electron, we can construct states for N electrons as products:

$$(4.17) \qquad |\Psi\rangle = |\alpha\rangle|\beta\rangle\cdots|\nu\rangle,$$

where, notionally, electron 1 is in the state $|\alpha\rangle$, electron 2 is in the state $|\beta\rangle$, etc. What is observed from the periodic table is that not all electrons can be in the state $|\alpha\rangle$. The Austrian physicist Wolfgang Pauli recognized that this can be accomplished mathematically by requiring the wavefunction to be antisymmetric under the exchange of particles.[14]

In equation 4.17, we have ordered the states according to electron number but, in practice, electrons are indistinguishable. There is no means by which we could select one electron from a pool of N electrons and identify it as the "third" one, according to our numbering scheme implied in equation 4.17. A mathematical mechanism for enforcing the Pauli principle is antisymmetrizing the wave function. Hence, we can write that two-electron electron wave functions must have the form

$$(4.18) \qquad |\alpha\beta\rangle = \frac{1}{\sqrt{2}} (|\alpha\rangle|\beta\rangle - |\beta\rangle|\alpha\rangle),$$

where the square root factor preserves the normalization. If we try to place both electrons into the same state $|\alpha\rangle$, the wavefunction vanishes and the sign of the wavefunction changes upon exchange: $|\beta\alpha\rangle = -|\alpha\beta\rangle$. More generally, the N-electron wavefunction can be obtained from the

---

[13] Fock published his "Konfigurationsraum und zweite Quantelung" in the *Zeitschrift für Physik* in 1932.
[14] Pauli published "Über den Zusammenhang des Abschlusses der Electronengruppen im Atom mit der Komplexstruktur der Spektren" in *Zeitschrift für Physik* in 1925. He was awarded the Nobel Prize in Physics in 1945 "for the discovery of the Exclusion Principle, also called the Pauli Principle."

determinant of an $N \times N$ matrix:

$$(4.19) \qquad |\alpha\beta\cdots\nu\rangle = \frac{1}{N!} \begin{vmatrix} \alpha(1) & \beta(1) & \cdots & \nu(1) \\ \alpha(2) & \beta(2) & \cdots & \nu(2) \\ \vdots & \vdots & \ddots & \vdots \\ \alpha(N) & \beta(N) & \cdots & \nu(N) \end{vmatrix}.$$

This construction is generally called the Slater determinant, after the American physicist John C. Slater, although the form had also been utilized by Heisenberg and Dirac previously.

EXERCISE 4.7. Suppose that an electron could occupy four different states: $|\alpha\rangle$, $|\beta\rangle$, $|\gamma\rangle$ or $|\delta\rangle$. What are the possible antisymmetric states available to two electrons?

EXERCISE 4.8. Create a 4×4 matrix with elements $a_1,\ldots,a_4$ through $d_1,\ldots,d_4$, representing the four states available to four electrons. Compute the determinant. Identify terms in which particles 1 and 4 are exchanged, e.g., $a_1\cdots d_4 \to a_4\cdots d_1$. Convince yourself that the wavefunction is antisymmetric.

We can obtain significant insights into the nature of quantum phenomena by studying the excited state spectrum of helium. We have noted that the duplexity problem required the introduction of a new quantum number that has been called (for better or worse) spin. The name itself evokes rotation, so we might ask if the spin of the electron is in any way an angular momentum? This is not to say that the electron is a small, rotating bead of some form. Rather, does the electron field entity possess intrinsic angular momentum?

We are constructing product states to describe multi-electron atoms. We can guess that they might look something like those we utilized to describe the single-electron hydrogen atom and, historically, this was the pathway followed. So, we have labelled the hydrogen states with the quantum numbers n, l, m and s, with the s arising from the subsequent need to explain the periodic table. What happens now when we try to combine two electrons into a state?

As it happens, there is a significant body of mathematics devoted to this question. It is related to addition theorems of the spherical harmonic functions that we mentioned previously but it can also be obtained through an understanding of group theory. If we ask what sort of states can be obtained by combining two spin $1/2$ things, it turns out that there are five possibilities:

(1) four spin singlets ($s = 0$),
(2) two spin doublets ($s = 1/2$),

(3) one doublet and two singlets,
(4) one triplet ($s = 1$) and one singlet or
(5) one quadruplet ($s = 3/2$).

If the spin is an angular momentum, then it will add to any orbital angular momentum present in the system. Traditionally, the total angular momentum is defined by the symbol **J** and the orbital angular momentum by the symbol **L**. If the spin **S** is also an angular momentum, then we would expect that the total angular momentum **J** = **L** + **S**, from Noether's theorem, would be a conserved quantity. What this means is that we would expect the operator **J** to commute with the Hamiltonian operator, giving rise to quantum states that can be labelled by j and the projection of **J** upon some axis m_j.

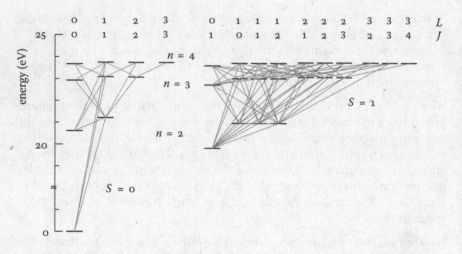

FIGURE 4.2. Energy levels in helium can be labelled by the total angular momentum J. Persistent transitions between levels are identified by the gray lines. The $S = 0$ and $S = 1$ manifolds of states are separated to improve the visibility of the diagram.

The spectrum of helium is significantly more complicated than that of hydrogen, as can be seen from figure 4.2, where the low-lying states have been assigned L, S and J values. These states can all be described by the promotion of a single electron from the 1s ground state into a hydrogen-like excited state. What we observe is that the first excited state (19.8 eV) has $(L, S, J) = (0, 1, 1)$. In spectroscopic notation, this would be denoted $^3S_1$. That is, it is a spin triplet with orbital angular momentum of zero and total angular momentum of one. Indeed, application of an external magnetic field can unambiguously determine that the state at 19.8 eV resolves into three components.

Moreover, the transitions between states also conform to the single-particle excitation model. For example, the first excited state $^3S_1$ cannot decay to the ground state $^1S_0$ via electric dipole radiation; the transition is only possible through a magnetic dipole mechanism. As a result, the 19.8 eV line is greatly reduced in intensity to those electric dipole transitions indicated in figure 4.2. In fact, early researchers were led to the conclusion that there might be two forms of helium: called orthohelium and parahelium, and we have emphasized the disjoint nature of the spectrum in our representation in figure 4.2.

EXERCISE 4.9.   The neutral helium (He I) spectral lines can be obtained through the NIST Atomic Spectral Database. What are the relative intensities of the transitions from the $|1s2p\rangle$ states to the $|1s1s\rangle$ ground state?

EXERCISE 4.10.   The helium spectrum indicates that the intrinsic angular momentum of the two electrons can be coupled either to produce states with no intrinsic angular momentum or an intrinsic angular momentum of $1\hbar$. These couple to the orbital angular momentum of the state, producing multiplets of total angular momentum J. Identify the configurations of the states illustrated in figure 4.2. What are the rules for coupling $S = 1$ intrinsic angular momentum to an $L = 2$ (D) state? That is, what values of J are available?

We have seen a glimpse of the difficulty in dealing with multi-electron atoms. There are no analytic solutions available, only approximations. A systematic scheme for dealing with the complexity was initially proposed by the English physicist Douglas Hartree in 1928.[15] Hartree called his approach the self-consistent field method. Essentially, one begins with (guesses) an initial configuration and then systematically computes updates until the process converges. Hartree's initial efforts were not terribly successful until the method was improved to utilize antisymmetrized states like those proposed by Slater. The method is now generally known as the Hartree-Fock method.

The complexity arises from the fact that the Hamiltonian for the N-electron problem contains N copies of the hydrogen Hamiltonian, plus the sum over all of the electron-electron Coulomb repulsion terms:

$$(4.20) \qquad \mathcal{H}_N = \sum_{i=1}^{N} \mathcal{H}_i + \frac{1}{4\pi\epsilon_0} \sum_{i<k} \frac{e^2}{|\mathbf{r}_i - \mathbf{r}_k|},$$

---

[15]Hartree published "The wave mechanics of an atom with a non-Coulomb central field" in two parts in the *Mathematical Proceedings of the Cambridge Philosophical Society* in 1928.

where the single-particle Hamiltonian, as defined by Schrödinger, is given by the following expression:

$$(4.21) \qquad \mathcal{H}_i = -\frac{\hbar^2}{2m}\nabla_i^2 - \frac{1}{4\pi\epsilon_0}\frac{Ze^2}{|\mathbf{r}_i|}.$$

The total N-electron wavefunction is a vector in the infinite-dimensional Fock space. Finite calculations require restrictions to a finite number of representative basis states. For hydrogen-like atoms (first column in the periodic table), the eigenstates of the multi-electron atoms are well described by a relatively few hydrogen eigenstates. As one moves towards the center of the periodic table, particularly for metallic atoms, the calculations become much more involved.

The advent of modern computers has given new life to the Hartree-Fock method. As one might imagine, iterating complex calculations until they converge is an activity in which humans do not excel. Fortunately, computers have no sense of tedium and are quite capable of repetitious calculations without algebraic error. A great deal of effort has gone into devising efficient numerical techniques and robust basis functions to ensure that the algorithm can converge to a solution. We shall discuss these calculations in somewhat more detail when we investigate molecular phenomena.

4.3. Density-Functional Theory

An alternative approach to understanding multi-electron atoms involves using the electron density instead of the wavefunction. We have asserted that the square of the wavefunction can be interpreted as the probability density for finding an electron at some position $\mathbf{r}$. Notionally, the electron density is just the integral over the squared wavefunction:

$$(4.22) \qquad \rho(\mathbf{r}_1) = \int d^3\mathbf{r}_2 \cdots \int d^3\mathbf{r}_N \, |\Psi(\mathbf{r}_1,\ldots,\mathbf{r}_N)|^2,$$

where the density is subject to the following normalization:

$$(4.23) \qquad N = \int d^3\mathbf{r}_1 \, \rho(\mathbf{r}_1),$$

where N is the total number of electrons. As a matter of practice, the function ρ is a function of three spatial dimensions where the wavefunctions are functions of $3N$ dimensions, suggesting that the computational effort associated with finding eigenstates could be greatly reduced.

The typical assumption in molecular physics is that one can simplify the problem by separating the electronic and nuclear degrees of freedom. This assumption rests on the notion that the nuclei are vastly heavier than the electrons and, thus, the electron density will equilibrate orders of magnitude more rapidly than any apparent motion of the nuclei. Thus, even if you want to study the impact of xenon atoms on a gold target, you could successfully study this problem by a series of stepwise increments of the xenon-gold nuclear distance and solving for the electron density at each (fixed) location of the nuclei. This approach is known as the Born-Oppenheimer approximation.

For molecules, we need to modify equation 4.21 to include the separate nuclear contributions to the potential:

$$(4.24) \qquad \mathcal{H} = -\sum_i^N \frac{\hbar^2}{2m}\nabla_i^2 + \frac{1}{4\pi\epsilon_0}\sum_i^N\sum_k^N \frac{Z_k e^2}{|\mathbf{r}_i - \mathbf{R}_k|} + \frac{1}{4\pi\epsilon_0}\sum_{i<k}\frac{e^2}{|\mathbf{r}_i - \mathbf{r}_k|},$$

where $\mathbf{R}$ represents the nuclear coordinates. The three terms on the right-hand side are termed the kinetic $\mathcal{T}$, electron-nuclear potential $\mathcal{V}_{en}$ and electron-electron potential $\mathcal{V}_{nn}$, respectively. Equation 4.24 describes the electron energy $\mathcal{E}$, the total energy of the system must include the nuclear-nuclear repulsion term $\mathcal{V}_{nn}$:

$$(4.25) \qquad W = \mathcal{H} + \frac{1}{4\pi\epsilon_0}\sum_{i<k}\frac{Z_i Z_k e^2}{|\mathbf{R}_i - \mathbf{R}_k|}.$$

Note that there is no kinetic term due to the nuclear centers in equation 4.25; they are fixed in space in the Born-Oppenheimer approximation. In 1939, Feynman proved what is known as his electrostatic theorem:

$$(4.26) \qquad \frac{dW}{d\mathbf{R}_i} = -\frac{1}{4\pi\epsilon_0}\sum_{i\neq k}\frac{Z_i Z_k e^2(\mathbf{R}_i - \mathbf{R}_k)}{|\mathbf{R}_i - \mathbf{R}_k|^3} - \frac{Z_i e^2}{4\pi\epsilon_0}\int d^3\mathbf{r}_1 \, \rho(\mathbf{r}_1)\frac{\mathbf{r}_1 - \mathbf{R}_i}{|\mathbf{r}_1 - \mathbf{R}_i|^3}.$$

This is precisely the result that one would obtain from a *classical* calculation of the electrostatic potentials due to the nuclear point charges and the electron density, lending further credence to the applicability of the Born-Oppenheimer approximation.

Initial models based on the idea of utilizing the electron density were developed by Llewellyn Thomas and, independently, by Enrico Fermi in the late 1920s but the idea was rejuvenated, particularly for molecular physics, by the Austrian-American physicist Walther Kohn in the 1960s.[16] On sabbatical in Paris at the École Normale Supérieure, Kohn proved two remarkable theorems, together with Pierre Hohenberg. The first theorem states that the electron density $\rho(\mathbf{r})$ defines the potential V in the Hamiltonian up to an additive constant. The second theorem states that the

---

[16]Walther Kohn was awarded the Nobel Prize in Chemistry in 1998 "for his development of the density-functional theory," and shared the prize with John Pople, who was cited "for his development of computational methods in quantum chemistry."

ground state electron density is defined by the minimum with respect to variation of the density. Upon returning to his home institution in San Diego, Kohn and his postdoctoral assistant Lu Sham then worked out the details of how to construct the density-functional theory equations.

EXERCISE 4.11. Physicists tire rapidly when forced to carry factors of $1/4\pi\epsilon_0$ around everywhere. Define the dimensionless quantity $\zeta = r/a_0$, where a_0 is the Bohr radius:

$$a_0 = \frac{4\pi\epsilon_0\hbar^2}{m_e e^2}.$$

Rewrite equation 4.24, substituting the dimensionless variables everywhere. Show that the equation takes on a simpler form.

Kohn and Sham demonstrated that the ground state of an N-electron system could be determined from the energy as a functional of the density:

$$(4.27) \qquad \mathcal{E}[\rho] = \int d^3\mathbf{r}_1\,\rho(\mathbf{r}_1)v(\mathbf{r}_1) + T[\rho] + \mathcal{V}_{ee}[\rho],$$

where v is an effective potential, that can include external potentials. The ground state density will be the density that minimizes the functional derivative:

$$(4.28) \qquad \lambda = v(\mathbf{r}_1) + \frac{\delta(T[\rho] + \mathcal{V}_{ee}[\rho])}{\delta\rho(\mathbf{r}_1)}$$

subject to the constraining equation 4.23. The λ in equation 4.28 is the Lagrange multiplier used to enforce the constraint.

EXERCISE 4.12. Suppose that we seek the minimum of the function $f(x,y)$ subject to the constraint $g(x,y) = 0$. This can be rewritten as a function $h(x,y,\lambda) = f(x,y) - \lambda g(x,y)$. Minimization of the unconstrained function h with respect to x, y and λ will minimize f and enforce the constraint g. Consider the following script:

```
f1[x_,y_]:= x y^2 + y
g1[x_,y_]:= x^2+y^2 - 2
delf1=Grad[f1[x,y],{x,y}]
delg1=Grad[g1[x,y],{x,y}]
soln=Solve[{delf1==lam delg1, g1[x,y]==0},{x,y,lam}]
ans=Table[f1[x,y]/.soln[[i]],{i,1,Length[soln]}]
mina=Min[ans]
locmin=Position[ans,mina]
```

Plot the functions f1 and g1. Compute the minimum. Is the constraint satisfied?

Kohn and Sham begin by first solving the problem with the electron-electron Coulomb interactions removed. That is, by solving for the wavefunctions of the following equation:

$$(4.29) \qquad [-\nabla^2 + v_{\text{eff}}]\psi_i = \varepsilon_i \psi_i,$$

where v_{eff} is the effective potential. The density for this approximation is constructed from the ψ_i:

$$(4.30) \qquad \rho(\mathbf{r}_2) = \sum_{i,s} |\psi(\mathbf{r}_i, s_i)|^2,$$

where we sum over both spatial and spin coordinates. Then, the effective potential can be updated:

$$(4.31) \qquad v_{\text{eff}}(\mathbf{r}_2) = v(\mathbf{r}_2) + \int d^3\mathbf{r}_1 \, \frac{\rho(\mathbf{r}_1)}{|\mathbf{r}_2 - \mathbf{r}_1|} + v_{xc}(\mathbf{r}_2).$$

Note here that the sum over electron-electron interactions has been replaced by the double integral over the electron density. The energy can be obtained as follows:

$$(4.32) \quad \mathcal{E} = T_0[\rho] + \int d^3\mathbf{r}_1 \, v_{\text{eff}}(\mathbf{r}_1)\rho(\mathbf{r}_1) - \frac{1}{2} \int d^3\mathbf{r}_1 \int d^3\mathbf{r}_2 \, \frac{\rho(\mathbf{r}_1)\rho(\mathbf{r}_2)}{|\mathbf{r}_2 - \mathbf{r}_1|} +$$
$$\mathcal{E}_{xc} - \int d^3\mathbf{r}_1 \, v_{xc}(\mathbf{r}_1)\rho(\mathbf{r}_1).$$

Remarkably, the density functional approach can utilize much of the machinery already in place for computing wavefunctions of multi-electron atoms. The key to its success resides in the mysterious exchange-correlation terms.

Initially, physicists proposed using a local density approximation (LDA) that includes a $\rho^{4/3}$ behavior. Subsequently, a generalized gradient approach (GGA) that admits non-local behaviors has been introduced. The form of the exchange correlation potential is a topic of current research interest but the recent popularity of the DFT methodology stems from work by Axel Becke, who developed an empirical relationship based on a hybrid functional that includes both LDA and GGA components.[17] Becke's approach provided the ability to compute results with accuracies comparable to high order quantum methods (scales like N^8) for essentially the cost of a low order Hartree-Fock calculation (scales like N^3). Numerous authors have expanded and refined the functions and the search for better exchange/correlation functionals continues. We'll take up performing DFT calculations in a subsequent chapter.

---

[17]Becke published "A new mixing of Hartree-Fock and local density-functional theories" in the *Journal of Chemical Physics* in 1993.

FIGURE 4.3. The performance of several DFT methods is substantially better than Hartree-Fock (HF) but worse than the high-order quantum G2 method. The comparisons are for enthalpies of formation (ΔH_f), ionization potentials (IP), electron affinities (EA), proton affinities (PA) and transition metal excited state energies (TM). Results represent mean absolute deviations from experimental values.

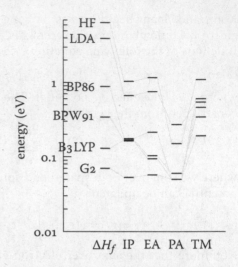

Typical results for a collection of about 150 atoms and small molecules are depicted in figure 4.3. The G2 method is an high-order quantum method that was state of the art in about 2000. It has been supplanted by G3 and G4 methods that are somewhat better. The Hartree-Fock method (HF) has mean absolute deviations from experimental values of nearly 6.5 eV for enthalpies of formation of small molecules. This is exceptionally poor. The simple local density approximation has been improved by a number of hybrid methods inspired by Becke's work. The most successful has been his three-point hybrid method that utilizes the Lee-Yang-Parr correlation function (B3LYP). This provides accuracies only somewhat worse than the G2 method but with computational times that scale like Hartree-Fock calculations.

Before continuing our journey, we should take a moment to discuss one of the key elements of density functional theory: the electron-electron interaction. This is the double integral of the squared density. While we often write such things without bothering to notice whether or not they are meaningful, let us consider the fact that the denominator vanishes when $r_1 = r_2$.

EXERCISE 4.13. Plot the function Exp[-x]Exp[-y]/Norm[x-y] over the domain $0 \le x \le 5$ and $0 \le y \le 5$.

In addition to dealing with potentially divergent terms, one also must compute integrals like the following:

$$\int d^3 \mathbf{r}\, v(\mathbf{r})\rho(\mathbf{r})$$

where the potential is not analytically known. This means that the integrals must be evaluated numerically.

A great deal of work has been performed toward the efficient numerical computation of integrals. A standard approach for bounded intervals utilizes what is known as Gaussian quadrature. Here, the integral is approximated by a finite sum:

$$(4.33) \qquad \int_{-1}^{1} dx\, f(x) = \sum_{i=1}^{N} w_i f(x_i),$$

where the w_i are weights and the x_i are points where the function is evaluated. These are typically the zeros of some special function.

In the density functional calculations, we are typically dealing with infinite integrals and the densities are generally falling exponentially from the nuclear centers. An extension of Gaussian quadrature to such integrals is known as Gauss-Laguerre quadrature. Here, we seek to approximate integrals of the following form:

$$(4.34) \qquad \int_{0}^{\infty} dx\, f(x) e^{-x} = \sum_{i=1}^{N} w_i f(x_i)$$

For the Nth order approximation, the x_i are obtained from the zeros of the Laguerre polynomial $L_N(x)$. The weights are then defined as follows:

$$(4.35) \qquad w_i = \frac{x_i}{(N+1)^2 [L_{N+1}(x_i)]^2}.$$

If the function f is not known explicitly, then we can rewrite equation 4.34 as follows:

$$(4.36) \qquad \int_{0}^{\infty} dx\, g(x) = \sum_{i}^{N} w_i g(x_i) e^{x_i}.$$

This places the positive exponential in the sum, cancelling the exponential behavior of g.

EXERCISE 4.14. Consider the function $f(x) = \sin \pi x/3$ or $g(x) = f(x)e^{-x}$. Compute the tenth order approximation of the integral. The x_i can be obtained from the NSolve function, using the LaguerreL function. Compute the weights and then compute the integral exactly using the Integrate function. How close is the approximation?

EXERCISE 4.15. The Gauss-Hermite quadrature is used for integrals of the form

$$\int_{-\infty}^{\infty} dx\, f(x) e^{-x^2} = \sum_{i}^{N} w_i f(x_i).$$

Here, the x_i are obtained from zeros of the Hermite polynomials and the weights are defined as follows:

$$w_i = \frac{2^{N-1}N!\sqrt{\pi}}{N^2[H_{N-1}(x_i)]^2}.$$

Compute the integral of the function $f(x) = 3x^4 + 5x^3$, using the $N = 8$ approximation. How does it compare to the exact answer?

4.4. Heavy Atoms

Students may have noticed that we have abandoned all pretext of ten-. sor notation and relativistic invariance in our discussions of the density-functional theory. This reflects the historical approach to multi-electron atoms. Coping with multiple electrons is difficult enough without carrying along all of the relativistic accoutrements. There would need to be a justification from experiment to incorporate the additional complexity.

In figure 4.4, we illustrate a typical spectrum from an x-ray source with a tungsten anode. Here, electrons are accelerated through a 35 kV potential and allowed to strike a tungsten plate. The smooth background is due to bremsstrahlung radiation caused by inelastic collisions of the electrons with the tungsten nuclei. The spectrum is dominated by large peaks; the largest arises from the $n = 2$ to $n = 1$ transition, known as the $K_{\alpha 1}$ transition.

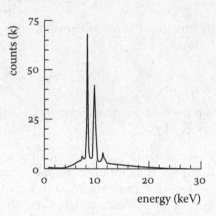

FIGURE 4.4. A typical spectrum obtained from high-voltage (35 keV) electrons impacting a tungsten anode includes a smooth background due to bremsstrahlung and characteristic peaks associated with the tungsten K-shell electrons.

As a result of this characteristic behavior, this simple x-ray source can be thought of as nearly monochromatic. This is fortuitous, as diffraction depends on the frequency. Consequently, no diffraction patterns would arise from a wide-band x-ray source. The Braggs would not have identified the atomic lattice without this feature.

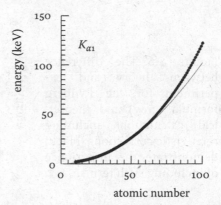

FIGURE 4.5. The $K_{\alpha 1}$ x-ray experimental energies obtained from the NIST database increase significantly as a function of atomic number. The Rydberg estimate (gray curve) underestimates the observed energies for large Z.

In figure 4.5, we plot the experimental $K_{\alpha 1}$ energies, along with the expected value from the Rydberg formula. Note that the energies at large atomic numbers deviate from the simple, non-relativistic formula. As a result, we should investigate how to include relativistic effects into our calculations. We note that the x-ray energies exceed 100 keV and that the rest mass of the electron is 511 keV. This is a hint that relativistic effects might be important.

EXERCISE 4.16. What is γ for a 100 keV electron? What is its velocity?

As a practical matter, both Hartree-Fock and DFT methods solve for electrons in the presence of some potential $v(\mathbf{r})$. So, it is reasonably straightforward to adapt the codes to include additional terms in the potential. For example, the electron-nuclear interactions can be modified to include the finite nuclear size:

$$(4.37) \qquad v_{ne}(r) = \begin{cases} -\dfrac{Ze^2}{4\pi\epsilon_o}\dfrac{r^2}{R^3} & r < R \\[2mm] -\dfrac{Ze^2}{4\pi\epsilon_o}\dfrac{1}{r} & r > R \end{cases},$$

where R is the nuclear radius. We'll discuss this in more detail subsequently. Relativistic corrections can be added similarly:

$$(4.38) \qquad \mathcal{H} = \mathcal{H}_o - \frac{p^4}{8c^2} + \frac{\Delta v}{8c^2} + \frac{1}{2c^2}\frac{1}{r}\frac{dv}{dr}\mathbf{l}\cdot\mathbf{s}.$$

Here the Hamiltonian $\mathcal{H}$ is the sum of the non-relativistic $\mathcal{H}_o$ and additional terms. The first additional term incorporates the relativistic mass-velocity relationship ($m = \gamma m_o$) to first order. The second term provides a non-local interaction between the electron and Coulomb fields and the final term incorporates spin-orbit coupling, due to the interaction of the electron's intrinsic magnetic field in the field of the nucleus.

Figure 4.6. The difference between theory and experiment for the Rydberg formula (gray) and theoretical calculations including relativistic corrections (black) demonstrate the importance of including such terms.

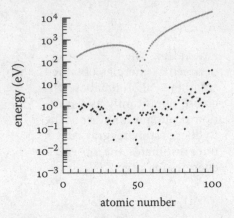

A sketch of the absolute difference between predicted and experimental energies of the K-shell levels is shown in figure 4.6. The black dots represent the best current theoretical approaches and the gray dots represent the Rydberg formula. As can be seen from the figure, the results improve by orders of magnitude but there is still room for improvement. This will be left as an exercise for the students.

Exercise 4.17. The NIST x-ray transitions energy database contains x-ray data beyond just the $K_{\alpha 1}$ information presented. Plot all of the x-rays for molybdenum and erbium. Plot the difference between experimental results and theoretical predictions.

V

On the Nature of the Nucleus

Ernest Rutherford's recognition that the high-angle scattering of alpha particles from gold films could be explained if the atom possessed a nuclear structure revolutionized our understanding of the microscopic world. In some sense, the electromagnetic force could be understood as the "glue" that held atoms together: the negatively charged electrons swirling about the positively charged nuclei. As the Coulomb force was, macroscopically, an inverse-square force, visions of tiny planetary systems were immediately evoked. Such initial enthusiasm, as we have seen, was rapidly dissipated as developing a mathematical description of the atom proved more arduous than originally envisioned.

In addition to resolving the questions surrounding the nature of the atom, physicists also began to investigate the nature of the atomic nucleus. The principal tool at their disposal was scattering but the technical sophistication of the experiments drastically improved over the simple lead-selenide scintillator/human eyeball apparatus used by Rutherford and his students. While scintillation materials remain an essential component of the experimental toolbox, detection devices are now electronic and, thus, vastly more precise, sensitive and reliable than humans. Rutherford's original experiments utilized beams of alpha particles obtained by collimating the output of naturally radioactive materials. Physicists subsequently learned to devise more intense beams and to accelerate them to higher and more tightly controlled energies.

The discerning student may have noticed that in the previous discussion of the atom that we have treated the nucleus as a point charge Ze, as did Rutherford in his initial discussions. How then, can we discern if this is the case or if the nucleus possesses a finite size? We can conduct scattering experiments and utilize theoretical calculations to interpret the data.

5.1. Electron Scattering

For the case of simple, electron-electron scattering, we have depicted one Feynman diagram for the process in figure 5.1. We identify the initial and

© Mark A. Cunningham 2018
M.A. Cunningham, *Beyond Classical Physics*,
Undergraduate Lecture Notes in Physics,
https://doi.org/10.1007/978-3-319-63160-8_5

final states by their momentum values. In most laboratory experiments, this is indeed how initial states are prepared. Electrons are accelerated to some energy and then directed towards a stationary target or counterpropagating beam. The majority of electrons in the initial beam do not interact with the target atoms and pass through to a beam stop. To improve the detection signal to noise ratio, detectors are placed at some angle from the initial beam direction. Depending on the detection system employed, either the momentum of the exiting particles is measured directly (through the curvature of the trajectory in a magnetic field) or the total energy of the particle is measured and the momentum inferred from the Einstein kinetic formula: $E^2 = p^2c^2 + m^2c^4$. Depending upon the experiment, there may or may not be any particular alignment or detection of the electron spin in initial or final states.

FIGURE 5.1. Electron scattering Feynman diagram. The electrons have momenta p_1 and p_2 in the initial state, and possess momenta p_3 and p_4 in the final state, exchanging a photon of momentum q.

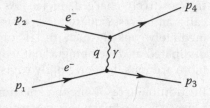

Complicating the interpretation of the measurements are the realities associated with measurements in general: finite precision and potential biases that can affect accuracy. The targets are not individual protons but gas cells containing molecular hydrogen; one must account for (or dismiss) the contributions of electron-electron scattering from any conclusions about electron-nucleus scattering. We also know that detectors subtend finite solid angles from the interaction point, meaning that momentum measurements have finite precision. Additionally, it is not possible to vary the transfer momentum q systematically. Instead, one must accumulate information in a statistical fashion, so that the precision will depend upon the square root of the total number of scattering events. Thus, in looking for events with a small probability of occurrence, one may have to wade through a very large number of uninteresting results.

In any case, the output of a scattering experiment is a determination of the cross section, which is related to the theoretical scattering amplitude. Computing the scattering amplitude generally requires some form of approximation. Initial work utilized non-relativistic forms of the matrix elements and then corrected for relativistic effects. Rutherford's original formula for the cross-section was

(5.1)
$$\left(\frac{d\sigma}{d\Omega}\right)_R = \left(\frac{Ze^2}{8\pi\epsilon_o E}\right)^2 \frac{1}{\sin^4\theta/2},$$

where θ is the angle between the incident beam and the measured electron. This formula assumed a point source for the nuclear charge Ze. An improved formula was produced by Mott, which accounted for electron spin and nuclear recoil. Mott's modification to the Rutherford formula is a multiplicative factor:

(5.2)
$$\left(\frac{d\sigma}{d\Omega}\right)_M = \left(\frac{d\sigma}{d\Omega}\right)_R \frac{\cos^2\theta/2}{1+(2\mathcal{E}/Mc^2)\sin^2\theta/2},$$

where M is the nuclear mass.[1] More generally, the interaction current that generates the scattering matrix element can be shown to have the following form:

(5.3)
$$j^k = \gamma^k F_1(q^2) + \sum_l \frac{i\kappa\sigma^{kl}q_l}{2Mc^2}F_2(q^2),$$

where q_k is the momentum transfer, γ^k is a Dirac gamma matrix and κ is a constant. The term $\sigma^{kl} = i[\gamma^k, \gamma^l]/2$ uses a shorthand notation for the commutator: $[\gamma^k, \gamma^l] = \gamma^k\gamma^l - \gamma^l\gamma^k$. The functions F_1 and F_2 are known as form factors. In the first-order approximation, the form factor F_1 is obtained from the Fourier transform of a charge density[2]:

(5.4)
$$F_1(\mathbf{q}) = \int d^3\mathbf{r}\,\rho(\mathbf{r})e^{i\mathbf{q}\cdot\mathbf{r}},.$$

Without loss of generality, we can choose the coordinate system to align $\mathbf{q}$ along the z-axis. Then the angular integrals can be performed, with the following result:

(5.5)
$$F_1(q) = 4\pi\int_0^\infty dr\,r^2\rho(r)\frac{\sin qr}{qr},$$

where q and r are the magnitudes of $\mathbf{q}$ and $\mathbf{r}$, respectively.

> EXERCISE 5.1. Assume that $\mathbf{q} = (0,0,q)$ and that θ represents the angle between the vectors $\mathbf{q}$ and $\mathbf{r}$. Write the integral in equation 5.4 in spherical coordinates. If the charge density has no angular dependence, show that equation 5.5 can be obtained by performing the angular integrals.

---

[1] In the current context, convenient values for the constants are $e^2/4\pi\epsilon_0 = 1.43996$ MeV·fm and $\hbar c = 197.3269788$ MeV·fm.

[2] Note: here we are following traditional usage, where there is a missing factor of $(2\pi)^{-3/2}$ in the Fourier transform. We shall account for it in the definition of the inverse transform.

EXERCISE 5.2.   One possible choice for the charge density is a Gaussian distribution:

$$\rho(r) = \frac{1}{(2\pi a^2)^{3/2}}\, e^{-r^2/2a^2},$$

where a represents the nominal width of the distribution. What is $F(q)$ for the Gaussian distribution?

EXERCISE 5.3.   Another possible choice for the charge density is an exponential $\exp(-r/a)$, where a is the nominal width. What is the required normalization factor for an exponential distribution? (We require $\int d^3\mathbf{r}\,\rho = 1$.) What is the form factor $F(q)$ for an exponential distribution?

EXERCISE 5.4.   Suppose that the charge inside the nucleus is uniformly distributed over a spherical volume of radius a. What is the required normalization factor for the density? What is $F(q)$ for a uniform distribution?

FIGURE 5.2. Form factors for different charge distributions. A gaussian distribution (black) is similar to the exponential distribution (gray). The form factor for a uniform distribution (lightgray) is oscillatory at large values of q.

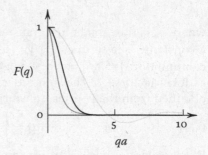

From figure 5.2, we can see that the consequence of a nuclear charge distribution other than a point charge is to reduce the scattering amplitude at large values of q. (Recall that the Fourier transform of the delta function is one, independent of q.) So, we should be able to discern the existence of a charge distribution from scattering experiments, particularly at higher beam energies. Just such experiments were conducted by Richard Hofstadter and his students at the electron accelerator at Stanford.[3] Different values of q can be found at different angles. We find that, for a given beam energy $\mathcal{E}$, that the transfer momentum is given by the following expression:

$$(5.6) \qquad q = \frac{2\mathcal{E}}{\hbar c}\frac{\sin\theta/2}{\left[1 + (2\mathcal{E}/Mc^2)\sin^2\theta/2\right]^{1/2}},$$

[3]Hofstadter was awarded the Nobel Prize in Physics in 1961 "for his pioneering studies of electron scattering in atomic nuclei and for his thereby achieved discoveries concerning the structure of the nucleons." He shared the prize with Rudolph Mössbauer, who was cited "for his researches concerning the resonance absorption of gamma radiation and his discovery in this connection of the effect which bears his name."

where M is the nuclear mass.

> EXERCISE 5.5.   Plot the value of q (in units of fm^{-1}) for electron-proton scattering. Consider angles between 20° and 150°, with electron beam energies of 100, 200 and 500 MeV.

FIGURE 5.3. Electrons with a beam energy of 550 MeV were scattered from a hydrogen gas cell. The scattered electron flux was measured as a function of angle from the beam axis. The results are in good agreement with an exponential model (black curve) of the proton charge distribution and not the Rutherford (gray) or Mott (light gray) models.

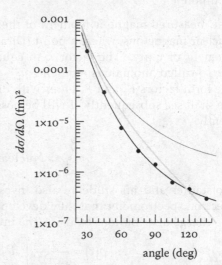

The second term on the right-hand side of equation 5.3 represents scattering from the proton's magnetic moment, which actually dominates at very high energies. An expression for the differential cross-section of electron-proton elastic scattering can be written as follows[4]:

(5.7)
$$\frac{d\sigma}{d\Omega} = \left(\frac{d\sigma}{d\Omega}\right)_M \left\{ F_1(q^2) + \left(\frac{\hbar c q}{2Mc^2}\right)^2 \left[2\big(F_1(q^2) + \kappa F_2(q^2)\big)^2 \tan^2 \theta/2 + \kappa^2 F_2^2(q^2) \right] \right\}$$

The magnetic form factor F_2 is obtained from the integral over the proton's magnetic moment distribution:

(5.8)
$$F_2 = \int_0^\infty dr\, r^2 \mu(r) \frac{\sin qr}{qr}.$$

Initially, Hofstadter and his students found reasonable agreement with models that utilized the same distribution function for both form factors, with an effective radius of about $a = 0.7$ fm. An example is depicted in figure 5.3.

> EXERCISE 5.6.   Plot the Rutherford, Mott and Rosenbluth (equation 5.7) cross sections for beam energies of 200 and 500 MeV. Use

---

[4]American physicist Marshall Rosenbluth published his "High energy elastic scattering of electrons on protons," in the *Physical Review* in 1950.

the distribution $F_1 = F_2 = 1/(1 + a^2 q^2)^2$ for the charge and magnetic moment distributions, with $\kappa = 1.79$. How does the Rosenbluth result vary as the effective radius a varies from 0 to 1 fm? How does the result vary if the magnetic radius is set to zero (point magnetic dipole)?

The measured magnetic moment of the proton is $\mu_p = 2.7928473508(85)$ nuclear magnetons, where a point (Dirac) proton should have a magnetic moment of unity. The factor κ in equation 5.7 is typically taken to be the so-called anomalous moment: $\mu_p - 1$. As we note from equation 5.7, the form factors F_1 and F_2 enter with different dependences on q and, as we shall see subsequently, it will be possible to determine them independently.

5.2. Nuclear Size

Hofstadter and his students also investigated heavier nuclei using the electron spectrometer that they developed. One can, in principle, recover the charge distributions by using the inverse Fourier transform:

$$(5.9) \qquad \rho(\mathbf{r}) = \frac{1}{(2\pi)^3} \int d^3\mathbf{q}\, F_1(\mathbf{q}) e^{-i\mathbf{q}\cdot\mathbf{r}}.$$

If we again assume no angular dependence in the charge distribution, this leads directly to the following formula for the charge distribution:

$$(5.10) \qquad \rho(r) = \frac{1}{2\pi^2} \int_0^\infty dq\, q^2 F_1(q) \frac{\sin qr}{qr}.$$

As a practical matter, the cross sections depend upon the square of the form factor and have a limited range of the transfer momentum q. This limits the ability to precisely define the charge distributions.

Inversion of the scattering data represents another example of what spectroscopists call the phase problem. In general, the form factor will be a complex function. In order for the inversion to be unique, one must obtain both the magnitude and the phase of the form factor but the cross section is proportional to the magnitude (squared). Consequently, the phase has to be obtained through modelling or other means. The fact that experimentally accessible momentum transfers are finite limits the spatial resolution of the reconstructed charge distribution: higher frequencies are equivalent to shorter wavelengths.

Beyond just the technical difficulties of arranging to accelerate electrons to ever greater energies, the analysis presented to this point has assumed elastic scattering: that is, kinetic energy of the electron is converted into kinetic energy of the nucleus but the total kinetic energy is conserved.

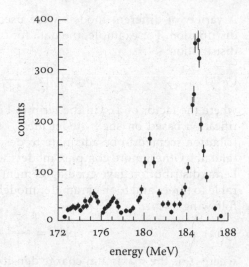

FIGURE 5.4. Electrons with energy 188 MeV were scattered at 80° from a carbon foil target. The peak at about 185 MeV corresponds to elastic scattering. The peaks at lower energies represent evidence of excited states of the carbon nucleus.

As we can see from figure 5.4, this is not the only possible outcome for energetic electrons. In this instance, Hofstadter and his students directed 188 MeV electrons at a carbon foil target. Looking at an angle of 80° from the beam direction, they were able to measure the energies of scattered electrons. The large peak at 185 MeV corresponds to elastically scattered electrons. Subsequent peaks at lower energies represent the transfer of initial kinetic energy of the electron into internal degrees of freedom of the nucleus. This is an extraordinarily important observation. Nuclei, that we believe to be composed of protons and neutrons, have excited states that can be excited electromagnetically.

We have already suggested a model for the nucleus in which the atomic number corresponds to the number of protons and the atomic mass corresponds to the sum of the number of protons and neutrons. Such a model accounts for the elements in the periodic table and the curious distribution of elemental masses, that arises from different isotopic content for each chemical element. What we observe from figure 5.4 is that nuclei appear to also exist in quantized states. A vast amount of experimental data has now been accumulated on the properties of nuclear matter. These data are generally accessible through curated databases, like the National Nuclear Data Center (NNDC) at Brookhaven National Laboratory (www.nndc.bnl.gov).

EXERCISE 5.7. Use the NuDat utility from the NNDC to obtain the level structure for ^{12}C. What are the three lowest excited states? Are they visible in the cross sectional data presented in figure 5.4?

A variety of different models were used to describe the nuclear charge distribution. For example, the data for carbon were best described by the distribution:

$$(5.11) \qquad \rho_C = \rho_0\left[1 + \frac{4r^2}{3a^2}\right]e^{-r^2/a^2},$$

where the factor of 4/3 in the term in brackets had some theoretical justification based on shell model ideas. For heavier nuclei, a Fermi distribution seemed to be adequate to describe the bulk of the experimental data. Other, more complex models were tried but the two-parameter Fermi distribution gave good agreement with experimental data, comparable to three- and four-parameter models. The Fermi distribution has the following form:

$$(5.12) \qquad \rho_F(r) = \frac{\rho_0}{1 + e^{(r-c)/a}},$$

where ρ_0 is the maximum charge density, c defines the nominal width of the distribution and a defines the sharpness.

EXERCISE 5.8. Plot the Fermi distribution and the modified Gaussian distribution, defined by the following expression:

$$\rho_G(r) = \frac{\rho_0}{1 + e^{(r^2-c^2)/a^2}}.$$

Note that the parameters a and c may have different values for the modified Gaussian distribution.

TABLE 5.1. Parameters for nuclear charge distributions.[*]

	Z	a (fm)	c (fm)		Z	a (fm)	c (fm)
^{12}C	6	0.71	—	^{115}In	49	0.523	5.24
^{40}Ca	20	0.568	3.64	^{122}Sb	51	0.568	5.32
^{51}V	23	0.5	3.98	^{197}Au	79	0.523	6.38
^{59}Co	59	0.568	4.09	^{209}Bi	83	0.614	6.47

[*]From Hofstadter's "Electron scattering and nuclear structure," published in the *Reviews of Modern Physics* in 1956. Note that carbon follows a Gaussian distribution; the other nuclei are described by the Fermi distribution.

EXERCISE 5.9. The normalization factor ρ_0 for the Fermi distribution is defined by the following requirement:

$$4\pi \int_0^\infty dr\, r^2 \rho_F(r) = Ze,$$

where Z is the number of protons in the nucleus. Using the values from table 5.1, determine ρ_o for the elements listed.

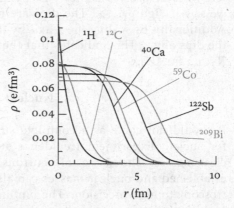

FIGURE 5.5. The best fits of nuclear charge distributions for heavy nuclei follow a nominal Fermi distribution.

As can be seen from figure 5.5, the charge distributions for heavier nuclei display a constant region at small radius before rolling smoothly to zero. Other models that reduce the charge density at $r = 0$ are also consistent with the data but the Fermi distribution has some modest theoretical support. Note that the peak central density ρ_o decreases with increasing atomic number A.

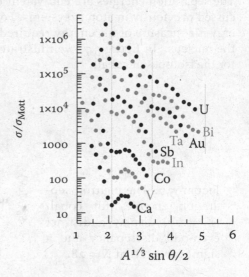

FIGURE 5.6. Scattering of 183 MeV electrons from nuclei display diffraction effects (dips in scattering cross section). The dips align when plotted as a function of $A^{1/3}$ (vertical light gray lines).

Electron-nucleus scattering also displays systematic effects. For example, in figure 5.6, we display Hofstadter's data for the scattering of 183 MeV electrons from a variety of different nuclei. To emphasize the effects, we have plotted the ratio of the experimental and Mott cross sections *versus* the somewhat curious abscissa $A^{1/3} \sin \theta /2$. The cross sections display dips that can be interpreted as arising from diffraction effects: electrons interfere with themselves as they pass by the nuclei, analogous to the patterns we saw in figure 1.12. The dips are more prominent for lighter nuclei. Additionally, by scaling the x-axis by the cube root of the atomic number, the dips align. This indicates that the nominal nuclear size R scales like $R \approx 1.2 \text{ fm} A^{1/3}$.

5.3. Nuclear Structure

It should not come as a complete surprise that there is a shell model for nuclei. Experimental evidence arises from observations of the nuclear masses. The masses of neutrons m_n and protons m_p are now well-established and nuclear masses can also be obtained through mass spectroscopy to high precision. The binding energy for a nucleus is obtained from the following simple formula:

$$BE(Z,N) = Zm_p c^2 + N m_n c^2 - Mc^2,$$

where M is the nuclear mass. From the binding energies, one can define the neutron S_n and proton S_p separation energies:

(5.13) $S_n = BE(Z,N) - BE(Z,N-1)$ and $S_p = BE(Z,N) - BE(Z-1,N)$.

The separation energies are equivalent to the ionization energies we discussed previously in atomic systems. For each isotope, the separation energy is a measure of the energy required to remove a single nucleon from the nucleus. In figure 5.7, we illustrate the neutron separation energies for the isotopes of cobalt.

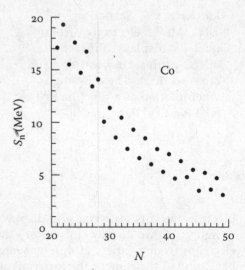

FIGURE 5.7. The neutron separation energies for cobalt ($Z = 27$) display a distinct even-odd discrepancy and a significant shift at $N = 28$.

What we observe from the figure is a distinct even-odd disparity. The separation energy is roughly 2 MeV greater in nuclei with even numbers of neutrons than it is in nuclei with an odd number of neutrons. From this, we can infer that the nuclear force that binds nucleons into the small volume of the nucleus includes a component that prefers pairs of nucleons. This pairing force is an emergent property of the nuclear force. Indeed, the ground states of all nuclei that have even numbers of protons and even numbers of neutrons have total angular momentum zero and are positive parity (0^+).

From figure 5.7, we also observe that there is a distinct shift in the separation energy at $N = 28$. Like the shifts in the ionization energy that heralded a shell structure for electrons in atoms, the shift in separation energy implies that nucleons in the nucleus also possess a shell structure. For nucleons, the magic numbers that determine shell closures are 2, 8, 20, 28, 50, 82 and 126. These are not the same magic numbers that we encountered in atomic systems.

> EXERCISE 5.10.   Use the NuDat utility from the NNDC website to obtain S_n values for iron and copper isotopes. Is the feature at $N = 28$ visible in these data?

> EXERCISE 5.11.   Use the NuDat utility from the NNDC website to obtain S_p values for isotopes with $N = 24$ and $N = 31$. Do you observe a pairing force for protons? Is there a feature at $Z = 28$?

A model that explained the observed shell features in the nuclear separation energies was independently developed by the German-American physicist Marie Goeppert-Mayer and the German physicist Hans Jensen in 1949.[5] One could obtain the observed magic numbers if the nuclear force favored high-spin states. This is in direct contrast to the electromagnetic force that binds electrons to atoms, where the energetically favored states possess low angular momentum.

A nominal level diagram is illustrated in figure 5.8. We note that labelling of the states in the nuclear shell model differs from the practice in atomic systems. Here, the first occurrence of each angular momentum level is denoted with the index 1 and the angular momentum states are indicated with lower-case letters. The nucleons (protons and/or neutrons) possess half-integral spin and couple to eigenstates of the Hamiltonian that possess total angular momentum J that is also half-integral. For example, the

---

[5]Goeppert-Mayer and Jensen shared half of the 1963 Nobel Prize in Physics "for their discoveries concerning nuclear shell structure." They shared the prize with the Hungarian physicist Eugene Wigner, who was cited "for his contributions to the theory of the atomic nucleus and the elementary particles, particularly through the discovery and application of fundamental symmetry principles."

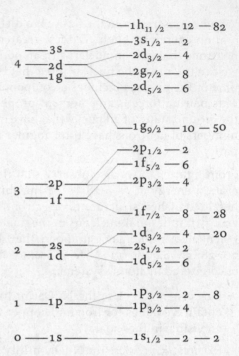

FIGURE 5.8. Magic numbers in the neutron and proton separation energies can be explained if nucleons preferentially populate high-spin levels.

first shell above the ground state is defined by the $1p_{3/2}$ and $1p_{1/2}$ levels, where the $1p_{3/2}$ level is lower in energy. We can see that these assignments agree with the experimental evidence from the fact that the ground state of ^{9}Be has $j = 3/2$, where the ground state of ^{13}C has $j = 1/2$.

EXERCISE 5.12.   Use the NuDat utility to examine the spin assignments for odd-mass nuclei adjacent $(Z + 1, N)$ and $(Z, N + 1)$ to nuclei with magic numbers for both Z and N. Do these agree with the assignments from figure 5.8?

Another emergent property of the nuclear force is the quadrupole nature of most nuclear excitations. In even-even nuclei, those with an even number of both protons and neutrons, not only does the ground state have $j = 0$ but the first excited state has $j = 2$. These observations suggest that the classical shape of larger nuclei is not spherical but spheroidal, particularly for high-spin excited states. One obtains just such behavior in liquids: a spinning liquid drop will flatten into a spheroidal shape. When spinning even faster, the liquid may neck down into more of a dumbbell shape to increase its moment of inertia. A number of models based on semiclassical ideas were used to describe nuclear phenomena, such as the liquid drop model of Aage Bohr, Ben Mottelson and Leo Rainwater, but we will instead discuss a modern variant, known as the Interacting Boson Model (IBM).

> EXERCISE 5.13. Use the `ParametricPlot3D` function to plot the surface defined by $(\cos u \sin v, \sin u \sin v, a \cos v)$, where $0.5 \leq a \leq 2$, $0 \leq u \leq 2\pi$ and $0 \leq v \leq \pi$. When $a < 1$, the spheroid is known as oblate. When $a > 1$, the spheroid is known as prolate.

The IBM was introduced by the Japanese physicist Akito Arima and the Italian physicist Franco Iachello in 1975.[6] Rather than taking the approach of previous groups, Arima and Iachello developed an algebraic approach to nuclear structure. The IBM considers the nucleus to consist of a static core and valence nucleons paired into either spin 0 (s) or spin 2 (d) bosons. States are defined as $\psi = |n_s, n_d, J, M\rangle$, where the quantum numbers n_s and n_d count the number of s- and d-bosons, respectively. The total angular momentum J and z-projection M are also preserved by the Hamiltonian. The most general Hamiltonian describing such a system that includes two-body interactions can be written as follows:

$$(5.14) \qquad \mathcal{H} = \varepsilon_s s^\dagger s + \varepsilon_d \sum_k d_k^\dagger d_k + \sum_{l_1 l_2 l_3 l_4 L} a_{l_1 l_2 l_3 l_4}^L (b_{l_1}^\dagger \times b_{l_2}^\dagger)^L \cdot (b_{l_3} \times b_{l_4})^L,$$

where $s^\dagger$ and s are the creation and annihilation operators for s-bosons and $d_k^\dagger$ and d_k are the creation and annihilation operators for d-bosons with $M = k$. The first two terms in the Hamiltonian are the number operators for s- and d-bosons, respectively. The last term in the Hamiltonian sums over all boson operators (denoted by b), coupled to total angular momentum L, with numerical weights $a_{l_1 l_2 l_3 l_4}^L$.

The initial model (known historically as IBM-1) did not differentiate between neutrons and protons, although subsequent treatments became increasingly sophisticated: treating neutrons and protons separately (IBM-2) and then as part of isospin multiplets (IBM-3 and IBM-4). The notion of isospin was originally introduced by Heisenberg in 1932 to explain symmetries observed in nuclei and was then systematized by Wigner in 1937.[7] Just as electrons have two possible internal states that we have labelled spin; Wigner recognized that the nucleons that make up nuclei can also be considered to be two components of another internal property that has come to be called isospin. The isospin symmetry is not exact but appears to be an approximate symmetry of the nuclear system.

---

[6]Arima and Iachello published a series of papers on the "Interacting boson model of collective states" in the *Annals of Physics* beginning in 1976.

[7]Heisenberg's "Über den Bau der Atomkerne" appeared in the *Zeitschrift für Physik* and Wigner's "On the consequences of the symmetry of the nuclear Hamiltonian on the spectroscopy of nuclei" in the *Physical Review*.

In any case we shall restrict our discussion to the initial form of the IBM-1 Hamiltonian, as indicated in equation 5.14. The creation and annihilation operators associated with the IBM-1 also possess a symmetry, that of the Lie algebra u(6). The Lie algebra is associated with the unitary Lie group U(6). A representation of the unitary Lie groups can be obtained with unitary matrices U, where $U^\dagger U = I$. The Norwegian mathematician Marius Sophus Lie found in the 1890s that continuous transformations could be studied most productively by considering infinitesimal operations; this strategy enabled a linearization of the mathematics and discarding of terms of higher order. The group operator became the commutator that we have introduced previously. Hermann Weyl introduced Lie's results into the physics literature in the 1920s.

EXERCISE 5.14.   The (mathematical) algebra defined by the IBM Hamiltonian is specified by the generators:

$$(s^\dagger \times s)^0, \quad (s^\dagger \times d)_k^2, \quad (d^\dagger \times s)_k^2 \quad \text{and} \quad (d^\dagger \times d)_k^L,$$

where the superscripts denote the total angular momentum L and k represents the z-component. The last operator has five different possible angular momentum couplings, from $L = 0$ to $L = 4$. The unitary group U(N) has N^2 generators. How many generators exist for the IBM?

EXERCISE 5.15.   The rotation group SO(3) can be represented by orthogonal matrices O, where $O^T O = I$. One basis for the algebra $so(3)$ is given by the following matrices:

$$(5.15) \quad L_x = \begin{bmatrix} 0 & 0 & 0 \\ 0 & 0 & -1 \\ 0 & 1 & 0 \end{bmatrix}, \quad L_y = \begin{bmatrix} 0 & 0 & 1 \\ 0 & 0 & 0 \\ -1 & 0 & 0 \end{bmatrix} \quad \text{and} \quad L_z = \begin{bmatrix} 0 & -1 & 0 \\ 1 & 0 & 0 \\ 0 & 0 & 0 \end{bmatrix}.$$

Show that the following relation holds:

$$[L_x, L_y] = L_z$$

and that the relation also holds for cyclic permutations of $\{x, y, z\}$.

EXERCISE 5.16.   Elements in the SO(3) group can be obtained from the generators by exponentiation:

$$(5.16) \qquad R_i = \exp(\theta_i L_i) = I + \sum_{k=1}^{\infty} \frac{\theta_i^k}{k!} L_i^k,$$

where I is the 3×3 identity matrix and L_i^k represents the matrix product of k copies of L_i. Show that equation 5.16 leads to the usual definition of the rotation matrix:

$$R_z(\theta_3) = \begin{bmatrix} \cos\theta_3 & -\sin\theta_3 & 0 \\ \sin\theta_3 & \cos\theta_3 & 0 \\ 0 & 0 & 1 \end{bmatrix}.$$

Hint: compute the first eight terms in the series explicitly. What pattern arises?

The coupling of two angular momenta into a final, total angular momentum is a problem that was solved by the German mathematicians Alfred Clebsch and Paul Gordan in their studies of invariants in group theory. When we encounter product spaces of Hilbert spaces, the Fock spaces we have discussed previously, it is the total angular momentum that is the physical observable, not the individual angular momenta of the different components. Using Dirac's notation, two bodies in states specified by angular momentum j_i and z-component m_i can be combined into a single state of angular momentum via the following formula:

$$(5.17) \qquad |j\,m\rangle = \sum_{m_1=-j_1}^{j_1} \sum_{m_2=-j_2}^{j_2} |j_1\,m_1\,j_2\,m_2\rangle\langle j_1\,m_1\,j_2\,m_2|j\,m\rangle,$$

which is the definition of the Clebsch-Gordan coefficient $\langle j_1\,m_1\,j_2\,m_2|j\,m\rangle$.

EXERCISE 5.17. Use the `ClebschGordan` function to identify the components of the $(d^\dagger \times d)^1_k$ generators for $k = -1,0,1$ in terms of the operators $(d^\dagger)^2_m$ and $(d)^2_m$, where $m = -2,-1,0,1,2$.

Of particular interest in the IBM are what are known as dynamical symmetries. We know that nuclear states can be labelled by their angular momentum, which indicates that the angular momentum operators $L^2 = L_x^2+L_y^2+L_z^2$ and L_z commute with the Hamiltonian. So, we are interested in determining subgroups of U(6) that also contain the angular momentum group SO(3). The IBM generators $(d^\dagger \times d)^1_k$ form a basis of the $\mathfrak{o}(3)$ algebra and, correspondingly the SO(3) group. So, we are interested in subgroups of U(6) that contains these generators. It turns out that there are three different decompositions of the original U(6) symmetry that contain SO(3) as a subgroup.

First, the d-boson generators alone form an algebra that is $\mathfrak{u}(5)$, with associated Lie group U(5). The chain of subgroups can be written as follows:

$$U(6) \supset U(5) \supset O(5) \supset O(3).$$

This group clearly contains the $\mathfrak{o}(3)$ generators. Hence, for the choice of Hamiltonian parameters in which only those multiplying $(d^\dagger \times d)^L$ are nonzero, we will find a U(5) symmetry. This is an example of a so-called hidden symmetry, where the Hamiltonian supports a larger (U(6)) symmetry than would be observed through experimental measurements (U(5)).

The $\mathfrak{u}(6)$ algebra also supports two other subalgebras that contain $\mathfrak{o}(3)$. One is spanned by the basis $\{(d^\dagger \times d)^1_k, (s^\dagger \times d + d^\dagger \times s)^2_k, (d^\dagger \times d)^3_k\}$, which defines the algebra $\mathfrak{o}(6)$. The group chain is the following:

$$U(6) \supset O(6) \supset O(5) \supset O(3).$$

The final subalgebra is spanned by the basis $\{(d^\dagger \times d)^1_k, [s^\dagger \times d + d^\dagger \times s - \sqrt{7/4}(d^\dagger \times d)^2_k]\}$, which defines the algebra $\mathfrak{su}(3)$. The group chain can be written as follows:

$$U(6) \supset U(3) \supset O(3).$$

> **EXERCISE 5.18.** The Lie algebra $\mathfrak{o}(n)$ has $n(n-1)/2$ generators. Show that the generators of the IBM subalgebra $\mathfrak{o}(6)$ specified above at least have the correct dimension.

In each of the different dynamical symmetries, the eigenstates take on a particularly simple form, as do the transition amplitudes between the states. We shall focus on the last dynamical symmetry, where the Hamiltonian is invariant under transformations involving the Lie group SU(3). This case will ultimately prove to provide a description of what were known historically as rotational nuclei. A consequence of the SU(3) symmetry is that the eigenvalues of the Hamiltonian can be written explicitly in terms of the invariant operators of the group, leading to an analytic expression for the energies. The invariant operators are also known as *Casimir operators* after the Dutch physicist Hendrik Càsimir. In this particular case, the energies of states in the SU(3) limit of the IBM take the following form:

$$(5.18) \qquad \mathcal{E}_{SU(3)} = \varepsilon_1 N + \varepsilon_2 N^2 + \kappa_1 L(L+1) + \kappa_2[\lambda^2 + \mu^2 + \lambda\mu + 3(\lambda + \mu)]$$

where the ε_i and κ_i are free parameters, L is the total angular momentum and the non-negative integers (λ, μ) characterize the SU(3) eigenstate. The limits on λ and μ depend upon the number of bosons N in the following fashion:

$$(5.19) \qquad \mu = 0, 2, 4, \ldots \quad \text{and} \quad \lambda = 2N - 6l - 2\mu,$$

where $l = 0, 1, 2, \ldots$ The possible values of L are also dependent upon λ and μ. The SU(3) states are labelled by an order number K but the energies do not depend explicitly upon K, which can take values $K = 0, 2, 4, \ldots, \min(\lambda, \mu)$. The angular momentum can then take values

$$(5.20) \qquad L = \begin{cases} 0, 2, 4, \ldots, \max(\lambda, \mu) & K = 0 \\ K, K+1, K+2, \ldots, K+\max(\lambda, \mu) & K > 0 \end{cases}.$$

EXERCISE 5.19. For the case of $N = 1$ bosons, $\lambda = 2 - 6l - 2\mu$. The requirement that λ be non-negative implies that we must have $l = 0$ and $\mu = 0$. Consequently, $K = 0$ and the only values of L that are possible are $L = 0, 2$. What are the possible values of the quantum numbers λ, μ, l, K and L for $N = 2, 3$ and 4? Using equation 5.18, what would be the energy levels for $N = 2$?

The nucleus ^{158}Gd has been identified as a good example of the SU(3) symmetry in nuclei, which are most often nuclei with many bosons, with the shells approximately half occupied. Gadolinium nuclei possess 64 protons and this particular isotope contains 94 neutrons. The closest magic numbers are 50 and 82, so the number of bosons required to describe this nucleus is $N = (64 - 50)/2 + (94 - 82)/2 = 13$.

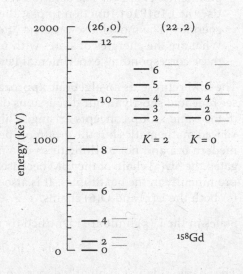

FIGURE 5.9. Low-lying (positive parity) states in ^{158}Gd can be identified with eigenstates of the SU(3) Hamiltonian. The bands are identified by their (λ, μ) characteristics as well as by the total angular momentum. Black bands represent the experimental levels and the gray bands are values obtained from the SU(3) approximation with $\varepsilon_1 = 435.146$ eV, $\kappa_1 = 11.9669$ eV and $\kappa_2 = -7.447368$ eV.

For a single nucleus, there is no means to separately determine both parameters ε_1 and ε_2, so we have chosen $\varepsilon_2 = 0$. The results of a fit to the low-lying, positive-parity energy levels are depicted in figure 5.9. Generally, the structure of the levels is well described with a three-parameter fit. Of course, there are some places where this simple model fails. For example, the SU(3) model Hamiltonian does not depend upon the order parameter K. If we look at the experimental results for the $K = 2$ and $K = 0$ submanifolds of the $(\lambda, \mu) = (22, 2)$ states, we see that the energies of the $L = 2, 4$ and 6 states are not degenerate, as predicted by the model. We can interpret these results to indicate that the actual Hamiltonian must contain terms that break the SU(3) symmetry, or that the SU(3) symmetry is only approximate. It is, of course, possible to improve the fits by using the complete IBM Hamiltonian and by, subsequently, incorporating more physics into the model: differentiating neutron and proton bosons, etc. Additionally, by coupling single-particle states to the boson core states, it is also possible to extend the IBM into an Interacting Boson-Fermion Model, thus permitting studies of odd-mass nuclei. Such strategies permit the systematic study of nuclei across the periodic table but are beyond the scope of this text.

> EXERCISE 5.20. Use the NNDC database to obtain the energy levels of ^{170}Er for the $(\lambda, \mu) = (34, 0)$ and $(30, 2)$ bands. Use the Least-Squares function to find optimal values of ε_1, κ_1 and κ_2. The matrix **M** required for the fit will have three columns: $\{17, L(L+1), \lambda^2 + \mu^2 + \lambda\mu + 3(\lambda+\mu)\}$ and the solution vector **b** will contain the associated experimental energies. The solution vector will then be $\mathbf{x} = \{\varepsilon_1, \kappa_1, \kappa_2\}$. Use the ListPlot function to plot the experimental and theoretical energies. The model also predicts $(26, 4)$ bands with $K = 0, 2$ and 4. What are the energies of states with $L = 0, \ldots, 6$ for these bands? Are there corresponding experimental levels?

The IBM represents an algebraic approach to nuclear physics that we will see repeated shortly in our discussions of high-energy physics, with many of the same group concepts arising. This may seem to rely on rather advanced mathematics but the group properties are well documented in this modern era and not as difficult as one might first fear. We only investigated the SU(3) chain of the IBM because, as we shall see, SU(3) will figure prominently in the next topics. It is also possible to find similar examples for both the U(5) and O(6) chains.

States in the U(5) limit are identified by the quantum numbers

$$|N, n_d, v, n_\delta, L, M\rangle.$$

The energies do not explicitly depend on the values of N, n_δ or M:

$$(5.21) \qquad \mathcal{E}_{U(5)} = \epsilon n_d + \kappa_1 n_d(n_d - 1) + \kappa_2 v(v + 3) + \kappa_3 L(L+1).$$

The allowed values for the quantum numbers are obtained from the following relations:

$$
\begin{aligned}
& N \geq n_d \geq v, \\
(5.22) \qquad & \lambda = (v - 3n_\delta) \geq 0 \quad \text{and} \\
& 2\lambda \geq L \geq \lambda,
\end{aligned}
$$

where $L = 2\lambda - 1$ is excluded.

States in the O(6) limit are identified by the quantum numbers

$$|N, \sigma, v, n_\delta, L, M\rangle.$$

Energies are given by the following formula:

$$(5.23) \qquad \mathcal{E}_{O(6)} = \kappa_1 (N - \sigma)(N + \sigma + 4) + \kappa_2 v(v + 3) + \kappa_3 L(L + 1).$$

Examples of both U(5)-like and O(6)-like nuclei have been found and transitions between the different dynamical symmetries have been investigated.

Trying to tie the bosons directly to single-particle states in the nuclear shell model has proven quite challenging. The quadrupole nature of the nuclear force is an emergent property, one that arises from the interactions of many nucleons. As a result, programs that have sought to derive the IBM from first principles have not found much success. This is yet another example of the difficulties encountered when dealing with many-body problems. So, while the IBM can be dismissed as empirical, it nevertheless provides a framework within which it is possible to study nuclear properties. Indirectly, then, it may shed some light on the nature of the nuclear force, although for our purposes, it serves as an introduction to an algebraic approach to theoretical physics.

5.4. Four Forces

In 1898, Ernest Rutherford left the formidable Cavendish Laboratory of Trinity College, Cambridge, where he was studying with J.J. Thomson, because the University refused him a promotion. Anxious to earn enough money to wed and support his fiancé, Rutherford accepted an appointment as the Chair of a new Department of Physics at McGill University in Montreal. Far from the epicenter of Thomson's discovery of the electron, Rutherford found himself without the equipment necessary to continue those studies and launched instead into a study of uranium and thorium salts. Rutherford's initial experiments amounted to measuring the ionic currents emitted from the radioactive elements, which he performed simply by placing the samples between two metallic plates and attaching the plates to an electroscope, as illustrated in figure 5.10. Inserting thin foils between the samples and the upper plate resulted in stepwise reductions in the current. From these results, Rutherford inferred that the radiation consisted of at least two components, that he named α and β. Subsequent investigations demonstrated that α particles were ionized ^{4}He and that β particles were electrons.[8]

Rutherford had discovered that the transmutation of elements from one into another was possible, although the energy scales were such that chemical mechanisms were not possible. The characteristic energy scale for

---

[8]Rutherford was awarded the 1908 Nobel Prize in Chemistry ""for his investigations into the disintegration of the elements, and the chemistry of radioactive substances."

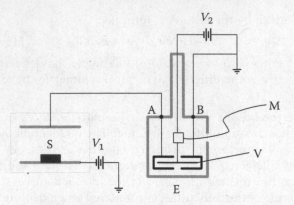

FIGURE 5.10. Rutherford's experiments utilized an electrometer E that was constructed with a fine wire to which was affixed a mirror M and which suspended a vane V. A constant voltage bias was applied to the vane and two sectors of a quartered electrode B. The current from the source S accumulated as a voltage on the remaining two sectors of the quadrant electrode A, producing a torsion of the wire and a deflection of the mirror.

chemical reactions is given by kT, where $k = 8.617 \times 10^{-5}$ eV/K is the Boltzmann constant and T is the absolute temperature (in Kelvin). Even for reactions conducted in furnaces at 1000 K, the characteristic energy is about 86 meV. For nuclear states, the characteristic energy is of order 1 MeV, about seven or eight orders of magnitude greater. No amount of fire or secret ingredients would enable alchemists to transmute lead into gold.

We have discussed some of the emergent behavior observed in heavy nuclei and note that in lighter nuclei, there is evidence that, to some extent, the nuclear wave function possesses a significant component that behaves like α particle clusters. Particularly for nuclei like ^{12}C and ^{16}O, we can observe states that resemble the behavior of three or four, respectively, α particles in a cluster.

When physicists began to investigate β decay processes in more detail, they found a further complication in our understanding of nuclear matter. The discovery of the uncharged neutron provided a ready explanation for the existence of multiple isotopes of the elements: the element is defined by the number of protons and the number of neutrons defines the isotope. The process of β decay seemed to indicate that, on occasion, a neutron could decay into a proton and electron. We now know that the rest masses of the electron and proton are less than the mass of the neutron, making

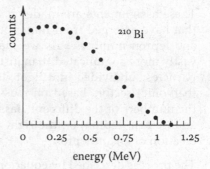

FIGURE 5.11. The measured energy dependence of β decay in ^{210}Bi produces a continuous distribution of emitted electrons.

such a decay energetically feasible. The problem arises when we try to predict the behavior of such a decay.

If, for simplicity, we consider a neutron to be simply a bound electron and proton, then conservation of momentum and energy dictates that the electron and proton emerge in opposite directions and that the bulk of the kinetic energy resides with the emitted electron. Yet, when one measures the kinetic energy of emitted β particles, like those observed in the decay of ^{210}Bi, illustrated in figure 5.11, one finds a continuous distribution of electron energies, with a peak in the spectrum at low energies. Such a distribution apparently contradicts the established principle of momentum conservation.

> EXERCISE 5.21. Consider the (relativistic) dynamics of two-particle decay in the rest frame of a free neutron. What are the energies of the emitted electron and proton? Convince yourself that the electron carries more kinetic energy. Recall that the relativistic kinetic energy is given by $\mathcal{E} - mc^2$.

Indeed, some physicists entertained the notion that, in microscopic systems, momentum might not be conserved or was conserved in a statistical sense. The dilemma was resolved in 1930 by Wolfgang Pauli, who observed that the continuous distribution of electron energies could be explained if the process was a three-body interaction in which one of the decay products was unobserved. Pauli's suggestion was rather fanciful: he postulated the existence of a new particle that did not interact via the electromagnetic or nuclear forces and whose sole purpose was to rescue momentum conservation.

$$(5.24) \qquad\qquad n \longrightarrow p^+ + e^- + \overline{\nu}_e$$

Today, the process is believed to occur via the path indicated in equation 5.24: a neutron decays into a positively charged proton, a negatively charged electron and a neutral (anti-)neutrino.

These assignments are made on the basis of two new conservation laws that have emerged from numerous experiments: conservation of baryon and lepton numbers. As we shall see shortly, the subnuclear world is vastly more complicated than just neutrons and protons, with the various particles subdivided into light (leptonic), medium (mesonic) and heavy (baryonic) sectors based on mass. It seems necessary that, like charge, the numbers of the different classes be conserved. Hence, the creation of an electron (lepton) on the right-hand side of equation 5.24 requires the creation of an antilepton.

The process described by equation 5.24 is one that is common in so-called neutron-rich nuclei. In proton-rich nuclei, one can also observe the reverse reaction:

$$(5.25) \qquad\qquad p^+ + e^- \longrightarrow n + \nu_e,$$

which is known as electron capture or even

$$(5.26) \qquad\qquad p^+ \longrightarrow n + e^+ + \nu_e,$$

where a positron is emitted in the final state. We note that equations 5.24–5.26 all conserve both baryon number and lepton number. Remarkably, the nucleus ^{64}Cu admits all three possibilities, as illustrated in figure 5.12.

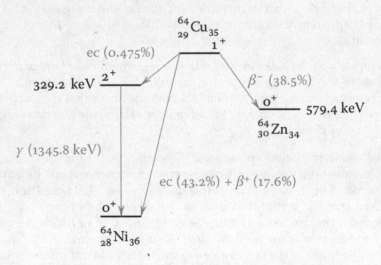

FIGURE 5.12. The nucleus ^{64}Cu decays via β^-, β^+ and electron capture (ec).

The fraction of positron emission can be determined experimentally by detection of the characteristic 511 keV γ rays arising from electron-positron annihilation events. About half of one percent of the time, the decay proceeds through the 2^+ first excited state of ^{64}Ni; these events are characterized by the observation of a characteristic 1346 keV γ ray.

EXERCISE 5.22. Use the NuDat utility of the NNDC website to study the properties of the nucleus ^{37}Ar. In particular, how does the nucleus decay?

In none of the experiments conducted to determine these decay probabilities were the neutrinos directly detected. This is not to say that the weakly interacting neutrinos are invisible but the cross-section for their detection is very small. Nevertheless, neutrinos have been observed and an observatory was constructed to utilize this weakly interacting property. The American physicist Raymond Davis utilized a permutation of the electron capture reaction in ^{37}Ar as a probe of neutrinos generated in fusion reactions in the sun. Nominally, we would have the radioactive ^{37}Ar decay into the energetically favorable ^{37}Cl, as follows:

$$(5.27) \qquad\qquad ^{37}\text{Ar} + e^- \longrightarrow {}^{37}\text{Cl} + \nu_e,$$

where some 814 keV of energy are released in the process. This is just equation 5.25, with some additional nucleons. What Davis recognized is that the reaction will also run in the opposite direction, provided that the incoming neutrinos had an energy above 814 keV.

Davis and his students set up shop in the Homestake Gold Mine in South Dakota, using the 1480 m depth to shield his experiment from cosmic rays other than neutrinos. The detectors were large tanks of perchloroethylene (C_2Cl_4) through which Davis bubbled small amounts of helium gas, to assist the migration of any ^{37}Ar produced. Some 24% of chlorine consists of the ^{37}Cl isotope, with the remainder being ^{35}Cl. The argon product has a half-life of 35 days, so Davis collected the gas emanating from the tanks and simply counted the decays to determine how many ^{37}Ar atoms had been produced and, thereby, the solar neutrino flux.

EXERCISE 5.23. Using the NNDC website, why did Davis not utilize the more common ^{35}Cl isotope in his studies?

Davis worked closely with American physicist John Bahcall, who had calculated the flux. Nominally, the fusion reactions that power the sun were assumed to be defined by the so-called pp-chain. The first step in the chain is given by the following:

$$(5.28) \qquad\qquad p^+ + p^+ \to {}^2\text{He}^* \to d^+ + e^+ + \nu_e,$$

where the unstable diproton generally decays back into two protons but occasionally decays into a deuteron, positron and neutrino. Once deuterium (^{2}H) is formed, there are a number of different pathways in the pp-chain that result in the formation of αs (^{4}He). The predominant branch is expected to be the following:

(5.29) $$d^+ + p^+ \rightarrow {}^3\text{He} + \gamma$$

(5.30) $${}^3\text{He} + {}^3\text{He} \rightarrow {}^4\text{He} + 2p^+.$$

There are other branches but the remarkable feature of neutrinos is that they interact so weakly that the majority leave the center of the sun where they are produced and arrive 1480 m below ground in the Homestake mine in relatively pristine condition.

Because the nuclear reactions that define the various pp-chain steps are so well understood, it came as something of a surprise to Davis and Bahcall that Davis systematically found only about a third the number of neutrinos that Bahcall predicted. Davis's experiments were revised and reconfigured, using bigger tanks and better detectors. They studied potentially abnormal argon chemistries to account for lower than expected detection efficiencies and still the problem persisted.

Other detectors were constructed, notably the Kamiokande and Super-Kamiokande experiments that utilized large water tanks and Čerenkov radiation to identify neutrino interactions.[9] The measurements from these other instruments confirmed Davis's results from the Homestake experiment: there are fewer solar neutrinos than nuclear physics predicts.[10] This was a puzzle that remained unsolved for 30 years.

---

[9]Russian physicists Pavel Alekseyevich Čerenkov, Il'ja Mikhailovich Frank and Igor Yevgenyevich Tamm were awarded the 1958 Nobel Prize in Physics "for the discovery and the interpretation of the Čerenkov effect."

[10]Davis and Japanese physicist Masatoshi Koshiba shared half of the 2002 Nobel Prize in Physics "for pioneering contributions to astrophysics, in particular for the detection of cosmic neutrinos." The Italian-American physicist Riccardo Giacconi was awarded the other half "for pioneering contributions to astrophysics, which have led to the discovery of cosmic X-ray sources."

Toward a Theory of Everything

We can continue our march towards the smallest things by simply continuing the process of scattering energetic electrons from nuclei, or nucleons. We've already noted that initial results of electron scattering demonstrated that protons have internal structure, visible through resonances
in the scattering data. With increased energy, though, analysis becomes
more complicated. For example, the final state of the e^-,p^+ collision is
no longer just e^-,p^+; there are other constituents, so the mass M used in
equation 5.7 will no longer be the proton mass. Instead, M becomes a
dynamical variable, defining the mass of the final hadronic state, usually
denoted by W.

Operationally, the transfer momentum q can be determined from the following expression:

$$(6.1) \qquad q^2 = -4\mathcal{E}\mathcal{E}' \sin^2 \theta/2,$$

where $\mathcal{E}$ is the incident electron energy, $\mathcal{E}'$ is the scattered electron energy and θ is the laboratory scattering angle. The relativistically invariant
parameter W now becomes

$$(6.2) \qquad Wc^2 = \left[M_p^2 c^4 + 2M_p c^2 (\mathcal{E} - \mathcal{E}') - q^2 c^2 \right]^{1/2},$$

where M_p is the proton mass.

In 1969, a collaboration headed by the American physicists Jerome Friedman and Richard Taylor at the Stanford Linear Accelerator Center (SLAC)
and Henry Kendall from MIT repeated the electron scattering experiments with 7 GeV electrons from the newly commissioned machine at
Stanford. What they observed was remarkable.[1] As illustrated in figure 6.1,
the ratio of the measured differential cross-section to the Mott cross-section

---

[1] Friedman, Kendall and Taylor shared the 1990 Nobel Prize in Physics "for their pioneering
investigations concerning deep inelastic scattering of electrons on protons and bound neutrons, which have been of essential importance for the development of the quark model in
particle physics."

© Mark A. Cunningham 2018
M.A. Cunningham, *Beyond Classical Physics*,
Undergraduate Lecture Notes in Physics,
https://doi.org/10.1007/978-3-319-63160-8_6

would be expected to fall exponentially with increasing transfer momentum. This was the result obtained by Hofstadter at lower energies, indicating that the charge distribution of the proton had finite size.

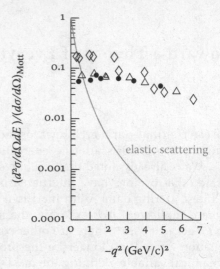

FIGURE 6.1. Scattering cross-sections for different values of Mc^2 show less variation as Mc^2 increases from 2 GeV ($\lozenge$) to 3 GeV ($\triangle$) to 3.5 GeV ($\bullet$).

What Kendall *et al.* observed was that the cross-section ratio became nearly constant at large values of M. Recall that for point charges, the form factor is a constant. So, the results of the deep inelastic scattering experiments suggest that, at short distances, the proton appears to be composed of point-like scattering centers.

6.1. Quarks

The discovery of the nucleus, and its component nucleons: protons and neutrons, ushered in a new era in physics. There had to be a force acting upon the nucleons that kept them restrained to an extraordinarily small volume. Moreover, the force had to be of short range; otherwise, the nuclei would all collapse into a single large nucleus—a result in direct conflict with observation of discrete atoms. An early suggestion on the nature of the nuclear force was provided by the Japanese physicist Hideki Yukawa, who suggested that a potential of the form

$$(6.3) \qquad V(r) = -\frac{g^2}{4\pi} \frac{e^{-\mu r}}{r},$$

where g is the strength of the interaction (analogous to the electric charge) and μ determines the range, would have the requisite properties.[2]

---

[2]Yukawa's "On the interaction of elementary particles. I." was published in the *Proceedings of the Physico-Mathematical Society of Japan* in 1935. Yukawa was awarded the 1949 Nobel

EXERCISE 6.1.   What is the (three-dimensional) Fourier transform of the Yukawa potential? If the transform variable is $\mathbf{k}$, then the spatial component of momentum is given by $\mathbf{p} = \hbar\mathbf{k}$. In the "natural" units preferred by high-energy theorists, where $\hbar = c = 1$, convince yourself that μ can be thought of as a mass.

Shortly after Yukawa's publication, a negatively charged particle with mass between that of the electron and proton was observed by Carl Anderson and Seth Neddermeyer at Caltech in their cosmic ray studies and also by Jabez Street and Edward Stevenson at Harvard.[3] Anderson dubbed the particle a mesotron but it subsequently became known as the muon, as expectations were that this was the mesonic particle Yukawa proposed to mediate the nuclear force. Somewhat unfortunately, the muon was discovered to decay solely to electrons and does not interact through the strong nuclear force, as indicated in the following equations:

(6.4) $$\mu^+ \rightarrow e^+ + \overline{\nu}_\mu + \nu_e \quad \text{and} \quad \mu^- \rightarrow e^- + \nu_\mu + \overline{\nu}_e.$$

Muons decay in approximately 2.2 μs, independent of charge. The three-body nature of the muon decay process was identified from the kinematics of the final-state electrons.[4] A number of years later, in a definitive experiment at the Alternating Gradient Synchrotron at Brookhaven National Laboratory, the distinct neutrino types were established.

Leon Lederman, Melvin Schwartz and Jack Steinberger and their coworkers conducted an experiment in which 13 GeV protons impacted a beryllium target, producing showers of Yukawa's mesons, which subsequently decayed, predominantly into muons and, supposedly, muon neutrinos. The debris from these collisions was passed through some 40 m of iron, repurposed from dismantled navy warships, which served to screen out everything but the neutrinos. These then passed through a 10-ton spark chamber constructed of 25 mm-thick sheets of aluminum, separated by 10 mm gaps. High voltage pulses were applied across the plates, resulting in sparks leaping between the plates where charged particles had left ionization trails in their wake; these were recorded photographically. A large, static magnetic field applied across the spark chamber permitted the determination of the particle momentum. After eight months of data collection, the experiment produced something like 10^{14} neutrinos and 51 spark chamber events. All of these particles were determined

---

Prize in Physics "for his prediction of the existence of mesons on the basis of theoretical work on nuclear forces."

[3]Neddermeyer and Anderson's "Note on the nature of cosmic-ray particles" appeared in the *Physical Review* in 1937, as did Street and Stevenson's subsequent "New evidence for the existence of a particle of mass intermediate between the proton and electron."

[4]Jack Steinberger's "On the range of the electrons in meson decay" was published in the *Physical Review* in 1949.

to be muons, produced by reactions analogous to those defined in equations 5.24–5.26.[5] The Brookhaven experiment demonstrated that there were two distinct classes of neutrinos: those associated with electrons and those associated with muons.

Yukawa's mesons were eventually discovered in photographic emulsions pioneered by the British physicist Cecil Powell in 1947.[6] Powell and his coworkers, notably Donald Perkins and Giuseppe Occhialini, exposed emulsions at altitude—in airplanes or on mountain tops—to mitigate attenuation of the cosmic rays by the atmosphere and then spent days poring over the developed films. They found several cases where the primary π particle stopped in the emulsion, giving rise to a secondary meson μ and, presumably, a neutral particle.

As large accelerators came online and experimenters no longer relied on random cosmic ray events, the constituents of the subatomic world became rather more numerous than anyone expected. Not only were there pions, the supposed Yukawa mesons, but an assortment of other strongly interacting particles: ρs, Ks, ηs together with numerous excitations of nuclear matter: Δs and Ns. There arose a consortium of high-energy physicists: the Particle Data Group (www-pdg.lbl.gov) who undertook the daunting task of collating and organizing the vast amount of data from all of the different experimental groups. Unlike transitions between atomic states, that are connected solely via electromagnetic interactions (photons), transitions between eigenstates of the nuclear force have multiple pathways. Identifying the selection rules associated with decay pathways provided physicists with a means for organizing and understanding the relationships amongst the states.

EXERCISE 6.2.   Use the PDG Summary Tables to identify the decay modes of the $\rho(770)$ meson and the $N(1440)$ baryon.

A decade after the discovery of the pion, physicists found themselves with an abundance of "elementary" particles and embroiled in attempts to find a means of classifying them. The picture that emerged from these attempts is now known as the quark model. In a first step, in 1961, American physicist Murray Gell-Mann and, independently, the Israeli physicist Yuval Ne'eman suggested a classification scheme of octets of mesons that

---

[5]Lederman, Schwartz and Steinberger were awarded the 1988 Nobel Prize in Physics "for the neutrino beam method and the demonstration of the doublet structure of the leptons through the discovery of the muon neutrino."

[6]Powell was awarded the Nobel Prize in Physics in 1950 "for his development of the photographic method of studying nuclear processes and his discoveries regarding mesons made with this method."

Gell-Mann called the Eightfold Way, in reference to the eponymous Buddhist principles. Among the predictions of the Eightfold Way was the existence of a previously unseen baryon: the Ω^-, that was subsequently observed experimentally.[7] In 1964 Gell-Mann and, independently, George Zweig suggested that the classification of particles could be understood if the baryons and mesons were themselves composites, formed from even more elementary particles that Gell-Mann named quarks, a name taken from the James Joyce novel *Finnegans Wake*. The multiplets occupied by the different particles arose from an underlying SU(3) group symmetry.

In this initial model, baryons were constructed from triplets of quarks and mesons from quark-antiquark pairs. The SU(3) symmetry arose from the three types of quarks, labelled u, d and s. The curious names arose from an initial analogy to nuclear isospin, where protons and neutrons were transformed into one another just as different spin states can be transformed into one another: both transformations are described by an $su(2)$ algebra. Hence, one might consider the proton to be the isospin "up" state and the neutron the isospin "down" state but for $su(3)$ there was need of a third "sideways" direction. Studies of the kaon particles, originally discovered by George Rochester and Clifford Butler in 1947 in cloud-chamber photographs, proved that their behavior was dissimilar to that of other particles. The "strange" behavior evinced by K^0 decays, which had two different lifetimes, provided an alternative name for the third quark, which has subsequently been adopted in most discussions.

TABLE 6.1. Quark quantum numbers. All quarks have half-integral spin.

q	B	Q	I_3	S	$\bar{q}$	B	Q	I_3	S
u	$1/3$	$2/3$	$1/2$	0	$\bar{u}$	$-1/3$	$-2/3$	$-1/2$	0
d	$1/3$	$-1/3$	$-1/2$	0	$\bar{d}$	$-1/3$	$1/3$	$1/2$	0
s	$1/3$	$-1/3$	0	-1	$\bar{s}$	$-1/3$	$1/3$	0	1

The quantum numbers associated with the quarks are listed in table 6.1. Each quark has an intrinsic spin $1/2\hbar$, like the electron, and a baryon number B of $1/3$. The quarks themselves possess third-integral charges Q but, magically, all composite particles will have integral charges. The isospin I algebra is also embedded in the $su(3)$ algebra, and the z-component I_3 appears for the u and d quarks. The s quark carries another quantum

---

[7]Gell-Mann was awarded the Nobel Prize in Physics in 1969 "for his contributions and discoveries concerning the classification of elementary particles and their interactions."

number, called strangeness S, that for the oddity of historical precedence is unity for the antiquark.

We can now explain the multiplicity of mesons observed experimentally as due to the different combinations of quark-antiquark pairs that can be obtained from the elements of table 6.1. This is a combinatorial problem that falls into the mathematical domain of group theory and, as a result, there is a large body of mathematical tools at one's disposal to help understand the results. If we look at mesons, we can construct nine different combinations from the quarks listed in the table. These actually form in the $su(3)$ algebra an octet and a singlet. The singlet state is the symmetric $u\bar{u} + d\bar{d} + s\bar{s}$ combination.

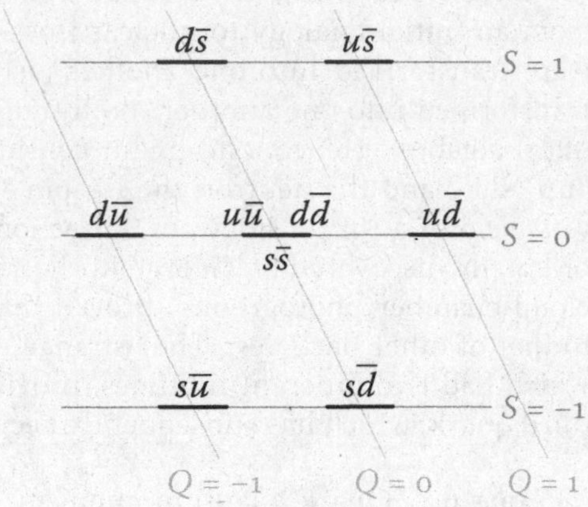

FIGURE 6.2. The nine possible combinations of quarks lead to different values of charge and strangeness.

We can assign the four K mesons to the strange states ($S \neq 0$) of the octet and the three pions, and the eta meson to the $S = 0$ states of the octet. The singlet state can be assigned to the η' meson.

EXERCISE 6.3.   Use the PDG Summary Tables to obtain the masses of the π, K, η and η' mesons. Assign each to a state defined in figure 6.2.

There are 27 different possibilities to generate baryons from triplets of quarks, which separate into a singlet, two octets and a decuplet. The proton and neutron can be assigned to one of the octets, as indicated in figure 6.3. These are all particles with total angular momentum $1/2\hbar$ and includes the Σs and Λs along with the cascade hyperons (Ξ). Of course, with three quarks in a baryon, it is also possible to obtain particles with total angular momentum of $3/2\hbar$.

The combinatorial aspects of the quark model can be illuminated through the use of a device known as the Young tableaux. These representations of the permutation group can be visualized graphically as a two-dimensional

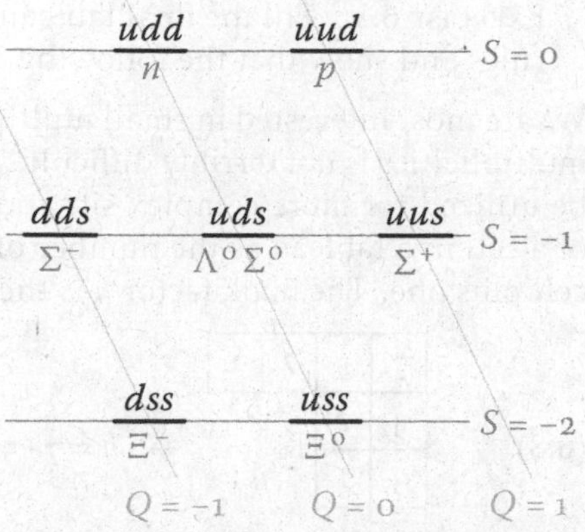

$$
\begin{array}{cc}
\dfrac{udd}{n} & \dfrac{uud}{p}
\end{array} \quad S = 0
$$

FIGURE 6.3. One octet of the possible combinations of three quarks contains the proton and neutron.

$$
\begin{array}{ccc}
\dfrac{dds}{\Sigma^-} & \dfrac{uds}{\Lambda^0 \Sigma^0} & \dfrac{uus}{\Sigma^+}
\end{array} \quad S = -1
$$

$$
\begin{array}{cc}
\dfrac{dss}{\Xi^-} & \dfrac{uss}{\Xi^0}
\end{array} \quad S = -2
$$

$$
Q = -1 \qquad Q = 0 \qquad Q = 1
$$

array of boxes, or tableaux, with (in standard form) decreasing numbers of boxes as you step down the columns. A tableaux with M boxes represents a tensor product of M particles. So, for SU(3), the representation of a single particle is just a single box:

$$
(6.5) \qquad 3 = \boxed{} = \boxed{1} + \boxed{2} + \boxed{3} .
$$

There are three possibilities for the contents of the box: 1, 2, 3 or u, d, s, if we are talking about quarks. Hence the single box in SU(3) is a triplet 3. To construct the adjoint representation, used for antiparticles, one replaces each column of the representation with a column of height $N-c$ and then rotates the resulting diagram around the vertical axis. For SU(3), the adjoint of the triplet is the following:

$$
(6.6) \qquad \overline{3} = \begin{array}{|c|}\hline \\ \hline \\ \hline\end{array} = \begin{array}{|c|}\hline 1 \\ \hline 2 \\ \hline\end{array} + \begin{array}{|c|}\hline 1 \\ \hline 3 \\ \hline\end{array} + \begin{array}{|c|}\hline 2 \\ \hline 3 \\ \hline\end{array} .
$$

Rows of boxes are symmetric products and columns are antisymmetric. Hence, when filling tableaux, the rule is that the index cannot decrease when filling rows but must always increase when filling columns. For the adjoint representation of the triplet, illustrated in equation 6.6, we also find a triplet but it will be designated by $\overline{3}$, to distinguish it. These tableaux correspond to the $d(1,0)$ and $d(0,1)$ irreducible representations of SU(3).

Constructing tensor products can be reduced to stacking boxes. For example, the quark-antiquark products are defined by the following:

$$
(6.7) \qquad 3 \otimes \overline{3} = \boxed{} \otimes \begin{array}{|c|}\hline \\ \hline \\ \hline\end{array} = \begin{array}{|c|}\hline \\ \hline \\ \hline\end{array} \oplus \begin{array}{|c|c|}\hline & \\ \hline & \\ \hline\end{array} .
$$

EXERCISE 6.4. Fill the final tableaux in equation 6.7 according to the rules and show that the following is true: $3 \otimes \bar{3} = 1 \oplus 8$.

We are most interested in small multiplets like $q\bar{q}$ and qqq, so counting the multiplicities is not terribly difficult. There is, however, a formula that can be utilized for more complex situations. We first define the *hook number* of a cell in a tableau as the number of boxes below and to the right of the cell plus one. The hook factor h is the product of all the hook numbers:

(6.8)

7	5	2	1
4	2		
3	1		
1			

$\rightarrow \; h = 7 \cdot 5 \cdot 2 \cdot 1 \cdot 4 \cdot 2 \cdot 3 \cdot 1 \cdot 1 = 1680.$

To determine the multiplicity of the SU(N) irreducible representation, we need to calculate a second factor F. This is obtained by entering N in the first cell and incrementing along the rows and decrementing down the columns. For the tableau used in equation 6.8, we would obtain the following:

(6.9)

N	N+1	N+2	N+3
N-1	N		
N-2	N-1		
N-3			

$\rightarrow \; F = N(N+1)(N+2)(N+3)(N-1)N(N-2)(N-1)(N-3).$

The multiplicity is then $m = F/h$. Note that for this particular tableau that N must be four or larger. Otherwise, the factor F will vanish.

EXERCISE 6.5. Construct the tableaux for the baryon representation $3 \otimes 3 \otimes 3$ (qqq) of SU(3). Note that the construction is associative, so that you can first construct the representations for $3 \otimes 3$ and then apply the final $\otimes 3$ to that result. Use the equations 6.8 and 6.9 to obtain the multiplicities.

The development of the quark model provided a framework for understanding the multitude of different particles that had been uncovered experimentally: nucleons had components and, thereby, could occupy different excited states. There were limitations to the model, though. The SU(3) symmetry was not exact; the meson masses were not close to the same. In the common vernacular, this is a *broken symmetry*. Moreover, now that the pions were recognized not to be Yukawa's mediators of the nuclear force, there was still no resolution to the original problem of nuclear confinement.

6.2. Electroweak Unification

We have previously discussed the rôle of spectroscopy in determining the nature of atomic matter. The lines observed at particular wavelengths indicate the differences in energy between different atomic states. The relative intensity of those lines provides further information about what are termed selection rules. Electric dipole (E1) transitions are the most efficient but states can also be connected through electric quadrupole (E2) or magnetic dipole (M1) transitions at reduced rates. As troublesome as the atomic structure proved to be, the situation for nuclear spectroscopy is more complex.

The energies associated with the excited states of nuclear matter are large enough that particle creation events can occur, in addition to the emission of electromagnetic energy. As a result, the final states of scattering events can generally include a multitude of charged and uncharged particles in addition to photons. Early spectrometers, like those employed by Hofstadter and his students in their studies of inelastic scattering of electrons, relied on the kinematic relationship between scattering angle θ and final state energy $\mathcal{E}'$, so there was no need to directly measure the energy of the final state electron. One could simply integrate the electron current present at a particular angle to provide an experimental estimate of the cross-section. The nuclear final states were not directly observed.

The invention of cloud chambers and, subsequently, bubble chambers provided experimenters with additional tools that enabled the direct visualization of charged particle trajectories through their ionization trails.[8] Analysis of bubble chamber photographs was a tedious enterprise. As physicists sought to understand the behavior of events with small total cross-sections, experimenters had to sift through millions of photographs for a few dozen "interesting" events. Physicists generally made poor scanners, as they tended to find interesting features in every photograph. The initial scanning duties were usually performed by non-physicists, who were told to find specific features. Photographs with those features were then submitted for detailed analysis.

An example of a strange interaction is depicted in figure 6.4, taken at the Lawrence Berkely Lab bubble chamber on March 26, 1959. The light

---

[8]Patrick Maynard Stuart Blackett was awarded the Nobel Prize in Physics in 1948 "for his development of the Wilson cloud chamber method, and his discoveries therewith in the fields of nuclear physics and cosmic radiation." Donald Arthur Glaser was awarded the Nobel Prize in Physics in 1960 "for the invention of the bubble chamber." Luis Walter Alvarez was awarded the Nobel Prize in Physics in 1968 "for his decisive contributions to elementary particle physics, in particular the discovery of a large number of resonance states, made possible through his development of the technique of using hydrogen bubble chamber and data analysis."

FIGURE 6.4. A negative pion π^- enters from the bottom of the image and interacts with a proton in the chamber to form a Σ^0 and K^0. The Σ^0 decays immediately to a Λ^0, which subsequently decays into a π^- and a proton (p). The K^0 eventually decays into a positron (e^+) and a π^-, along with an unobserved neutrino. Image ©2010 The Regents of the University of California, through the Lawrence Berkeley National Laboratory.

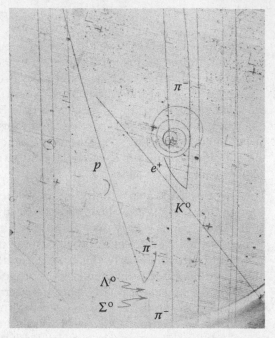

π^- track entering from the bottom of the image disappears. Subsequently, two vertices are observed higher in the frame and both can be traced kinematically back to the end of the π^- track. Another interesting feature of the image is the spiral just above the higher vertex. From the curvature, it is a negatively charged particle, with decreasing radius. It is the more energetic of two electrons scattered by a different incident π^- and not part of the strange interaction.

EXERCISE 6.6.   Use the quark model to describe the events depicted in figure 6.4. Why is the interpretation that a Σ^0 was produced initially, followed by a Λ^0? Hint: consult the PDG particle tables. The K^0 momentum is aligned with the end of the initial π^- track and the $e^+ \pi^-$ vertex. In which direction must the neutrino be emitted?

EXERCISE 6.7.   Consider the decay of a Σ^0 to a Λ^0 and a photon, in the rest frame of the Σ^0. What is the photon four-momentum in the final state, if the Λ^0 is emitted in the positive z-direction?

The interpretation of the photograph is as follows:

$$\pi^- + p \to \Sigma^0 + K^0 \to \pi^- + e^+ + \nu_e$$
(6.10)
$$\downarrow$$
$$\gamma + \Lambda^0 \to p + \pi^-.$$

TABLE 6.2. Properties of light mesons.

	mass (MeV/c²)	lifetime (s)	decay (fraction)
$\pi^\pm$	139.570	2.603×10^{-8}	$\mu^\pm \nu_\mu$
π^0	134.976	8.52×10^{-17}	2γ
$K^\pm$	493.677	1.238×10^{-8}	$\mu^\pm \nu_\mu \,(0.63),\ \pi^\pm \pi^0 \,(0.21),$ $\pi^\pm \pi^+ \pi^- \,(0.06),\ \pi^0 e^\pm \nu_e \,(0.05),$ $\pi^0 \mu^\mp \nu_\mu \,(0.03),\ \pi^\pm \pi^0 \pi^0 (0.02)$
K_S^0	497.611	8.956×10^{-10}	$\pi^+ \pi^- \,(0.69),\ 2\pi^0 (0.31)$
K_L^0	497.611	5.116×10^{-8}	$\pi^\pm e^\mp \nu_e \,(0.41),\ \pi^\pm \mu^\mp \nu_\mu \,(0.27),$ $3\pi^0 \,(0.19),\ \pi^+ \pi^- \pi^0 \,(0.12),$ $\pi^+ \pi^- \,(0.002),\ 2\pi^0 \,(0.0008)$

Analysis of the interaction begins with careful measurement of the tracks to determine particle momenta and, if the track curvature changes, the rate of momentum changes. Several possible kinematic models are then compared to identify which best satisfies the constraints of momentum conservation.

What made the strange interactions so strange was, in large measure, the fact that the decays pictured in figure 6.4 could be observed at all. A rough estimate of the timescales associated with nuclear events can be obtained by dividing the nuclear size by the velocity of light. This suggests that nuclear events should be characterized by times of the order of 10^{-23} s. Yet, the strange particles leave tracks that are centimeters in length, suggesting lifetimes on the order of 10^{-10} s. Lifetimes and decay pathways for the π and K mesons are listed in table 6.2.

EXERCISE 6.8. From the lifetimes listed in table 6.2, determine the characteristic path length $c\tau$. Could all be visualized in bubble chamber photographs?

The K mesons were originally thought to be part of an isospin triplet, like the pions but it was quickly observed that there were two neutral K-mesons, with significantly different lifetimes and decay pathways but possessing (nearly) the same mass. The K_S^0 has the shorter lifetime and decays predominantly to two pions. The K_L^0 has the longer lifetime and decays in a variety of pathways involving pions and leptons or three pions. The two pion decays noted at the end of the table were wholly unexpected. Cronin and Fitch conducted an experiment at the AGS at Brookhaven in which neutral K mesons were passed down a 17.5 m pipe before encountering the detectors. Presumably, the K_S^0 mesons had all decayed by this point, so only K_L^0 mesons should have survived. What Cronin and Fitch observed was that about one in every five hundred events was characteristic of a two-pion decay that should be attributable to K_S^o decay. The conclusion that Cronin and Fitch painstakingly demonstrated was that the forbidden two-pion decay of the K_L^o was not, in fact, forbidden.[9]

The Cronin-Fitch experiment fell close on the heels of another experiment conducted by the Chinese-American nuclear physicist Chien-Shiung Wu in collaboration with low-temperature physicists at the NIST laboratories in Washington DC. The theorists Chen Ning Yang and Tsung-Dao Lee had revisited the experimental support for the CPT theorem that was widely held to guide the behavior of particle interactions. The ideas underlying the CPT theorem were that the equations of motion would be invariant to charge conjugation (swapping particles for anti-particles), parity (space inversions) and time reversals. Lee and Yang found little experimental evidence for the ideas and suggested to Wu that she investigate. To obtain sufficient signal, it was necessary to conduct the experiment at cryogenic temperatures. Wu and her colleagues quickly determined that parity was violated in the β decays of ^{60}Co. The Cronin-Fitch experiment demonstrated that the product CP is also not conserved.[10]

The experimental results forced a retrenchment for particle theorists, as the most popular ideas did not incorporate parity and charge conjugation violations. We shall not try to disentangle the exact chronology of events or the primacy of ideas but shall rather observe that the present state of affairs is based on three fundamental ideas.

The first idea is that particles can be described by a non-Abelian gauge field theory. Recall that quantum electrodynamics is an Abelian gauge field theory: the gauge group is U(1) and the group action amounts to multiplication by a complex number of unit magnitude. This can be represented by the following transformation:

$$\psi' = e^{i\phi}\psi.$$

Application of two such transformations is independent of the order:

$$\psi'' = e^{i\phi}e^{i\theta}\psi = e^{i(\phi+\theta)}\psi = e^{i(\theta+\phi)}\psi = e^{i\theta}e^{i\phi}\psi.$$

This would not necessarily be the case for other groups, in particular the SU(N) groups, where the group elements do not, in general, commute. This problem was studied by Chen Ning Yang and Robert Mills in 1954 under the context of isospin symmetry.

---

[9]James Watson Cronin and Val Logsdon Fitch were awarded the Nobel Prize in Physics in 1980 "for the discovery of violations of fundamental symmetry principles in the decay of neutral K-mesons."

[10]Yang and Lee were awarded the Nobel Prize in Physics in 1957 "for their penetrating investigation of the so-called parity laws which has led to important discoveries regarding the elementary particles."

In modern terminology, the generators t^a of a Lie algebra satisfy the following commutation relations:

(6.11) $$[t^a, t^b] = i f^{abc} t^c,$$

where the coefficients f^{abc} are known as structure constants. The indices a, b and c in equation 6.11 run from one to the total number of generators and are placed as superscripts conventionally. These are not Lorentz indices and the superscript notation does not indicate a Lorentz contravariant. Yang and Mills demonstrated that one could define a field tensor F_{ik}^a as follows:

(6.12) $$F_{ik}^a = \frac{\partial}{\partial x^i} A_k^a - \frac{\partial}{\partial x^k} A_i^a + g \sum_{bc} f^{abc} A_i^b A_k^c,$$

where the A_i^a are gauge potentials, akin to the potentials of the electromagnetic field and g is a coupling constant, akin to the electric charge.

The equations of motion for the fields can be defined through the covariant derivative:

(6.13) $$D_i = \frac{\partial}{\partial x^i} + i\frac{e}{c} A_i + ig \sum_a t^a A_i^a.$$

This is just the minimal coupling that we described in equation 3.12, now expanded to incorporate more complicated algebras. Where previously we had noted that the minimal coupling was just the simplest way to couple particles and field, now this mechanism can be demonstrated to be uniquely determined. The equations of motion for electrons or quarks mimic the structure of the Dirac equation, with the addition of the new gauge group defined by the Lie algebra, instead of just the electromagnetic field. Indeed, the literature expressly utilizes notation like that in equation 6.12 to reinforce the analogy to quantum electrodynamics.

Exercise 6.9. Use the definition of the covariant derivative in equation 6.13 to write the equation for a spinor field ψ, as in equation 3.13.

The second idea is that vector fields that arise in the Yang-Mills equations obtain mass through coupling to a scalar field. This mechanism is often called the BEH mechanism after Robert Brout, François Englert and Peter Higgs.[11] The three modified an earlier suggestion by Jeffrey Goldstone that incorporated a previously unobserved complex scalar field $\varphi = (\varphi_1 + \varphi_2)/\sqrt{2}$. The first component is presumed to have a non-zero

[11]Englert and Higgs were awarded the Nobel Prize in Physics in 2013 "for the theoretical discovery of a mechanism that contributes to our understanding of the origin of mass of subatomic particles, and which recently was confirmed through the discovery of the predicted fundamental particle, by the ATLAS and CMS experiments at CERN's Large Hadron Collider." Brout had passed away in 2011 and was ineligible.

vacuum expectation value $\langle\varphi_1\rangle = v$ and to couple to the gauge field A_i^a through terms like the following:

$$(6.14) \qquad ig\sum_i \frac{\partial}{\partial x^i}\varphi^*\varphi A_i^a - g^2\varphi^*\varphi\sum_{ikbc} f^{abc}g^{ik}A_i^b A_k^c.$$

The second term behaves like a mass $(g^2 v^2)$ multiplying the gauge fields and the first term preserves gauge invariance. The BEH mechanism solves the Yukawa problem of needing massive bosons to mediate the nuclear force. The Yang-Mills fields would otherwise be massless, like the photon. Recent experimental measurements at the Large Hadron Collider at CERN identified a bosonic entity with a mass of roughly 125 GeV that can be interpreted as proof of the existence of the Higgs boson.

The final idea underpinning our current description of particle interactions is symmetry breaking, wherein the vacuum state does not possess all of the symmetry explicit in the Lagrangian. In the case of the weak interactions, we'll see this in more detail presently. Significant support to these ideas was provided by Gerardus t'Hooft and his thesis advisor Martinus Veltman when they were able to prove that Yang-Mills theories that obtained mass through the BEH mechanism are renormalizable.[12] Their result meant that there was the possibility of deriving a coherent theory in terms of gauge fields.

So, the picture of the weak interactions is that they are governed by the Lie algebra $\mathfrak{su}(2)$. There are three generators of $\mathfrak{su}(2)$ that give rise to three massive bosons through the gauge fields A_i^a. These are historically known as the $W^\pm$ and Z° bosons. A Feynman diagram for neutron decay is illustrated in figure 6.5.

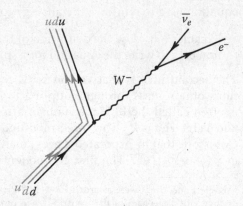

FIGURE 6.5. A d quark in the neutron emits a W^- boson and converts into a u quark. The W^- subsequently decays into an electron and neutrino.

---

[12]t'Hooft and Veltman were awarded the Nobel Prize in Physics in 1999 "for elucidating the quantum structure of electroweak interactions in physics."

Here, the neutron is depicted as an assembly of udd quarks, one of the d quarks is transformed into a u through the emission of a W^- boson. Unlike the photon of the electromagnetic field, which is uncharged, the bosons that mediate the weak interactions can carry electromagnetic charge. In this instance, the W^- boson decays into an electron and (anti-) neutrino.

One might naïvely then think that a composite theory of weak interactions coupled with electromagnetic interactions could be constructed from the direct product of the $su(2)$ algebra describing the weak interactions and the $u(1)$ algebra of electromagnetism. Indeed, the so-called electroweak interaction theory is based on an SU(2)×U(1) gauge group but the existence of CP-violating reactions means that the separation into weak and electromagnetic factors is more complex that one might initially guess.[13] We can suggest that the generators of SU(2) could be described by the triplet (W^-, W°, W^+), in what might be called weak-isospin. The generator of U(1) is another neutral boson B°. CP-violating reactions can be accommodated by mixing the neutral components of the direct product. This is accomplished as follows:

$$(6.15) \qquad \begin{bmatrix} \gamma \\ Z^\circ \end{bmatrix} = \begin{bmatrix} \cos\theta_W & \sin\theta_W \\ -\sin\theta_W & \cos\theta_W \end{bmatrix} \begin{bmatrix} B^\circ \\ W^\circ \end{bmatrix},$$

where θ_W is known as the Weinberg angle or weak mixing angle.

The physical photon that represents the electromagnetic interactions emerges as a composite of the weak isospin neutral boson (W°) and the so-called hypercharge boson (B°). The Weinberg angle can be established in a number of ways experimentally; the most obvious is the relationship of the Z° and $W^\pm$ masses: $\cos\theta_W = M_W/M_Z$, from which a nominal value of $\sin^2\theta_W = 0.23$ can be obtained. The use of the direct product SU(2)×U(1) gauge group also permits the use of the BEH mechanism to give mass to the $W^\pm$ and Z° bosons while leaving the photon massless.

6.3. Standard Model

The development of the electroweak theory was a significant achievement but there was still no resolution of the original problem that we set out to resolve: developing a model for the strong nuclear force. Yukawa's idea of massive mediators of forces to provide strong, short-distance effects seems

---

[13]The Nobel Prize in Physics in 1979 was awarded to Sheldon Glashow, Abdus Salam and Steven Weinberg "for their contributions to the theory of the unified weak and electromagnetic interaction between elementary particles, including, *inter alia*, the prediction of the weak neutral current." The Nobel Prize in Physics in 1984 was awarded to Carlos Rubbia and Simon van der Meer "for their decisive contributions to the large project, which led to the discovery of the field particles W and Z, communicators of weak interaction."

to have worked for the weak nuclear force: the massive mediators had been predicted and then observed experimentally. For the strong force, the situation was different.

The quark model had provided a simplified description of the multitude of baryons and mesons that had been observed but, unlike the massive mediators of the weak force, no massive mediators of the strong force had been observed and certainly no individual quarks had been observed. Initially, due to the fractional charges of the quarks, there was some resistance to the idea that quarks existed. Certainly no fractionally charged particles had ever been observed experimentally.

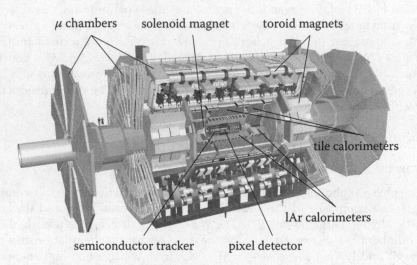

FIGURE 6.6. The ATLAS detector installed on the LHC at CERN extends 44 m between the end μ chambers and is 25 m in diameter. The equipment has a mass of 7000 Mg. Image ©2008 CERN, with permission.

There were significant improvements to experimental technology, driven in large measure by the advent of digital computing. Bubble chambers were supplanted by detectors that provided signals that could be directly interfaced to computers. Moreover, as experimenters began investigating interactions at higher energies, particle lifetimes decreased and multiplicities increased. The advent of the multiwire proportional chamber provided position-sensitive detection of charged particles and several placed sequentially could provide enough information for computer analysis to reconstruct the particle tracks.[14]

Detectors have now evolved into behemoths massing thousands of metric tons (Mg). The largest at present is the ATLAS detector at CERN, illustrated in figure 6.6. The detector is composed of an inner detector with a 2.1 m diameter and 6.2 m length composed of a number of high-resolution, position-sensitive detectors. The pixel detector provides 80 million channels of data, the semiconductor tracker provides another six million channels of information on the position of charged particles and photons in the immediate vicinity of the interaction region. The inner detector is enclosed in a large solenoidal magnet that provides a longitudinal field within the interaction region.

Beyond the inner detector is an array of liquid argon (lAr) and tile calorimeters that measure the energy deposited by electromagnetic and hadronic particles, respectively. The remaining particles leaving the detector are undoubtedly muons; these are identified by the μ chambers surrounding the calorimeters. The only undetected entities are neutrinos that generally do not interact with the detectors.

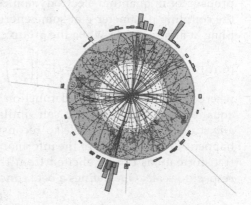

FIGURE 6.7. The ATLAS detector recorded this event on 14 Sep 2011. This z-projection of the central portion of the detector includes four muon tracks (light gray) and several hadron showers. Image ©2011 CERN, with permission.

An interesting event from the ATLAS detector is illustrated in figure 6.7. The final state includes four muons (light gray, long tracks) that are assumed to be the result of a pair of Z° bosons each decaying into $\mu^+\mu^-$ pairs. This is one possible decay pathway for the Higgs boson. The remaining particles dump their energy into the calorimeters. A histogram of energy deposited in each sector indicates the location of hadronic jets.

The Large Hadron Collider (LHC) circulates bunches of approximately 10^{11} protons in two beam pipes that intersect in four stations along the

---

[14]Georges Charpack was awarded the Nobel Prize in Physics in 1992 "for his invention and development of particle detectors, in particular the multiwire proportional chamber."

27 km circumference. At the ATLAS site, bunches cross 4×10^7 times per second, generating 20-25 proton-proton interactions in each crossing, or about 10^9 interaction events per second. Events are filtered in three stages before being recorded for further analysis. Even with a 200,000-fold reduction in event rate by the trigger system, the ATLAS collaboration records over 3 PB (1 PB=10^{15} bytes) annually. Even with this staggering event rate, production of Higgs bosons is estimated to occur about once every three hours. As yet, no fractionally charged particles have ever been observed.

As a result of numerous experiments, theorists came to the conclusion that quarks cannot be separated but it was not clear how to construct a model of the strong interactions. While there was a clear prejudice that the model would have to be a gauge field theory like the electroweak theory, there were some issues. No one knew if such a theory could be renormalizable or if a perturbation theory could be constructed that would be convergent.

One key idea that led to the modern theory is the concept of scale invariance. Gell-Man and Low used the idea to study the behavior of the photon propagator in quantum electrodynamics at high energy. They found that the coupling parameter g at some energy scale μ was related to the coupling at another scale M via the group equation:

$$(6.16) \qquad g(\mu) = G^{-1}\left[\left(\frac{\mu}{M}\right)^d G(g(M))\right],$$

where G is some unspecified function. The consequence of this group equation is that the theory is self-similar; if one can obtain a solution at *any* scale, it will be possible to reconstruct the theory at any other. It happens then that much of the information about the nature of the scaling transformations can be elicited from the variation of the coupling with respect to scale. One defines a beta function:

$$(6.17) \qquad \beta(g) = \frac{\partial g}{\partial \mu}$$

that enables a perturbation reconstruction of the function G, thereby defining the theory at all scales. Kenneth Wilson provided a computational pathway to implement the renormalization group and used the method to solve a longstanding problem in magnetic materials.[15] In the context of magnetic materials, one can envision an array of spins located at the lattice sites. At this most detailed scale, we consider the interaction of neighboring atoms, which provides a natural length scale for the system.

---

[15]Wilson was awarded the Nobel Prize in Physics in 1982 "for his theory for critical phenomena in connection with phase transitions."

For particle theory, there is no natural smallest length scale. This is the essential problem that we faced in quantum electrodynamics that Freeman Dyson resolved. What Wilson recognized was that there was a recursive means to transform between scales. At larger scales, we can consider the effective interaction of blocks of spins made from smaller blocks of spins.

The final key observation was made by David Gross and Frank Wilczek and, independently, David Politzer who found that non-Abelian gauge field theories could have the property that, in the limit as the energy scale goes to infinity, the beta function becomes negative and the effective coupling vanishes.[16] This is known not to be true for quantum electrodynamics, so it was quite surprising to find that *asymptotic freedom* would hold for the strong interactions.

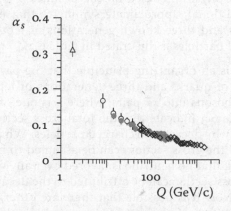

FIGURE 6.8. The strong coupling coefficient α_s decreases as the momentum transfer Q increases. Data are from τ decays ($\triangle$), DIS jets ($\circ$), e^+e^- jets (gray $\bullet$), and $p\overline{p}$ jets ($\diamond$). Error bars are comparable to the size of the points for the $p\overline{p}$ data.

Remarkably, we see evidence for asymptotic freedom in experimental data, as illustrated in figure 6.8. The effective strong coupling factor α_s, analogous to the fine structure constant α used in quantum electrodynamics, can be estimated from results of measurements of cross-sections at different energies.

If we look at the original Yukawa potential, we see that it is divergent at short distances. The picture that emerges from asymptotic freedom is that the quark couplings inside the proton essentially vanish as long as the quarks remain close. Green and Wilczek proposed that nature presents an exact SU(3) symmetry that they called the **color symmetry**; this symmetry is distinct from the approximate SU(3) symmetry of the eightfold way. All baryons are color triplets, composed of three quarks that are, individually, red, green and blue. The quarks exchange gluons, the mediators of the SU(3) color interaction. The gluons are themselves colored and,

---

[16]Gross, Politzer and Wilczek were awarded the Nobel Prize in Physics in 2004 "for the discovery of asymptotic freedom in the theory of the strong interaction."

Table 6.3. Fundamental particles in the Standard Model.

quarks	u	c	t	gauge bosons	g
	d	s	b	weak bosons	$W^{\pm}, Z$
leptons	e	μ	τ	photon	γ
	ν_e	ν_μ	ν_τ	scalar boson	H

so, the individual quark colors do not remain constant but the baryon it-self remains colorless. This theory is known as quantum chromodynamics (QCD).

Overall, the non-Abelian gauge field theory that seems to describe high-energy physics is based on the product of SU(3)×SU(2)×U(1). There is an additional, approximate symmetry that encompasses the six known quarks and three known generations of leptons.[17] The table of funda-mental particles is illustrated in table 6.3.

There is an organizing principle that the particles sort into three gener-ations of quarks and three generations of leptons. Studies of the decay of Z° bosons into $x\overline{x}$ pairs, where x is one of the fermionic constituents of table 6.3 indicate that the total cross section can be explained almost entirely by decays into charged particles. What remains, the invisible por-tion of the cross section, can be assumed to be decays into $\nu\overline{\nu}$ pairs. As-suming that no neutrino is favored over any other, the missing portion of the cross section can be attributed to the decays into three neutrino chan-nels. Now it is possible that there are other, massive neutrinos that are kinematically forbidden but repeated searches for such *sterile* neutrinos have not identified any evidence for their existence.

> Exercise 6.10. Use the PDG Summary Tables to identify the princi-pal decay modes of the Z°. What are the branching ratios for the $\ell\overline{\ell}$ decays for the three generations of leptons $\ell = (e, \mu, \tau)$?

There have been over three decades of physics experiments investigat-ing the nuances of the Standard Model, all of which agree to the percent level. All experiments conducted to date are in good accord with pre-dicted values obtained from the Standard Model. The only experimen-tal discrepancy of note involves the electromagnetic radius of the pro-ton. As we discussed earlier, the proton radius can be obtained through

---

[17]Burton Richter and Samuel Chao Chung Ting were awarded the Nobel Prize in Physics in 1976 "for their pioneering work in the discovery of a heavy elementary particle of a new kind." The Nobel Prize in Physics in 1995 was awarded "for pioneering experimental con-tributions to lepton physics" to Martin L. Perl "for the discovery of the tau lepton" and Frederick Reines "for the detection of the neutrino."

scattering measurements. It is also possible to investigate through precision measurements of the Lamb shift. Recent measurements of the Lamb shift in muonic hydrogen—atoms in which a negative muon replaces the electron—have obtained a distinctly different value.

EXERCISE 6.11. The present PDG value for the proton charge radius is $r_p = 0.8751(61)$ fm, where the uncertainty in the last two digits (standard deviation) is listed in parentheses. The result of Lamb shift measurements in muonic hydrogen is $r_p = 0.84087(39)$ fm. The difference is small but statistically significant. How many standard deviations separate the two?

6.4. Strings

The main pieces of the Standard Model fell into place roughly during the 1970s, with refinements taking place over the next few years. Developers of the theory and experimentalists who provided the data that defined the behavior of the elementary particles were frequently rewarded by the Nobel Prize committee members and one might conclude that there was a great deal of satisfaction within the community. Indeed, the recognition that a non-Abelian gauge field theory could be constructed that was renormalizable, that a host of technical details could be wrestled to ground and that calculations agreed with experimental measurements did provoke a large measure of satisfaction.

On the other hand, the fundamental symmetry of nature embodied in the Standard Model is $SU(3) \times SU(2) \times U(1)$, which seems a bit inelegant, as does the model itself. The concept of an atom composed of protons, neutrons and electrons is tidy. There are three constituents. There is one coupling constant α that needs to be determined. Where we stand now in our understanding of nuclear matter is that there are six kinds of quarks, three generations of electron-like leptons, massive vector bosons, gluons and a scalar Higgs field. This seems like a lot of fundamental entities. So, physicists have not yet decided to call their construction the Standard Theory, like Einstein's Theory of Relativity, as there are about twenty free parameters within the Standard Model that must be fixed by comparison with experiment. Despite its successes, there is no general consensus that the Standard Model has yet risen to the level of being called a theory.

Unsurprisingly, physicists have attempted to find further simplifications and several programs seeking unifications of the four known forces: strong, weak, electromagnetic and gravitational have been launched. In some sense, the construction of the Standard Model followed Einstein's logic in determining his General Theory of Relativity. If you define the symmetry you wish to impose through the choice of gauge group, define the covariant derivative (what mathematicians would call the connection) and pick the number of dimensions in which you want to live, then construction of the theory is a mathematically solved problem. Turning the mathematical crank to obtain the equations of motion is, essentially, straightforward. Solving those equations, of course, remains a nontrivial, often herculean, task.

The important problem physicists face is that there are often no experimental data to provide guidance. There are no accelerators capable of reaching the energies necessary to probe more deeply into what might lie beyond the Standard Model. There are a few very high energy cosmic rays that are observed from time to time but the data rate of interesting events is too low to provide answers within the academic lifetimes of faculty members. So, while there has been much speculation, there has been little progress.

An obvious place to begin is to revisit the $SU(3) \times SU(2) \times U(1)$ symmetry. In 1975, Howard Georgi and Sheldon Glashow recognized that this symmetry could be embedded in the larger $SU(5)$ group. In some sense, this would be more aesthetically pleasing but a consequence of the embedding is that the $SU(5)$ model ultimately predicts proton decay. This is due to the fact that, within the $SU(5)$ model, quarks and electrons are just different members of a particular representation and there exist generators that mix these states. A number of experiments have been conducted and the current limits on the proton lifetime exceed 10^{29} years for all channels and 10^{33} years for specific decays like $p \rightarrow e^{+}\pi^{0}$. Astrophysicists estimate the age of the universe to be on the order of 1.3×10^{10} years, so protons are stable on time scales that vastly exceed the age of the universe. Concisely, $SU(5)$ is not the gauge theory that defines elementary particles.

A number of other grand unified theories were proposed, based on other Lie groups that contained the Standard Model group as a subgroup. Inevitably, these theories predicted the existence of particles that have not yet been observed and transitions (like proton decay) that have also not been observed. Fixing these problems generally required some mathematical sleight of hand to push the masses of the unobserved particles beyond the energy range accessible to experiment.

An alternative approach that was subsequently abandoned was based on an idea originally proposed by Jogesh Pati and Abdus Salam in 1974.[18] Suppose all of the particles in table 6.3 are themselves composite. Pati and Salam coined the term **preons** to describe their "pre-quarks" and found

---

[18]Pati and Salam's "Lepton number as the fourth 'color' " was published in the *Physical Review D*.

that as few as two preon types could explain the part of table 6.3 that was known at the time. A number of subsequent studies also have postulated that the fundamental particles of the Standard Model are composite. The fundamental issue with these ideas is they require the existence of an additional, stronger force that is required to bind the preons into leptons and quarks. At the energies accessible to experimentalists, there has been no definitive signature of the compositeness of the elements in table 6.3.

Prior to the ascendance of the Standard Model, a number of physicists explored an approach to quantum theories known as the S-matrix. Originally proposed[19] by John Wheeler in 1937 as a means of describing atomic nuclei, Heisenberg proposed using the methodology as a basis for understanding particle behavior.[20] The key idea of the S-matrix theory was that the scattering amplitude between initial and final states could be defined, quite abstractly, in terms of a matrix S that was unitary, preserved relativity and was analytic except at poles. This last condition arises as a result of the properties of so-called analytic functions of a complex variable.

If one considers integrating a function $f(z)$ along some closed contour in the complex plane, the French mathematician Augustin-Louis Cauchy demonstrated in 1825 that, for analytic functions, the integral will vanish. Otherwise, the value of the integral is completely determined by singular points of the function.

$$(6.18) \qquad \oint_{\partial S} dz\, f(z) = 2\pi i \sum_j \text{Res}(z_j).$$

In equation 6.18, the sum includes all of the singular points z_j of the function f within the contour that encloses the boundary of some surface ∂S in the complex plane. The residues (Res) of a function can be determined from an analysis of the singularity. If the function f can be expressed as the ratio of two functions $f(z) = g(z)/h(z)$, then f is singular when h vanishes. In this case, we have the following definition of the residue:

$$\text{Res}(z_j) = \frac{g(z_j)}{(dh/dz)_{z_j}}.$$

In cases where the singularity is stronger than first order, the residues can be obtained from higher derivatives of f:

$$\text{Res}(z_j) = \frac{1}{(n-1)!} \lim_{z \to z_j} \frac{d^{n-1}}{dz^{n-1}} \left[(z - z_j)^n f(z) \right],$$

---

[19]Wheeler's "On the mathematical description of light nuclei by the method of resonating group structure" was published in the *Physical Review*.
[20]Heisenberg's "Die beobachtbaren Grössen in der Theorie der Elementarteilche" was published in the *Zeitschrift für Physik* in 1943.

is the result for a pole of order n.

The Italian physicist Tullio Regge developed an approach to solving for the S-matrix scattering amplitudes based, essentially, on the idea of allowing the angular momentum to become a complex number.[21] For a time, analyzing the Regge trajectories: behavior of poles in the complex plane was a singular focus in theoretical physics. Ultimately, physicists concluded that the S-matrix program wasn't providing any additional insights into the nature of the strong force and it was largely abandoned. In 1968, though, the Italian physicist Gabriele Veneziano discovered that Leonhard Euler's beta function could serve as a description of an analytic S-matrix, with an additional property of duality.[22] Veneziano's dual S-matrix captured the attention of a number of theorists.

EXERCISE 6.12. The β function is defined as follows:

$$\beta(x,y) = \frac{\Gamma(x)\Gamma(y)}{\Gamma(x+y)},$$

where Γ is the Euler Gamma function. Plot Gamma for the domain $-10 \leq x \leq 10$. Plot the real and imaginary parts of the complex function Beta[x + I y,-3.2] for $-10 \leq x \leq 10$ and $-5 \leq y \leq 5$.

By 1970, Yoichiro Nambu, Leonard Susskind and Holger Bech Nielsen had found a physical interpretation of Veneziano's theory: it was a quantum mechanical theory that corresponded to a classical system of vibrating strings. The strings could be open or closed, forming a loop upon themselves. A number of physicists contributed to the development of this initial string theory but it didn't lead to the desired grand unification. First, it was discovered that to make the theory consistent, it had to exist in a 25-dimensional space and it included solutions that travelled faster than light. These *tachyon* solutions are harbingers of death, as they destroy causality. While vastly popular in science fiction circles, they are decidedly unwanted in physical theories.

The original string theory only contained bosons but fermions were added to the theory by Pierre Ramond in 1970. To accomplish this, Ramond created an algebra unlike any he had seen previously. We have noted that the Lie algebras for bosons are defined by the commutation relations amongst the generators: $[a_i,a_j] \equiv a_i a_j - a_j a_i = 0$, for example. For fermions, the generators anticommute: $\{a_i,a_j\} \equiv a_i a_j + a_j a_i = 0$. Ramond's algebra had both commuting and anti-commuting operators. Today this is known as supersymmetry.

[21] Regge published "Introduction to complex orbital momenta" in *Il Nuovo Cimento* in 1959.
[22] Veneziano published "Construction of a crossing-symmetric Regge-behaved amplitude for linearly rising Regge trajectories" in *Nuovo Cimento*.

Ramond found that his strings could live peacefully in nine space dimensions and that his theory did not have tachyon solutions. Over the next several years, a number of key results were obtained. First, the theory was found to be relativistically correct. Then, in 1972, Andrei Neveu and Joël Scherk found that superstrings had loop states that correspond to gauge bosons. Finally, in 1974, Scherk, John Schwarz and, independently, Tamiaki Yonega found that some of the massless bosons of the supersymmetric theory could be interpreted as gauge bosons for the gravitational force. Hence, string theory is the fundamental theory that everyone sought.

The beauty that is often attributed to string theory arises from the natural way that gravitation arises from the theory that began as a statement of strong interactions. Open strings correspond to particle/anti-particle pairs and loops correspond to bosons. All interactions are described by the breaking and joining of loops.

Despite large interest in the new theory, there were still concerns that the theory might not work. It might not be renormalizable or have anomalies. These fears were abated by work by Michael Green and John Schwarz in 1984, who demonstrated that anomalies cancelled in a supersymmetric theory based on the group SO(32).[23] Their work spawned a resurgence of interest in string theory but a decade of work ended with the status of the theory still confused.

Physicists soon determined that there were five different superstring theories that were internally consistent. All required nine spatial dimensions for consistency, so six dimensions have to be hidden. The leading idea is that they are wrapped up into an infinitesimal size that is unobservable to us. This is an idea that dates back to an early discovery by the Finnish theorist Gunnar Nordström in 1914 and rediscovered by the German physicist Theodor Kaluza in 1921. Kaluza added a fifth dimension to Einstein's equations from his general theory of relativity.[24] Einstein's metric tensor is extended into five dimensions:

$$(6.19) \qquad \tilde{g}_{ik} = \begin{bmatrix} g_{ik} & \phi^2 A_i \\ \phi^2 A_k & \phi^2 \end{bmatrix}.$$

The Christoffel symbols become five dimensional as well. One can then follow Einstein's methodology and recover not only the theory of general relativity but the electromagnetic stress tensor and Maxwell's equations.

[23]Green and Schwartz published "Anomaly cancellations in supersymmetric D-10 gauge theory and superstring theory" in the *Physics Letters B* in 1984.
[24]Nordström published "Über die Möglichkeit, das Elektromagnetische Feld und das Gravitationsfeld zu vereiningen" in the *Physikalische Zeitschrift* in 1914. Klein published "Zum Unitätsproblem in der Physik" in the *Sitzungsberichte der Königlich Preußischen Akademie der Wissenschaften* in 1921.

In taking up the work, Oskar Klein made the additional assumption that the metric tensor was independent of the fifth degree of freedom[25]:

(6.20)
$$\frac{\partial \tilde{g}_{ik}}{\partial x^5} = 0.$$

This greatly simplified the resulting equations and, of course, made the fifth dimension invisible.

The Kaluza-Klein theory falls just short of miraculous: it fails because the *ad hoc* fifth dimension is not a dynamical variable like the four of space-time. It has to be frozen in order to extract Maxwell's equations independently from the gravitational equations. Allowing it to be dynamical results in processes that mix the electromagnetic and gravitational fields in ways that have not been observed.

In a similar sense, the superstring theories require some sort of mechanism that can render the extra six spatial dimensions invisible. In addition to supersymmetry, the six-dimensional space is assumed to be invariant to conformal transformations; these requirements can be satisfied if each point in the six-dimensional space is characterized by three complex numbers. Such spaces are known as Calabi-Yau spaces, after the mathematicians Eugenio Calabi, who first studied them, and Shing-Tung Yau, who proved Calabi's conjecture that the curvature of such spaces satisfies a particular constraint only if a topological invariant of the space vanishes.

There was initial hope that there might only be a few Calabi-Yau spaces; identifying the one required to make string theory work properly would be tedious but the task wouldn't take long. Unfortunately, there are many Calabi-Yau spaces, possibly an infinite number but this is, as yet, undetermined.

In 1995, physicist Ed Witten proposed that one could *derive* the five known superstring theories from a larger, 11-dimensional supermembrane theory. If you take one of the dimensions of the supermembrane theory to be a circle and wrap one dimension of a membrane around that circle, you now have what appears to be a one dimensional object (string) moving through space. Witten found that there were five ways to wrap membranes around the circle; each corresponded to one of the known superstring theories. He called his invention M-theory but without identifying what he intended M to mean. The strings, though in this representation, are an emergent property of a larger system.

Witten's conjecture spurred a so-called second revolution in string theory. Eleven dimensions turns out to be the largest number of dimensions in

---

[25]Klein published "Quantentheorie und fünfdimensionale Relativitätstheorie" in the *Zeitschrift für Physik* in 1926.

which one could create a theory of supergravity. There was much speculation that this latest conjecture would provide a pathway to the final theory. For a variety of reasons, this has not turned out to be the case. Twenty years later, string theorists now speak of the landscape of 10^{500} potential string theories that might be constructed without any particular means of identifying which one of those might be relevant.

All of the string theories require supersymmetry for mathematical stability but supersymmetry requires that there be a number of particles that are the supersymmetric partners of known particles. These have not been observed. So, if they exist, they must have energies beyond what we can generate in the largest accelerators. If there were a few dozen unexplained bumps in the cross-sections obtained at the LHC, they would provide guidance as to how to winnow the theories down to a tractable number. As it happens, most theorists believe that unification will happen somewhere around the Planck mass:

$$(6.21) \qquad m_\mathrm{P} = \left[\frac{\hbar c}{G}\right]^{1/2} \approx 1.2 \times 10^{19} \ \mathrm{GeV}/c^2.$$

Such energies have only existed at the beginning of the universe, just after the big bang.

Thus, theorists have turned to astronomical measurements to determine if there might be clues embedded in the structure of matter in the universe that will provide guidance. As we shall discuss subsequently, current astronomical ideas indicate that the universe is expanding and has large-scale structure that must have arisen due to early time fluctuations in the matter density. The expansion is potentially an indication that Einstein's cosmological constant is small and positive. As superstring theory includes gravitation, the existence of a positive cosmological constant might provide some direction but has not yet led to any resolution.

On a positive note, the struggles with superstring theory have brokered something of a rapprochement between theoretical physicists and mathematicians. Physicists use mathematics but can be untidy in their approach. Mathematician are exacting but generally unconcerned about applications of their efforts. Einstein required the tutelage of Grossmann to make sense of the dry mathematical derivations and, more recently, string theorists have sought out mathematicians like Michael Atiyah, who are interested in applications.

As we have mentioned, physicists would like to know if soliton solutions are supported by Maxwell's equations. How does one frame such a question mathematically? Well, mathematicians have developed many ideas about the nature of mappings on manifolds in higher dimensions. In particular, the Atiyah-Singer Index Theorem provides an answer to the question of how many solutions can exist for a particular set of differential equations. The fields of differential geometry and analysis contain tools to help understand such behaviors, as does the field of topology. Atiyah's other contributions to mathematics have led to illuminating the connections between such disparate fields and provide calculational tools that even physicists can learn to use. In return, physicists have provided nontrivial examples for mathematicians to consider; both fields can benefit from the interchange.

At this point, there is no earth shattering resolution to the problems besetting superstring theory. No one has found a means to construct a meaningful theory or prove that the entire approach is infeasible. Meanwhile, the Standard Model has held sway for thirty years. It is not elegant but there is no compelling reason to abandon it.

VII

On the Nature of the Chemical Bond

To this point, we have been considering ever smaller length scales in our efforts to understand the fundamental nature of matter. We shall now pivot and begin considering larger entities. Most of the world around us is not composed of individual atoms but is, instead, composed of molecules: specific combinations of atoms that are bound in specific arrangements. This is the world of chemistry.

In some sense, the fact that there is a scientific discipline called Chemistry is disappointing. One might well imagine that there should be a discipline called molecular physics where we simply use all of the quantum mechanics that has been developed this far to calculate all of the properties of molecules that we require. As a practical matter, we will find that calculating the properties of interest for chemical systems lies well beyond the capabilities of current technology.

Chemical systems represent an inherently many-body problem. This is problematic for anyone wishing to compute quantities like enthalpies of formation or reaction rates. As a result, chemistry remains a science dominated by experiment. For example, in 1985 Harold Kroto joined forces with Richard Smalley and Robert Curl. Smalley and Curl had developed a molecular beam spectrometer for use in their studies of metal clusters, illustrated in figure 7.1. After leaving the exit port B, the gas underwent free expansion, cooling the material and freezing the molecular constituents. After another meter of travel, the gas entered a mass spectrometer to identify the constituents.

Kroto was interested in carbon species in the atmospheres of red giant stars and, ultimately, convinced Smalley and Curl to abandon their primary research topics and spend some time investigating carbon. What they observed was wholly unexpected: instead of some sort of smooth distribution of carbon clusters, the group found their spectrum was dominated by a single peak at 60 carbon atoms, with a smaller peak at 70.[1]

---

[1]Kroto *et al.* published "C_{60}: Buckminsterfullerene" in *Nature*. Curl, Kroto and Smalley were awarded the Nobel Prize in Chemistry in 1996 "for their discovery of fullerenes."

© Mark A. Cunningham 2018
M.A. Cunningham, *Beyond Classical Physics*,
Undergraduate Lecture Notes in Physics,
https://doi.org/10.1007/978-3-319-63160-8_7

FIGURE 7.1. High pressure helium gas enters the apparatus
at A. Carbon is ablated from
a rotating disk D by laser
pulses and continues to react
through the tube until reaching point B, where the gas is
free to expand.

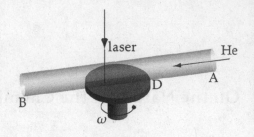

After several tense days of trying to understand why carbon should cluster in such a particular fashion, Smalley and coworkers finally established
the three-dimensional nature of their new molecule, that self-assembled
in the gas phase. The truncated icosahedral shape is reminiscent of a foot-
ball, with alternating panels of hexagons and pentagons, as pictured in
figure 7.2.

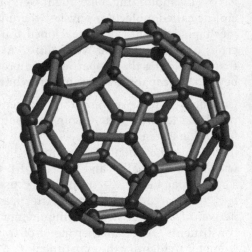

FIGURE 7.2. The C_{60} molecule
possesses icosahedral symmetry. Each atom (dark spheres)
sits at the vertex of a penta-
gon and two hexagons. Atoms
are connected via (gray) tubes
to indicate nearest neighbors.
All sixty atoms are equivalent.

Figure 7.2 represents one possible representation of the C_{60} molecule,
known as a ball-and-stick representation, that emphasizes the network
of interactions between neighboring atoms. There are a number of different representations of molecular structure—each has its own purpose—
but we need to recognize that, just as atoms are not clusters of grapes
surrounded by whirling electrons, molecules are not balls connected by
sticks. Chemists call these interactions **chemical bonds** and they form
the focus of our discussion.

Of course, Curl, Kroto and Smalley did not visualize tiny footballs. Their apparatus simply provided a large peak at the mass 720 Da, which corresponds to sixty copies of ^{12}C, and a smaller peak at 840 Da, corresponding to C_{70}. Confirmation of their proposed structure required accumulating enough C_{60} to conduct infrared spectra, X-ray diffraction data and NMR spectra. A team headed by physicists Wolfgang Krätschmer and Donald Huffman found a means to generate gram quantities of C_{60} and demonstrated convincingly in 1990 that the molecule possesses the three-dimensional structure illustrated in figure 7.2. In the same year, Kroto and his student Jonathan Hare succeeded in obtaining the NMR spectrum of ^{13}C and found it to contain just a single line, demonstrating the equivalence of each of the sixty carbon atoms that comprise the molecule.[2]

Unfortunately, theorists had very little to say about the fullerenes, even after their discovery. Sixty carbon atoms imply 360 electrons scattered around sixty centers. Computers of the day were simply not fast enough to provide any significant information, much less predict the existence of such an unusual structure. Not surprisingly, we have subsequently learned that three-dimensional structures of carbon are vastly more common that we could have imagined.

7.1. Electronic Structure

Chemistry began with the recognition by early researchers like Henry Boyle, Joseph Priestly and Antoine-Laurent Lavoisier that elements combined in particular combinations. Those combinations were related to the elements' positions within the periodic table, illustrated in figure 4.1. We now recognize that the combinations result from the participation of so-called valence electrons, those electrons that are not in closed shells. As a result, all of the atoms in vertical columns in the table behave in similar fashion. For example, all elements in the first column form molecules with a single hydrogen atom: H_2, LiH, NaH, KH, etc. All elements in the last column are chemically inert, and are known as the noble gases.

The combinations of elements that arise to form molecules can be deduced from the simple rule that all atoms seek closed shells, either by donating or accepting electrons. A representation of this behavior is known as the Lewis dot notation. Dots, representing valence electrons, are displayed around the elemental symbol. For second-row elements, the shell closes

---

[2]Krätschmer *et al.* published "C_{60}: A new solid form of carbon" in *Nature*. Kroto *et al.* published "Isolation, separation and characterisation of the fullerenes C60 and C70: the third form of carbon" in the *Journal of the Chemical Society, Chemical Communications*.

with eight electrons; so, we can obtain the explanation for the chemical formula for calcium chloride ($CaCl_2$) as follows:

(7.1) $\cdot Ca \cdot$ $\cdot \ddot{\underset{..}{C}l}:$ $\rightarrow$ $:\ddot{\underset{..}{C}l}: Ca :\ddot{\underset{..}{C}l}:$

Here, calcium contains two electrons beyond the closed shell and chlorine is one electron short of a closed shell. By combining two chlorine atoms with a single calcium atom, all of the atoms achieve closed shells. Calcium does so by donating its two electrons and reverting to the previous closed shell. Each chlorine atom accepts an electron to complete the current shell.

This simple idea can be used to explain many of the bonding patterns for the lighter elements but is of less value when discussing heavier isotopes. Shell closures involving eight electrons, as in equation 7.1, are graphically simple but, when discussing shell closures of 18 electrons, the graphical representation is not so simple. Metals can often have multiple oxidation states. Iron, for example, can exist in both +2 and +3 oxidation states, forming FeO and Fe_2O_3, respectively. Even for light atoms, though, behaviors can be complicated: nitrogen and oxygen can form multiple stable molecules with different stoichiometries.

> EXERCISE 7.1.   Draw the Lewis dot diagrams for carbon, nitrogen and oxygen. Use these to define the following molecules: CO_2, CO, N_2O, NO and NO_2.

While a useful heuristic device for the lighter elements, the Lewis dot method is not quantitative and not terribly predictive. Of course, the answer is to compute the electron density utilizing a quantum approach but such calculations are quite formidable. We have outlined the basic theoretical underpinnings of quantum chemistry calculations; the most accessible are those based upon density-functional theory. There are, of course, a vast number of technical details that we have omitted, so we shall not attempt to construct a density-functional code here.

In this text, we shall utilize the NWChem software package developed at Pacific Northwest Laboratories. It is freely available and was designed for parallel architectures.[3] There are a number of alternative codes but we shall focus on NWChem for its scalability. We will restrict ourselves to rather modest calculations but, given adequate computing resources, the code is capable of handling hundreds of atoms and running on thousands of processors.

---

[3]The code is available at www.nwchem-sw.org. M. Valiev, E.J. Bylaska, N. Govind, K. Kowalski, T.P. Straatsma, H.J.J. van Dam, D. Wang, J. Nieplocha, E. Apra, T.L. Windus, W.A. de Jong published "NWChem: a comprehensive and scalable open-source solution for large scale molecular simulations" in *Computational Physics Communications* in 2010.

Students should be cautious about interpreting results obtained from simulations. The NWChem code is relatively user-friendly, as far as computational chemistry codes go, but there are many ways to make errors in the input files that will lead to spurious results. Even then, as we have seen, even relatively monumental calculations achieve only modest accuracies; it is important to always refer to experimental data for ground truth. Check to ensure that your results are reasonable, whatever that means. Nevertheless, as we shall discover, computational results can be utilized to help interpret sometimes ambiguous experimental results.

The beginning point for calculations is establishing an initial geometry of the molecule you wish to investigate. If you put several carbon, oxygen and hydrogen atoms into a box and shake it, there are potentially dozens of different molecules that could be constructed. So, to start calculations on a particular molecule, you need to begin with a starting structure that is close to the actual structure. Otherwise, you may spend forever working on molecules other than the one of interest. Fortunately, in this modern era, there are several online databases that archive structural data. For small molecules, we will make use of the Crystallographic Open Database (COD), which provides a graphical interface for searching for structural data. This can be quite useful for physicists because chemical names can be quite baffling. If you can draw a picture of the molecule, translation into the chemical name is done automatically in the background. As an example, we shall investigate the properties of the carbon dioxide molecule (CO_2).

EXERCISE 7.2.   Enter the COD website and use the graphical search tool to construct a molecule with two double (=) bonds. Label the outer atoms as oxygens. The central atom will be unlabelled but it is assumed to be carbon. Searching for this structure will retrieve several crystallographic information files (.cif). Select the most recent and save it to your local computer.

As we have mentioned, crystallographers have discovered that there are a finite number of ways in which crystal structures can exist in a three-dimensional world. They are governed by what are known as *point groups* that define the rotational symmetries of the crystal. Any crystal can be decomposed into unit cells, with coordinate axes defined by the vectors **a**, **b** and **c**. The information that defines a particular molecule is written into a format that makes sense to crystallographers. These are crystallographic information files (.cif). The content of the .cif files includes the coordinates of the atoms within the asymmetric unit in terms of fractions of the axes vectors, which is a particularly compact way of specifying the crystal. The complete crystal can be recovered by replication of the asymmetric unit into the unit cell and thence into a larger assembly. At times, we must transform the .cif file into a more usable format. There are a number of possible solutions but we shall utilize the Mercury utility provided by the Cambridge Crystallographic Data Centre.[4]

EXERCISE 7.3. Enter the CCDC website and download the Mercury installer and install Mercury on your computer. Load the CO_2 .cif file. You should see two oxygen atoms and a carbon in the center. Save the molecule as a .xyz file. Mercury provides a number of analysis capabilities. What is the carbon-oxygen bond length? What is the angle formed by the three atoms, with carbon at the vertex?

We now have an initial set of coordinates for an NWChem calculation. One can proceed in a number of different ways depending upon the local NWChem installation. For example, there is a graphical interface called ECCE that enables molecular construction, job submission and analysis of results in a kinder, gentler fashion. For simplicity, we shall avoid that pathway as ECCE only runs on Unix-based (Linux) machines. NWChem is also happier on Linux architectures but Windows and Mac versions can be compiled. So, below is an input script that will run a simple NWChem job under Linux. The script file is named co2 and through some of the early magic in the script will create a variable called JOBNAME that is the name of the file. The script creates an input file named co2.nw and ultimately runs a multiprocessor job with mpirun. The initial module commands utilize the Linux module environment for path management. These are not necessary if the mpirun and nwchem executables are visible to the user through other means.

```
#!/bin/tcsh
#
# This script runs NWChem on a Linux machine
#
module load mpich
module load nwchem
#
setenv NWCHEM_PROCS 4
setenv NWCHEM_MEMORY "1200 mb"
setenv NWCHEM_SCRATCH /home/mark/nwchem/scratch
#
set procid=`echo $$`
setenv JOBNAME `ps -p $procid | grep $procid | awk 'print $4'`
setenv NWCHEM_ROOT `echo $JOBNAME | awk -F_ 'print $1'`
setenv NWCHEM_SUFFIX `echo $JOBNAME | awk -F_ 'print $2'`
mkdir $NWCHEM_SCRATCH/$procid
```

---

[4]The CCDC website is www.ccdc.cam.ac.uk

```
#
# Build the input file
#
cat « finis > $JOBNAME.nw
title "CO2 optimization"
start $NWCHEM_ROOT
permanent_dir $cwd
scratch_dir $NWCHEM_SCRATCH/$procid
memory $NWCHEM_MEMORY
print low
ecce_print $JOBNAME.ecce

charge 0
geometry units angstrom
  C 0.100000 0.000000 0.000000
  O1 0.608040 0.608040 0.608040
  O2 -0.608040 -0.608040 -0.608040
end

basis
  * library "6-31++G**"
end
dft
  xc b3lyp
  iterations 1000
  direct
  noio
end

task dft optimize
dplot
  vectors $NWCHEM_ROOT.movecs
  limitxyz units angstrom
  -3.0 3.0 100
  -3.0 3.0 100
  -3.0 3.0 100
  spin total
  gaussian
  output $NWCHEM_ROOT.cube
end

task dplot
finis
#
# Run the job
#
mpirun -np $NWCHEM_PROCS nwchem $JOBNAME.nw >& $JOBNAME.nwo
#
# Clean up scratch directory
rm -rf $NWCHEM_SCRATCH/$procid
```

The strategy of using a script to write the input files is a convenience that is useful when running many NWChem jobs, particularly sequential operations. Naming files co2_01, co2_02, co2_03, etc., will help the user at least remember the sequence of events when revisiting the calculations at a later date. The .nw file generated by the script is a text file and can be viewed directly. When NWChem executes, it generates a number of scratch files. These are stored in the directory specified by the scratch_dir directive, which the script creates and then removes when complete. The remaining output files will be placed into the directory specified by the permanent_dir directive. The main output of NWChem is directed into a file called co2.nwo.

EXERCISE 7.4. Run the co2 script. (On Linux machines, this can be accomplished via ./co2)

Note that the geometry is that obtained from the .cif file, with the modification that a small offset was made to the carbon atom. This avoids a technical issue with the conversion of the .xyz coordinates into the internal coordinates of the molecule utilized by NWChem. The script specifies a relatively large basis set and conducts a DFT calculation using the B3LYP exchange correlation. These choices are a bit cryptic but should produce a reasonably good description of the charge density without requiring extraordinary compute times. (On the author's laptop, this job runs in under 15 s.) The final task for the script is to utilize the dplot utility to write the three-dimensional charge density to a file called co2.cube.

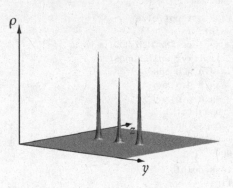

FIGURE 7.3. The charge density ρ for CO_2 on a plane through the nuclear centers is strongly peaked at the nuclear locations. The small asymmetry is due to sampling on a rectangular grid not aligned with the nuclear locations.

The charge density obtained from the NWChem calculation is illustrated in figure 7.3, in which a slice through the nuclear centers is displayed. The density provided by NWChem is positive, reflecting a suppression of the electron charge $q = -e$. The molecule, as can also be seen from the Mercury plots of the crystal structure, is linear. There is a small asymmetry present but this is due to the fact that the density is computed on a grid that is not aligned with the nuclear centers.

We note that the charge density is strongly peaked at the nuclear centers. This is rather different than the usual pictures of molecules presented elsewhere. As a sanity check, NWChem sums the values of the charge density and estimates the volume integral by multiplying by the volume element. In this instance, the integrated charge was $22.05e$, where we would nominally expect a value of $22e$, given the atomic numbers of oxygen (8) and carbon (6). From this result, we can conclude that the electron charge density is largely confined within the 6 Å cube in which the density was computed.

We also note that the C–O distance determined by the x-ray crystallography data is 1.053 Å, whereas the NWChem calculation returned a value of 1.169 Å. We should recognize that the NWChem calculation computes the distance in vacuum, where no other atoms interact with the molecule, whereas the x-ray diffraction data are obtained from solid-phase measurements. This sort of discrepancy may require close investigation to determine whether it is a shortcoming of the calculation that can be improved by using better methods or larger basis sets or whether it arises from crystal-packing effects.

EXERCISE 7.5. Repeat the CO_2 calculation but use the following basis sets: 3-21g*, 6-31g* and 6-311++g**. How does the geometry change with the choice of basis set. Note that the first two basis sets are smaller than we initially used and the last is significantly larger.

It is possible to use *Mathematica* routines to visualize the charge density. The following code will read the .cube file.

```
ReadCube[filename_] :=
Module[{atomList, atomZ, avec, bohrA, bvec, cdens, charge, coordList, cvec, dum,
  fileid, go, head1, head2, molxyz, nAtoms, na, nb, nc, nextline, ufactor, xvec},
bohrA = 0.52917721067;
(*Read the header*)
fileid = OpenRead[filename];
head1 = Read[fileid, String];
head2 = Read[fileid, String];
(*Read the number of atoms and grid origin*)
go = 0, 0, 0;
nextline = Read[fileid,String];
{nAtoms, go[[1]], go[[2]], go[[3]]} = ImportString[nextline,"Table"][[1]];
(*Read the a, b and c vectors and scale from bohrs, if necessary*)
avec = 0, 0, 0;
nextline = Read[fileid, String];
{na, avec[[1]], avec[[2]], avec[[3]]} = ImportString[nextline,"Table"][[1]];
ufactor = If[na > 0, bohrA, 1.];
bvec = 0, 0, 0;
```

```
nextline = Read[fileid,String];
{nb, bvec[[1]], bvec[[2]], bvec[[3]]} = ImportString[nextline,"Table"][[1]];
cvec - 0, 0, 0;
nextline = Read[fileid, String];
{nc, cvec[[1]], cvec[[2]], cvec[[3]]} = ImportString[nextline, "Table"][[1]];
go = ufactor go;
avec = ufactor avec;
bvec = ufactor bvec;
cvec = ufactor cvec;
xvec = 0, 0, 0;
(*Read the list of atoms:  Z, charge, x,y,z*)
atomList = {};
coordList = {};
Do[nextline = Read[fileid, String];
{atomZ, charge, xvec[[1]], xvec[[2]], xvec[[3]]} =
ImportString[nextline, "Table"][[1]];
If[atomZ>0,
atomList = Append[atomList, ElementData[atomZ, "Abbreviation"]];
coordList = Append[coordList, xvec];, dum=1], {nAtoms}];
(* Write a .xyz file into a string, scale the coordinates because the*)
(* default is pm.  We'll use Angstroms.*)
molxyz = ExportString[{atomList, 100. ufactor coordList},
{"XYZ", {"VertexTypes","VertexCoordinates"}}];
(* Read the density and scale to e/A^3*)
cdens = Partition[
Partition[ReadList[fileid, Number, na nb nc], nc], nb]/bohrA^3;
Close[fileid];
{na, avec, nb, bvec, nc, cvec, go, molxyz,cdens}];
```

The ReadCube function requires the name of the .cube file. Its usage is as follows:

```
{na,avec,nb,bvec,nc,cvec,go,molxyz,cdens} = ReadCube["co2.cube"];
```

EXERCISE 7.6.   Read the co2.cube file with the ReadCube function. The volume element is obtained from the triple product $dV = \mathbf{a} \cdot (\mathbf{b} \times \mathbf{c})$ and the sum of the density can be obtained with the Flatten and Total functions. Convince yourself that there are 22 electrons present. The *Mathematica* function Import produces a three-dimensional graphics element when the file type is a chemical .xyz file. The returned variable molxyz contains a string image of such a file. Use the ImportString function to visualize the molecule.

EXERCISE 7.7.   Visualizing three-dimensional densities is problematic. We can use the ListContourPlot3D function to contour the density and use the PlotRange directive to display half of the image.

```
dmax=Max[Flatten[cdens]]
gmax=go+(na-1)avec+(nb-1)bvec+(nc-1)cvec
grange = Table[{go[[i]],gmax[[i]]},{i,1,3}]
prange=grange
prange[[1,2]]=0
ListContourPlot3D[cdens,
Contours->{0.00005 dmax, 0.0005 dmax, 0.005 dmax, 0.05 dmax, 0.5 dmax},
Mesh->None,DataRange->grange,PlotRange->prange ]
```

The code above will plot contours of the charge density for values of half the peak, five percent of the peak, etc.

EXERCISE 7.8.   The cdens variable is a three-dimensional array of size na × nb × nc. Select a two-dimensional slice through the center and plot with the ListPlot3D function.

Select the line through the center of the two-dimensional slice and plot with the ListPlot function. What is the value of the density between the atoms, as a percentage of the peak value?

EXERCISE 7.9.   The integrated density should be just the number of electrons. The numerical sum of the entire density matrix is close to that value (when multiplied by the volume element). Subdivide the volume into three parts and estimate the charge associated with each atom. Do you find integral numbers of electrons on each nuclear center?

As we begin our studies into the nature of the chemical bond, we will make use of numerical simulation as a means of developing an understanding of the fundamental properties of chemical systems. To the extent that the simulations accurately reflect the actual properties of the systems, this affords us the opportunity to conduct studies that would be difficult otherwise.[5] As with other venues in physics, it is crucial to tie these calculations to experimental data to understand any systematic problems.

Using simulations, we can arrange to move the atoms in the simulation in any particular fashion that we desire and conduct constrained optimizations to study the energy dependence associated with different geometries. A particular complexity arises from the sheer size of the parameter space; there are many degrees of freedom in any collection of atoms but simulation enables us to vary them one at a time. The mechanism within NWChem for constrained optimizations relies on the Z-matrix. The geometry defined by the Z-matrix is illustrated in figure 7.4.

---

[5]Larry Curtis and coworkers have published a number of comparisons of DFT calculations, state-of-the-art quantum calculations and experimental data, most recently in 2005 in the *Journal of Chemical Physics*. Their "Assessment of Gaussian-3 and density-functional theories on the G3/05 test set of experimental energies" paints a somewhat dismal picture of the current capabilities of even the best methods.

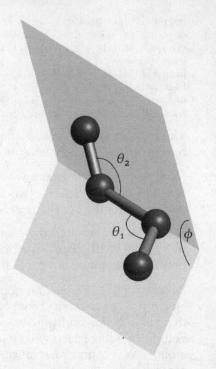

FIGURE 7.4. An alternative way to specify atom positions is the Z-matrix. The Z-matrix defines the atom distances, angles between three adjacent atoms (θ_1, θ_2) and dihedral angles (ϕ) defined as the angle between the two planes formed by four atoms.

The Z-matrix provides an alternative means of specifying the geometry. The geometry block in the previous script could be replaced by the following:

```
geometry
   zmatrix
     O1
     C1  O1  d1
     X   C1  one   O1  ninety
     O2  C1  doc   X   ninety   O1 t1
   variables
     d1  1.16941
     t1  179.99801
   constants
     doc   1.06936
     one   1.0
     ninety 90.
   end
end
```

This will force the distance between the carbon and the second oxygen to remain constant when the other variables change to optimize the geometry. We note the use of the dummy atom X to avoid an ambiguity that arises in specifying linear molecules. The dummy atom is placed an arbitrary distance (1 Å) away from the center atom at an angle of 90°. The O_1–C–O_2 angle is then specified as the dihedral angle.

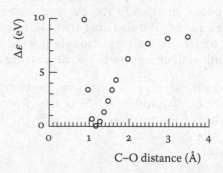

FIGURE 7.5. The difference in energy between the optimized molecule and the molecule with one of the C–O distances constrained shows a sharp minimum at the optimum distance.

One of the difficulties associated with using NWChem, or any other such code, is that it is not particularly user-friendly. In particular, the code uses atomic units internally for efficiency. Distance units are bohrs and energy units are Hartrees. These can be readily converted, as was done in the ReadCube function. Additionally, the authors of such codes tend to be verbose (even with the `print low` directive), so the output files can become quite large. To find some of the more important lines, one can search for the @ character. (On Linux systems, grep @ co2.nwo will search for all occurrences of @ within the file.) Final, optimized geometries can be found in the vicinity of the word "converged." It is possible to remedy some of the tediousness associated with ferreting relevant information from the output files via scripting but we will utilize brute force for now.

EXERCISE 7.10. Replace the geometry block in the original script file with the Z-matrix specified above. Replicate the geometry block and the `task dft optimize` directive several times. Change the value of the constrained distance doc to obtain the energy as a function of distance. See if you can replicate/extend figure 7.5.

What we observe from figure 7.5 is that there is an attractive interaction between the oxygen atom and the oxygen-carbon molecule. There is a sharp minimum in the energy at the distance of 1.17 Å and beyond about 3 Å the energy difference reaches a plateau. At this point, we can consider that the carbon dioxide molecule has separated into a carbon monoxide (CO) molecule and an oxygen atom. Hence, the attractive interaction has a limited range. We also observe that the energy scale is measured in eV, characteristic of chemical systems and six orders of magnitude smaller than the characteristic energies we observed in nuclear systems.

In a nutshell, this is why alchemists were doomed to failure in their quest to transmute lead into gold. It is, of course, possible to strip three protons from a lead nucleus to produce a gold isotope but the energy required to do so far outstrips the energy available in any furnace. Chemical reactions produce energies in the eV range, where nuclear reactions work in the MeV range. We also note that, despite the fact that the charge densities in figure 7.3 are very strongly peaked at the nuclear centers, there is a significant energy barrier to decreasing the internuclear separation.

EXERCISE 7.11.   Calculate the charge densities for the molecules N_2O, NO and NO_2. Note that for calculations with an odd number of electrons, one must add the following directives to the dft block:

```
dft
  odft
  mult 2
  ⋮
end
```

This will perform an open shell calculation with an unpaired electron. What are the intermolecular distances? Describe the molecular shapes. Plot the charge densities.

7.2. Emergent Behavior

A key observation is that molecules possess a three-dimensional structure. This is found in the crystal structures and replicated to a greater or lesser degree in the simulations. Of particular interest to chemists is the structure of the carbon-carbon bond. We illustrate the structures of some small hydrocarbon molecules in figure 7.6. Here, we see that in ethane (C_2H_6), the top structure in the figure, the hydrogen atoms are arrayed in a tetrahedral configuration around the carbon atoms.

In terms of Lewis dot diagrams, we can represent the molecules as follows:

$$
\text{(7.2)} \qquad
\begin{matrix} \text{H} & \text{H} \\ \text{H:} \ddot{\text{C}} \text{:} \ddot{\text{C}} \text{:H} \\ \text{H} & \text{H} \end{matrix}
\qquad
\begin{matrix} \text{H} & & \text{H} \\ \quad \ddot{\text{C}} \text{::} \ddot{\text{C}} \\ \text{H} & & \text{H} \end{matrix}
\qquad
\text{H:C:::C:H}
$$

but while this representation provides a reasonable explanation for the stoichiometries, it does not represent the three-dimensional structures. As we shall see, particularly in the case of biological molecules, the structure of a molecule is directly related to its function.

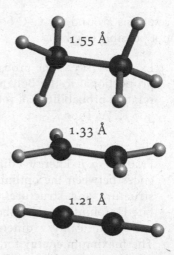

FIGURE 7.6. The carbon-carbon distance changes as more electrons participate in the binding interaction. The geometry of the hydrogen atoms (light gray) around the carbon atoms (dark gray) also changes from tetrahedral (top) to planar (center) to linear (bottom).

The C–C distance in ethane (from our simulations) is 1.55 Å. For the case of ethene (C_2H_4), that distance decreases to 1.33 Å. Chemists would describe ethene as possessing a double bond between the carbons. This is also evidenced by the planar nature of the atoms surrounding the carbon. A triply bonded carbon is depicted in the structure of ethyne (more commonly known as acetylene), in which the molecule is linear and the C–C distance is reduced to 1.21 Å.

EXERCISE 7.12. Calculate the structures for the three simple molecules CH_4, NH_3 and OH_2. What are the bond lengths? What are the angles H–X–H, where X is one of the heavy atoms?

Another characteristic of double bonds is that there is a significant barrier to rotation of the molecule around the double bond. This can be seen in figure 7.7, where we have plotted the differences between the energies of the optimized structures and structures where the molecule has been twisted around the C–C bond. For the ethene structure, the rotational energy displays a two-fold symmetry, minimal when the two planes defined by the two CH_2 groups are aligned and maximal when they are at 90° to one another. The maximum energy is comparable to molecular binding energies, as per figure 7.5. On the other hand, for the ethane structure, the rotational energy displays a three-fold symmetry that is maximal when the two CH_3 groups are aligned and minimal when they are rotated by 60°, as indicated in figure 7.6. The maximum energy is only about 0.12 eV. As we shall see subsequently, the thermal energy in a system is characterized by $k_B T$, where k_B is the Boltzmann constant and T is the absolute temperature. At 300 K, we find $k_B T = 25.8$ meV, suggesting that rotations around the C–C bond in ethane will be possible but generally not possible in ethene.

> EXERCISE 7.13.   The probability that a system can occupy a state is proportional to the Boltzmann factor $\exp[-\Delta\mathcal{E}/k_B T]$. What are the relative probabilities of rotating the ethane and ethene molecules at 300 K? At 1000 K?

FIGURE 7.7. Energy differences between the optimized structure and structures rotated around the C–C bond are significantly different. The maximum energy for the single bond is about 0.12 eV, suggesting that rotations around a single bond will be energetically permitted.

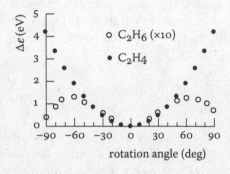

We have not considered rotations for the ethyne molecule (C_2H_2) as there is no rotational degree of freedom around the C–C bond. From a quantum perspective, there is no means for placing a dot on the side of one of the hydrogen atoms to determine whether or not the system is rotating around the axis defined by the C–C bond. It is possible to rotate the molecule around an axis perpendicular to the C–C bond. Those rotations are, of course, quantized.

Another characteristic of molecules is their vibrational spectrum. These vibrational modes can be correlated with the classical mechanical modes of a system of masses coupled by springs, although in a quantum system it is somewhat misleading to take the analogy too literally. In general, there are $3N-6$ vibrational modes in a molecule with N atoms. We subtract three for translations of the center of mass and another three for rotations about the center of mass. For linear molecules, this becomes $3N-5$ as the one rotational degree of freedom around the axis defined by the molecule is unobservable.

We've plotted the computed energies of the vibrational modes of the hydrocarbon molecules in figure 7.8, along with the experimental values. The root-mean-square deviations for ethane, ethene and ethyne are 8 meV, 3 meV and 11 meV, respectively. The general characteristics of the spectra are reasonably well produced but it is not the case that the results are as good as one might hope. Indeed, the calculations include a scaling factor

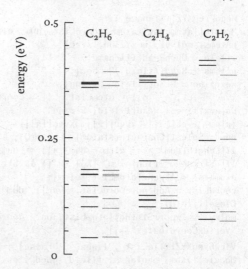

FIGURE 7.8. The experimental (black) and calculated (gray) vibrational levels for the three molecules are plotted. The theoretical values were scaled by 0.965.

of 0.965 which is generally accepted practice as the methodology systematically overestimates the vibrational frequencies.

This scaling of the frequencies is a shortcoming of the models that is known but difficult to remedy. Studies of many systems indicate that different combinations of theoretical model and basis sets will require different scaling factors but for any particular combination, a single scaling factor appears to suffice. This is not a particularly satisfactory state of affairs. Further work is clearly required.

We note that NWChem reports the frequencies in units of cm^{-1} but for consistency, we have converted these values into electron volts. The highest frequencies are associated with vibrations of the hydrogen atoms. All of the vibrational frequencies increase in energy as the bond type shifts from single to double to triple. The (greatly exaggerated) displacements can be visualized with use of the vibrational analysis module in NWChem. After optimization, one can add the following block to the input script:

```
freq
   animate
end
task dft freq
```

This analysis can be a lengthy one and the animate directive will lead to the generation of a host of .xyz files that represent the various modes. These can be visualized by utilizing the following *Mathematica* functions:

```
ReadFreqXYZ[filename_] :=
Module[{atomList, atom, coordList, dum, elem, fileid, frames,
nAtoms, molxyz, nextline, xvec},
fileid = OpenRead[filename];
nAtoms = Read[fileid, Number];
nextline = Read[fileid, String];
xvec = {0., 0., 0.};  atomList = {};  coordList = {};
Do[nextline = Read[fileid, String];
{atom, xvec[[1]], xvec[[2]], xvec[[3]]} = ImportString[nextline, "Table"][[1]];
dum = Select[Characters[atom], LetterQ];
If[Length[dum] > 1, elem = dum[[1]] <> dum[[2]], elem = dum[[1]]];
If[ StringPosition[elem, "X"] == {} && StringPosition[elem, "Bq"] == {},
atomList = Append[atomList, elem];
coordList = Append[coordList, xvec];, dum = 1], {nAtoms}];
Close[fileid];
molxyz = ExportString[{atomList,100. coordList}, {"XYZ", {"VertexTypes",
"VertexCoordinates"}}]];

ViewFreqXYZ[fileList_, linear_: False] :=
Module[{atom, contents, fileid, goodFiles, molframes, nActive,
nAtoms, nModes, nSteps, x, xmin, xmax, y, z},
fileid = OpenRead[ExpandFileName[fileList[[1]]]];
nAtoms = Read[fileid, Number];
Close[fileid];
contents = ImportString[ReadFreqXYZ[ExpandFileName[fileList[[1]]]], "Table"];
{nActive} = contents[[1]];
xmin = Infinity;  xmax = -Infinity;
Do[{atom, x, y, z} = contents[[2 + i]];
xmin = Min[xmin, x, y, z];
xmax = Max[xmax, x, y, z];, i, nActive];
If[linear, nModes = 3 nAtoms - 5, nModes = 3 nAtoms - 6];
goodFiles = Partition[fileList, Length[fileList]/(3*nAtoms)];
nSteps = Length[goodFiles[[1]]] - 1;
goodFiles = goodFiles[[-nModes ;; -1, 1 ;; nSteps]];
molframes =
Table[ImportString[ReadFreqXYZ[ExpandFileName[goodFiles[[i, j]]]], "XYZ",
ViewCenter -> {0.5, 0.5, 0.5}, ViewPoint -> {1.3, -2.4, 2.},
DataRange -> 100 {{xmin, xmax}, {xmin, xmax}, {xmin, xmax}}],
{i, 1, nModes}, {j, 1, nSteps}];
Manipulate[ListAnimate[molframes[[i]]], {i, 1, nModes, 1}]]
```

Note that **ViewFreqXYZ** takes an optional logical argument. Set it to True
if the molecule is linear. The function also sets some viewpoint information to force the rendering engine to be consistent when rendering all of
the frames. These values can be adjusted to obtain different views.

EXERCISE 7.14.   Conduct a frequency analysis of the ethane molecule. The molecular vibrations can be visualized with the following
script:

```
SetDirectory["where you put the files"]
fileList=FileNames["freq.m-*"]
ViewFreqXYZ[fileList]
```

This will take some time to read the relevant files and produce a table of graphics images. These are sorted by mode number and rendered using the ListAnimate function.

Exercise 7.15. A significant example of emergent behavior is the structure of benzene (C_6H_6) that forms a six-membered ring. Construct a Z-matrix for benzene and optimize its structure. A common strategy for larger molecules is to first optimize the structure with a small basis set (3-21g) before optimizing with the target basis set (6-31++g**). (Run the optimization as two steps: first with the basis library 3-21g and then run a second job with the high-level basis library but starting from the geometry of the first run.) This strategy places the atoms in close proximity to their final positions and minimizes the number of times that a DFT calculation has to be performed with the full basis set. What is the structure of this aromatic compound? How is the electron density distributed?

For lighter elements, there are distinct patterns to chemical bonding, reflected by the success of the Lewis dot methodology. If one wants to study metals, the problems become more challenging. The DFT methods we have been utilizing scale approximately like N^3, where N is the total number of electrons, higher order methods are worse: scaling like N^6 or N^8. Consequently, moving from systems with 20–25 electrons to something like Fe_2O_3 with 76 electrons means that calculations will run nearly 30 times longer.[6] With the advent of massively parallel computers, the scope of computational chemistry has been broadened to include larger and larger molecules but a number of alternative approaches have also been considered. For example, it is possible to use effective core potentials and treat only the valence electrons. This reduces the number of active electrons in iron, for example, by 10.

Exercise 7.16. Consider the two molecules FeO and Fe_2O_3. (Note that Fe_2O_3 will be planar. It will be necessary to add dummy atoms out of the plane to avoid geometry issues in the optimization.) Use the 6-31++g** basis for oxygen and the 6-31g** basis along with the associated polarization basis for iron for the structures. What are the iron-oxygen distances? What are the angles defined by O–Fe–O and Fe–O–Fe?

---

[6]Part of the motivation for the initial development of NWChem was the requirement to be able to study plutonium chemistry, without the need for access to plutonium, which is a highly controlled substance. With 94 electrons, even a single plutonium atom in a model made computations nearly intractable.

```
basis
  O library "6-31++g**"
  Fe library "6-31g**"
  Fe library "6-31g** polarization"
end
```

We note that the calculation for iron (III) oxide will require significant time on a desktop or laptop computer. This is rather disappointing as we are generally going to be interested in the properties of bulk materials, not just the gas-phase dynamics of a single molecule. As we shall see going forward, it is possible to treat larger systems but only by sacrificing accuracy. Further approximations will be necessary to overcome the poor scaling properties of current methods.

7.3. Hydrogen Bonding

We have thus far concentrated on what chemists would call covalent bonds. As we have seen in CO_2, for example, the electron density does redistribute itself around the nuclear centers. The relatively mobile valence electrons effectively migrate to neighboring sites. There is a hierarchy of such bonds, the strongest of which are called ionic bonds in which the electron is considered to have completely migrated to the adjacent center. As it happens, there is another binding mode available, one that is key to the chemistry of biological processes: the hydrogen bond.

If one has access to large computing resources, it is reasonably straightforward to observe the phenomena by placing a couple of dozen or so water molecules in close proximity and optimizing the structure. Unfortunately, while that task is reasonably straightforward to state, it is rather formidable in practice. Understanding the properties of bulk water remains a current research topic, as numerical methods do not reach sufficient levels of accuracy and efficiency to supplant experimental data. Engineers have taken great pains to tabulate the properties of water in steam tables. No computational method can, as yet, replicate these data.

So, we shall drastically simplify the problem and consider just two water molecules. If we turn these loose in an optimization, it will likely not converge as the molecules will rotate aimlessly. In order to constrain the optimization, we first optimize a single water molecule. Then, constraining the O–H distances within each molecule to be the optimized value and the H–O–H angles to also be the optimized value, we permit only the distance between molecules and the relative orientation to vary. The dihedral angle between the first water molecule (1 in figure 7.9) and the hydrogen of the second water molecule is fixed at 125.25°c. Further, the oxygen of molecule 2 is constrained to be collinear with the hydrogen and oxygen from molecule 1, as depicted in figure 7.9.

FIGURE 7.9. The geometry of the water duplex studied is depicted with the hydrogen atoms shaded light gray and oxygen atoms shaded dark gray. Only the separation d_{OH} and angle θ_{HOH} were varied in the optimization.

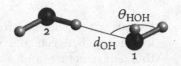

EXERCISE 7.17. Construct the Z-matrix for the water duplex problem. The optimized O–H distance is 0.96528 Å and the optimized angle is 105.72566°. The dihedral angle between the first water molecule and the hydrogen of the second is 125.25°. Compute the optimum distance d_{OH} and angle θ_{HOH} for the water duplex.

EXERCISE 7.18. Use the Z-matrix from the previous exercise to investigate the energy dependence on the dihedral angle. Hold the distance d_{OH} and angle θ_{HOH} fixed and vary the dihedral angle from 90° to 180°. What is the optimum angle?

Conventionally, one specifies the distance between the heavy atoms, even though the calculations constrained the O–H distance. In figure 7.10, we observe that the optimal distance is approximately 3 Å and the well depth is approximately 0.25 eV. Hydrogen bonds are an order of magnitude weaker than covalent bonds and have a restricted range; there is little interaction beyond about 4 Å. Nevertheless, hydrogen bonds play a significant rôle in the properties of water. Indeed, as one undoubtedly learned in elementary school, the formula for water is H_2O but this is true only in a macroscopic sense. In ice, each oxygen atom is tetrahedrally coordinated by four hydrogen atoms. Two hydrogens are covalently bound and two are hydrogen bonded. This crystal structure can be confirmed via x-ray diffraction.

The picture for liquid water is more complex. Using extended x-ray absorption fine structure (XAFS) measurements, one can ascertain that the coordination number of each oxygen in liquid water is 4.4, not 4, as would be expected in a tetrahedral structure. The current interpretation is that, as the solid melts into liquid, the crystal structure becomes disordered and an additional water molecule may invade the interstitial space between lattice water molecules. One suggestion is that liquid water may be characterized by ring structures more than a diffuse tetrahedral alignment but this remains an active area of research. Unfortunately, one requires precise (expensive) theoretical calculations of large numbers (poorly scaling) of water molecules. The precision and size of the calculations to date have not been able to reproduce the known properties.

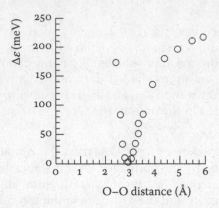

FIGURE 7.10. The difference in energy between the optimized structure and other structures displays a minimum at an oxygen-oxygen distance of about 3 Å. The depth of the potential well is less than 1 eV.

Ideally, one would like to put a large number of water molecules in a box to study the hydrogen-bonding properties but, as we have seen, the calculations scale poorly with increasing N. Moreover, there will be a large number of molecules on the surfaces of the box that do not interact in the same fashion as those in the bulk. So, to minimize the impact of these on the final result, there is an incentive to increase the box size further. Unfortunately, it is impossible to consider boxes of sufficient size that one can make direct contact with experiment. Even a drop of water contains something like 10^{19} water molecules and a calculation involving that number of molecules lies well beyond anything that is currently, or likely ever, feasible. One might even question if the properties of a single drop of water are representative of bulk water.

FIGURE 7.11. The hydronium ion 1 adopts a pyramidal geometry, as opposed to the planar water molecule 2. Oxygens are rendered as dark gray and hydrogens as light gray.

Things are not entirely hopeless, though. While not as quantitative as one would like, simulations do provide insight into the microscopic behavior of chemical systems. One property of water that chemists consider is the pH, notionally, the concentration of protons within the water. Actually, a better description is the concentration of hydronium ions H_3O^+, as protons are energetically favored to be bound to a water molecule rather than floating freely. The proton is, however, quite likely to migrate from one oxygen to another. The hydronium ion is illustrated in figure 7.11 in a model that also contains a water molecule that is aligned in a conformation where one hydrogen from hydronium is hydrogen-bonded to the water molecule.

FIGURE 7.12. In the model, the dihedral angle, the O–O distance and the distance d_{OH} were constrained but the remaining O–H distances and angles θ_1 and θ_2 were optimized. The energy difference peaks when the proton is at the midpoint.

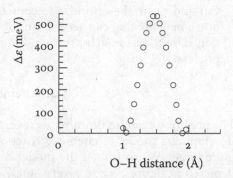

We can perform a series of calculations in which we vary the one O–H distance. In practice, performing an unconstrained optimization of two molecules will result in non-interesting, non-physical rotations. So, one must add constraints to maintain the geometry but allow some flexibility. In this instance, we note that the H–O–H angles for hydronium are somewhat larger than the corresponding angle in water. Additionally, the O–H separations are slightly larger. To accommodate this, we can permit those parameters to vary as the distance d_{OH} is increased. The results of this simulation are depicted in figure 7.12. Note that the energy required to surmount the barrier between the two states is only about 0.55 eV, much lower than would be expected for a covalent bond.

EXERCISE 7.19. Construct a Z-matrix for the hydronium-water interaction. Restrict the non-transferring O–H distances to be identical within each molecule but be different between the two molecules; i.e., there are two distances d_1 and d_2. Each molecule supports a distinct angle H–O–H θ_1 and θ_2. Maintain a constant dihedral angle of 132° between the planes of the water molecules and the transferring proton. Finally, constrain the O–O distance to 2.95 Å.

Run a series of optimizations from $d_{OH} = 1.0$ Å to 1.95 Å, essentially moving the proton from one oxygen to the other. See if you can recover the results of figure 7.12.

EXERCISE 7.20. It is important to visualize the results of the calculations, as it is quite easy to go astray when generating the input files for the simulation. Construct a *Mathematica* function to read the XYZ structure information written to the output .nwo file. Each line consists of the atom index, atom name, charge and position. The

ExportString function can write a proper .xyz file but the Import-String function requires only element names and cannot process numeric identifiers.

Cut and paste the optimized geometries from the output .nwo file into a new file that can serve as input to your new ReadNWGeom function. Use the ListAnimate function to examine the proton transfer process.

7.4. Chemistry

The previous exercise illuminates several of the problems that we face in developing a model of chemical reactions. First, the parameter space is quite large. We restricted the model calculations underlying the results depicted in figure 7.12 to a very small subset of the total space. For example, the intramolecular O–H distances were defined to be the same for both of the hydrogen atoms but these can clearly be different. Additionally, the O–O distance was fixed throughout but it seems likely that this distance would vary as the proton exchange takes place. As a result, the pathway indicated is just one of a multitude of possible pathways between the initial and final states. Additionally, the calculations represent the energetics of a gas-phase reaction whereas much of chemistry takes place in solution, in the condensed phase as it is called.

Nevertheless, the example illustrates an essential part of chemistry: we are interested in the pathways between initial and final states, not solely in the endpoints of the system. The simplest model we can construct for a chemical system is that depicted in figure 7.12, where the system passes from the initial state to the final state through an intermediate transition state. The concept of a transition state was first formulated by the Hungarian-British chemist Michael Polanyi and his British colleague Meredith Evans and, independently, by the Mexican-American chemist Henry Ehring in 1935.[7] This is a crude approximation of reality on many levels but does explain why reactive molecules do not spontaneously combine. For example, it is well known that the combination of hydrogen gas and oxygen gas can form water vapor, accompanied by a prodigious release of thermal energy. Yet, mixing the two in a container does not yield an explosion, at least at room temperature. This is due to the fact that there is an ensemble of states along the pathway from initial to final that are higher in energy than either initial or final states. As a result, the

---

[7]Evans and Polanyi published "Some applications of the transition state method to the calculation of reaction velocities, especially in solution" in the *Transactions of the Faraday Society*. Eyring published "The activated complex in chemical reactions" in the *Journal of Chemical Physics*.

probability of occupying such a state is reduced by the Boltzmann factor $\exp(-\Delta\mathcal{E}^{\ddagger}/k_B T)$, where $\mathcal{E}^{\ddagger}$ is the traditional notation for the energy of the transition state.

Normally, at this juncture, we would begin an exposition on tying the microscopic results that we have obtained thus far into a macroscopic description of the thermodynamics of chemical systems. Historically, this was motivated by the fact that chemical observables are inevitably averages over large numbers of molecules. Consequently, one was forced into a discussion of statistical thermodynamics and chemical potentials.

The situation changed in the late 1980s, when the Egyptian-American chemist Ahmed Zewail and his students began utilizing ultrashort laser pulses to study the femtosecond behavior of chemical bond reorganizations.[8] Their work was made possible by rapid advances in laser technology that permitted stable laser pulses with full width at half maximum intensity on the order of 50-100 fs. Zewail recognized that this time scale is comparable to time it takes to form or break chemical bonds and so provided a new window into studying transition states directly.

The Femtosecond Transition-state Spectroscopy (FTS) method pioneered by Zewail utilized a pump/probe technique, sketched in figure 7.13. The pump beam is generated via any of several steps of dye-lasers or nonlinear, second-harmonic generation of an initial, infrared laser beam. In the initial studies on iodine cyanide (ICN), the pump beam had a wavelength of 306 nm, which corresponds to a photon energy of 4.05 eV, which is sufficient to cause the dissociation of ICN into I and CN fragments. The progress of the dissociation reaction is monitored via a probe pulse, which in the case of ICN was resonant with CN at about 388 nm. The probe beam can also be generated through sum/difference mixing of different source beams, resulting in the ability to tune the beam across a relatively wide bandwidth. The relative polarizations could be defined independently and the arrival times of the two pulses was controlled by an actuator capable of submicron precision. The actuator adjusts the pathlength of one of the two beams, shown as the probe beam in figure 7.13.

EXERCISE 7.21. What is the relative change in path length required to obtain a 100 fs change in the arrival time of the laser pulse? Plot the waveform of a pulse with a wavelength of 306 nm and a width of 100 fs. (Assume a Gaussian profile for the amplitude.)

---

[8]Zewail and his students published "Femtosecond real-time probing of reactions" in several parts in the *Journal of Chemical Physics* beginning in late 1988. Zewail was awarded the Nobel Prize in Chemistry in 1999 "for his studies of the transition states of chemical reactions using femtosecond spectroscopy."

FIGURE 7.13. The pump and probe beams are generated from a single, initial laser pulse. Polarization can either be parallel or orthogonal (as shown). The probe pulse travels down an additional leg controlled by the actuator (A) that can adjust the relative arrival time. The beams are recombined through a half-silvered mirror (M4) and directed through the molecular beam (or gas cell). The detector (D) is placed out of the beam/laser plane to improve sensitivity. For ionization experiments, a set of voltage grids (V1,V2) provide a bias voltage to drive the ionic current.

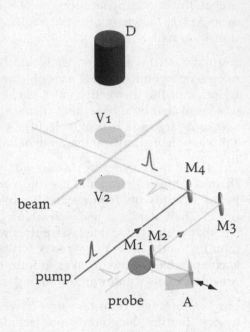

Autocorrelation experiments on the pump/probe beams demonstrated that the widths of the laser pulses was below 100 fm. In order to gauge the time resolution of their system, Zewail and his team utilized the multi-photon ionization of N,N-diethylanaline (DEA). The ionization of DEA happens promptly upon absorption of pump and probe photons. A set of voltage grids provides a bias potential that allows collection of the ion current in a photomultiplier tube. The gray points in figure 7.14 demonstrate the rise time of the ionization current, illustrating the instrument factor that we discussed earlier. The output signal is the convolution of essentially a step function in current with the instrument sensitivity.

After confirming their time resolution, Zewail and team examined the dissociation reaction of iodine cyanide (ICN) into iodine (I) and cyanide (CN) fragments. Detection of the cyanide fragment was performed by using a probe beam with a wavelength of about 388 Å, that caused laser-induced fluorescence that could be detected by the photomultiplier tube. Having determined time zero of their experiment to be the midpoint between the off- and on-levels of the DEA ionization curve, Zewail was able to determine that the dissociation reaction required 205±30 fs to complete. An example of one experimental ICN run is also illustrated in figure 7.14.

FIGURE 7.14. The autoioniza-
tion signal (gray dots) from
N,N-diethylanaline (gray figure) defines the rise time of
the system. The disasso-
ciation of ICN (black dots)
is delayed by approximately
200 fs.

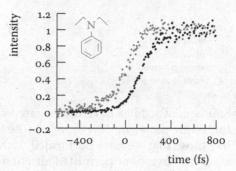

Moreover, by lengthening the wavelength of the probe signal, Zewail could
determine the progress of the reaction along a potential energy surface He
found that they could explain their results with a model based on treating the nuclear motions in a classical sense, thus validating the general
strategy employed in most electronic structure calculations. The Born-
Oppenheimer approximation presumes that the electronic structure equi-
librates instantaneously to accommodate nuclear motion. This is an oversimplification but Zewail's studies provide experimental support that, at
least at lower energies, this is a viable strategy.

FIGURE 7.15. The difference
between energies of the ICN
optimized structure (d_{IC} =
2.01 Å) and molecules with
the I–C distance constrained
to larger values (open circles)
suggests a dissociation energy
of about 3.5 eV. The triplet
state (gray dots) populated
by the laser pump pulse lies
above the dissociation energy.

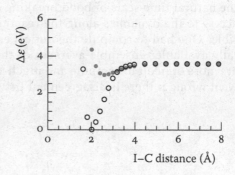

EXERCISE 7.22. Construct a model of the ICN system in NWChem.
Use the 6-311G* basis and the accompanying polarization basis for
the iodine atom. Use the 6-31++G** basis for carbon and nitrogen.
First compute the optimized structure and then control the I–C distance out to 7 Å. How does the electron density evolve along the
pathway? What is the total charge around the iodine nucleus? Now
compute the triplet (mult 3) excited state and then control the I–C
distance as before. Hint: tighten the convergence criteria with the
following directive:

```
driver
   gmax 0.00015
   grms 0.00010
   xmax 0.00060
   xrms 0.00040
end
```

Subsequent studies have investigated a number of chemical processes at the single-molecule level that can be directly compared to calculations like those that we have performed with NWChem. Improving laser technologies have now permitted discussion of attosecond-scale resolution of the bond-forming and bond-breaking processes that define chemistry. The ability to study systems like ICN that possess relatively few internal degrees of freedom has provided a tractable theoretical problem. We will shortly dive into the realm of intractable theoretical problems but can note that the chemical systems provide an arena in which quantum and classical ideas are often juxtaposed. The electron densities certainly require a quantum treatment but nuclear motions can often be treated via simple Newtonian mechanics.

What is compelling about the work of Zewail and his students is that they provided experimental data that could be compared more directly with theoretical results than was previously possible. They were able to effectively watch single molecules evolve at a time scale that corresponded to the natural time scale of bond breaking and bond formation. Previously, access to the dynamics along the reaction pathways was simply inaccessible. One had to compute theoretical estimates of what was experimentally available: ensemble averages of thermodynamic properties. These are more difficult to compute and much more difficult to understand what went wrong if there is disagreement between theory and experiment.

On Solids

Given the difficulties that we have encountered in our studies of molecular physics, it might seem that investigations of macroscopic objects like crystals would be completely intractable. After all Avogadro's number is so large that even a piece of material the size of a grain of rice contains something like 10^{19} atoms. Coping with numbers of that size is clearly beyond the capability of any computer in existence or planned for the foreseeable future. Nevertheless, it is quite possible to deal with such systems by exploiting the symmetries available in crystals.

Here, we are discussing the spatial symmetry demonstrated by the atoms in a crystal structure, not the invariance of equations in some more abstract vector space. As the Braggs discovered in their early diffraction studies, atoms in crystals sit at lattice sites that repeat endlessly. The number of different possibilities for crystal lattices was determined to be 14 by the French physicist Auguste Bravais in 1848.[1] This number arises from the requirement for translational invariance.

As depicted in figure 8.1, one can define lattice vectors $\mathbf{a}$, $\mathbf{b}$ and $\mathbf{c}$ that serve as a basis for the lattice. All lattice points can be constructed from integral multiples of the basis vectors. That is, we must have the following relation:

$$(8.1) \qquad \mathbf{x} = N_a \mathbf{a} + N_b \mathbf{b} + N_c \mathbf{c},$$

where $\mathbf{x}$ is any lattice site and N_a, N_b and N_c are integers. One can also describe the lattice in terms of the magnitudes of the basis vectors: a, b and c, respectively and the angles between the vectors α, β and γ, as indicated in the figure.

> EXERCISE 8.1. Use the `LatticeData` function to plot the different Bravais lattices. What are the lattice vectors corresponding to each type?

---

[1]Bravais published "Mémoire sur les systèmes formés par des points distribués regulièrement sur un plan ou dans l'espace" in the *Journal de L'Ecole Polytechnique*.

© Mark A. Cunningham 2018
M.A. Cunningham, *Beyond Classical Physics*,
Undergraduate Lecture Notes in Physics,
https://doi.org/10.1007/978-3-319-63160-8_8

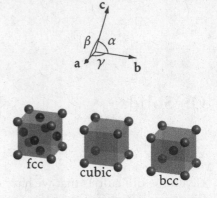

FIGURE 8.1. The basis vectors **a**, **b** and **c** define the lattice sites. For the cubic lattice, there are two additional Bravais lattices: face-centered (fcc) and body-centered (bcc).

The Bravais lattices are subdivided into seven subgroups. These are listed in table 8.1. The first three entries in the table all have orthogonal lattice vectors but differ in the relative lengths. Trigonal crystals have constant lattice vector lengths and equal, but not orthogonal, angles between the vectors. The remaining types have lattice vectors that differ in length and angles. All of the types, though, are prisms, with varying degrees of asymmetry and skewness.

TABLE 8.1. Crystal lattices are segregated into classes depending upon the relationships of the lattice vectors.

cubic	$a = b = c$	$\alpha = \beta = \gamma = \pi/2$
tetragonal	$a = b \neq c$	$\alpha = \beta = \gamma = \pi/2$
orthorhombic	$a \neq b \neq c$	$\alpha = \beta = \gamma = \pi/2$
trigonal	$a = b = c$	$\pi/2 < \alpha = \beta = \gamma < 2\pi/3$
hexagonal	$a = b \neq c$	$\alpha = \beta = \pi/2, \gamma = 2\pi/3$
monoclinic	$a \neq b \neq c$	$\alpha = \beta = \pi/2 \neq \gamma$
triclinic	$a \neq b \neq c$	$\alpha \neq \beta \neq \gamma$

In dealing now with molecular crystals, we can envision replacing the small spheres in figure 8.1 with replicas of the molecules. So, in addition to the basic lattice structures, there are potentially a number of other symmetry operations that can be applied. There are, in fact, 230 different crystallographic point groups. We shall not bother to list them all here.

The point groups arise from different symmetry operations like rotations and reflections. For example, the simple cubic lattice is invariant under rotations by $\pi/2$ around any axis. This is a four-fold rotational symmetry.

The molecular symmetry will affect the overall crystal symmetry. In figure 8.2, we illustrate a situation where there are molecules located at the

FIGURE 8.2. A simple molecule is placed at hexagonal lattice points but alternates in orientation. There is a three-fold rotational symmetry.

vertices of a hexagon, which has six-fold symmetry. The molecules are oriented alternately around the hexagon. Thus, the six-fold symmetry is reduced to a three-fold symmetry. Rotating by $\pi/3$ produces a lattice that is inverted along the vertical axis, not the same lattice.

There are 32 different point groups but the total number of possible crystal symmetries is not 32×14 but 230, due to degeneracies. Our main point is that the number is finite. Crystals only exist in a relatively small number of configurations despite being constructed from vastly larger numbers of atoms and molecules. There is a complex nomenclature for the crystal structures but we shall not focus on those details here. In practice, the Bravais lattice defines what is known as the unit cell. This is the translationally invariant part of the crystal lattice. The unit cell, in turn, can be reconstructed from the asymmetric unit and copies of the asymmetric unit made by the generators of the symmetry group. The asymmetric unit is not defined uniquely; any of the symmetry-generated copies could also serve as the asymmetric unit.

EXERCISE 8.2. Retrieve some .cif files from the Crystallography Open Database. (COD ID 1507553 and 5000108 are reasonable first choices.) Load the .cif files into the Mercury program. This provides the ability to examine the unit cell by selecting the Packing option in the Display, or the Asymmetric Unit (default). Describe the symmetry transformations that define the unit cell.

The list of symmetry groups includes cyclic groups of order 1, 2, 3, 4 and 6. The groups of order five and orders larger than six cannot be used to generate unit cells that are translationally invariant. So, it seemed the problem was solved. Crystallographers improved their instrumentation and gained access to digital computers but all of the structures that they observed fell neatly into one of the 230 possible crystal groups. Remarkably, in 1974, British mathematician Roger Penrose found a means of producing objects with five-fold symmetry but which did not possess translational invariance.[2]. These objects can be constructed from simple primitives that give rise to a long-range five-fold rotational symmetry but the pattern does not simply repeat.

EXERCISE 8.3.   Download the "Penrose Tiles" notebook from the Wolfram Demonstrations Project website. This provides a simple demonstration of the construction of the quasicrystal. The quasicrystals do not have translational invariance but do demonstrate self-similarity: the quasicrystal looks the same at different length scales.

Even more remarkably, the Israeli material scientist Dan Shechtman discovered materials with five-fold (actually icosahedral) symmetry in 1982.[3] While on sabbatical at the National Bureau of Standards laboratories (now the National Institute of Standards and Technology (NIST)), Shechtman observed ten-fold symmetry in a diffraction pattern from a sample of an aluminum-manganese alloy, as depicted in figure 8.3. In particular, note the pentagonal placement of the diffraction points. Other two-fold and three-fold patterns from the same crystal led Shechtman to conclude that the material possessed icosahedral symmetry. The icosahedron was a geometric solid known to the Greeks but one cannot produce a crystal by stacking icosahedra without leaving gaps.

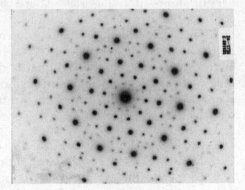

FIGURE   8.3. The   electron diffraction pattern of a sample of an Al-Mn alloy displays ten-fold symmetry.   Image provided courtesy of Dan Shechtman, who acquired it on April 8, 1982, while on a guest appointment at the NIST laboratories.

---

[2]Penrose published "The rôle of aesthetics in pure and applied mathematics" in the *Bulletin of the Institute of Mathematics and its Applications*
[3]Shechtman and his collaborators published "Metallic phase with long range orientational order and no translation symmetry," in the *Physical Review Letters* in 1984. Shechtman was awarded the Nobel Prize in Chemistry in 2011 "for his discovery of quasicrystals."

Shechtman's results were fiercely debated for years and publication of his initial discovery was delayed for two years. A range of notable scientists, including Nobel Prize winning chemist Linus Pauling, were quite vocal about their disbelief of Shechtman's discovery. Due to the relatively simple proof that crystals cannot have five-fold symmetry, Shechtman's results were attributed to crystal defects, twinning, poor experimental controls and fantasy. Nevertheless, Shechtman persisted and other groups have subsequently produced icosahedral quasicrystals. The existence of solid-phase materials with long-range order that possesses five-fold symmetry, but is not translationally invariant, is now well established, as can be seen in figure 8.4.

FIGURE 8.4. A quasicrystal of a scandium-zinc alloy displays icosahedral symmetry. Image courtesy of Paul C. Canfield at Ames Laboratory and Department of Physics and Astronomy, Iowa State University.

The quasicrystal is an alloy of scandium and zinc, that is obviously macroscopic. It was obtained by Paul Canfield, Alan Goldman and coworkers at Ames Laboratory, who were investigating rare-earth binary systems. The icosahedral form, when illuminated with x-rays, produces diffraction patterns like those depicted in figure 8.3. This is not an amorphous material shaped into an icosahedron by grinding or other processing. It is another example of Shechtman's impossible quasicrystalline material.

8.1. Bulk Properties

Moving back to our original discussion of solid crystalline materials, the Kohn-Sham equations that define the wavefunction or electron density must now satisfy the additional constraint that the density be periodic. (We shall avoid the complications that arise in quasicrystals in this discussion.) Each unit cell is a replica of all of the other unit cells; this feature will permit us to describe the properties of macroscopic-scale systems. In our discussions to this point, the basis functions for the calculations have been gaussians centered on the nuclear sites. These have proven to be reliable and robust for molecular simulations. They also have the property that the electron density vanishes as the distance tends to infinity.

In crystals, this last property is not necessary. All that is required is that the density be continuous at the boundaries of the unit cell. This suggests a Fourier transform approach (also called the plane-wave approximation) may be successful in describing the electron density. Such a method was proposed by the Swiss physicist Felix Bloch in 1929.[4] The wave functions can then be expanded as follows:

$$(8.2) \qquad \psi(\mathbf{r}) = \sum_n e^{i\mathbf{k}_n \cdot \mathbf{r}} u(\mathbf{r}),$$

subject to the additional periodicity constraint:

$$(8.3) \qquad \psi(\mathbf{r}) = \psi(\mathbf{r} + N_a \mathbf{a} + N_b \mathbf{b} + N_c \mathbf{c}).$$

The periodicity constraint is naturally satisfied if the wavenumbers $\mathbf{k}_n$ are equal to 2π times the reciprocal vectors $\tilde{\mathbf{a}}$, $\tilde{\mathbf{b}}$ and $\tilde{\mathbf{c}}$.

As a result, solutions of the infinite crystal can be determined quite readily, particularly for simple systems. The program NWChem is capable of such calculations. Before entertaining such calculations, we need to define some vocabulary. Wigner and Seitz provided a mechanism for defining the volume occupied by a particular atom in the lattice, which corresponds to determining the Voronoi cell.[5] One selects a particular atom in the lattice and then draws a plane perpendicular to the line joining that atom and one of its neighbors, at the midpoint of the line. Repeating the process for all of the neighbors ultimately results in a polygonal volume element that contains the selected atom.

One can also do the same in the reciprocal lattice, as indicated in figure 8.5. The volume here is known as the *Brillouin zone*, after the French physicist Léon Brillouin. Using the Bloch formalism, solutions can be determined solely from their behavior in a single Brillouin zone. The particular polygon obtained depends upon the lattice vectors. For an fcc crystal, the Brillouin zone is a truncated octahedron. Historically, the center of the Brillouin zone is denoted Γ. There are eight points L at the centers of hexagonal faces and six points X at the centers of square faces. There are 24 points W at the vertices and 24 points U at the midpoints of the edges of the square faces. Finally, there are 12 points K at the midpoints of the shared edges of two hexagonal faces.

EXERCISE 8.4. The lattice vectors for an fcc lattice are $\mathbf{a} = (0, \frac{1}{2}, \frac{1}{2})$, $\mathbf{b} = (\frac{1}{2}, 0, \frac{1}{2})$ and $\mathbf{c} = (\frac{1}{2}, \frac{1}{2}, 0)$. Calculate the reciprocal vectors. Use

---

[4]Bloch published "Über die Quantenmechanik der Elektronen in Kristallgittern" in the *Zeitschrift für Physik*.
[5]Eugene Wigner and Frederick Seitz published "On the constitution of metallic sodium" in the *Physical Review* in 1933.

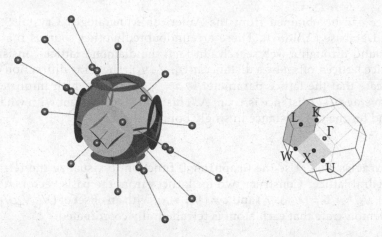

FIGURE 8.5. The reciprocal lattice of an fcc crystal is depicted at left. A series of planes dividing the lines joining other atoms in the lattice at their midpoints defines a volume known as the Brillouin zone, a truncated octahedron. The center of the Brillouin zone is known as Γ and other symmetry points are also identified at right.

the Graphics3D function to display an array of wave vectors:

$$\mathbf{k}_{hkl} = 2\pi h \tilde{\mathbf{a}} + 2\pi k \tilde{\mathbf{b}} + 2\pi l \tilde{\mathbf{c}},$$

where $-1 \leq h, k, l \leq 1$. Use the InfinitePlane function to produce a plane that intersects the line joining the origin (in reciprocal space) and each lattice point, at the midpoint of the line. Set the PlotRange to clip the top half of the resulting plot, to visualize the interior. Can you see half of the truncated octahedron depicted in figure 8.5?

EXERCISE 8.5. The lattice vectors for a bcc lattice are $\mathbf{a} = (\frac{1}{2}, \frac{1}{2}, -\frac{1}{2})$, $\mathbf{b} = (-\frac{1}{2}, \frac{1}{2}, \frac{1}{2})$ and $\mathbf{c} = (\frac{1}{2}, -\frac{1}{2}, \frac{1}{2})$. Compute the reciprocal lattice vectors and determine the Brillouin zone, as in the previous exercise.

The different crystal lattices all have symmetry points like those displayed in figure 8.5 but the names of the points differ. We shall not dwell on the different names; those can be obtained readily through the crystallo-graphic databases. Instead, we want to focus on the emergent behaviors that arise due to the many-body interactions present in crystals. Solid-state physics possesses a multitude of such behaviors so we will necessarily restrict the discussion.

As a first step, let us consider the structure of diamond, which forms a tetrahedral lattice (space group Fd$\bar{3}$m). A .cif file defining the diamond lattice can be obtained from the American Mineralogist Crystal Structure Database (AMCSD). There are undoubtedly other sources that can be found through a web search. In fact, the diamond lattice consists of two fcc lattices, offset by a distance $a(\frac{1}{4}, \frac{1}{4}, \frac{1}{4})$. The x-ray diffraction data indicate that the lattice parameter is $a = 3.56679$ Å, which implies that the average C–C distance is 1.544 Å. This value is consistent with what we found for the C–C distance in singly bound ethane.

Exercise 8.6. Use the `Graphics3D` function to visualize the tetrahedral lattice. Construct two fcc lattices from the basis vectors $\mathbf{a} = (0, \frac{1}{2}, \frac{1}{2})$, $\mathbf{b} = (\frac{1}{2}, 0, \frac{1}{2})$ and $\mathbf{c} = (\frac{1}{2}, \frac{1}{2}, 0)$ with an offset of $(\frac{1}{4}, \frac{1}{4}, \frac{1}{4})$. Demonstrate that each atom is tetrahedrally coordinated.

We can utilize the NWChem software to explore solid state systems, as it contains an implementation of a pseudopotential plane wave method, in which the valence electrons are treated as discussed above and the core electrons are represented by a static potential. We should make a cautionary note before proceeding that solid state calculations can be quite time-consuming, both in terms of computational time and personal time. As computational time scales like N^3 or N^4, even exploiting symmetry will lead to large calculations. Additionally, finding things like an optimized structure can be quite tricky and are at the forefront of research efforts. Consequently, it is quite likely that students will find that they can attempt a simple, theoretical calculation that turns out to lock up their computers for hours on end. Access to more capable computational facilities will be advantageous but not even the most powerful machines can address all problems. Brute force can be helpful at times but one must think seriously about what questions to ask.

The following excerpt from an NWChem script file demonstrates a calculation of the optimal lattice parameters for diamond.

```
geometry center noautosym noautoz print
    system crystal
        lat_a 3.56
        lat_b 3.56
        lat_c 3.56
        alpha 90.0
        beta  90.0
        gamma 90.0
    end
    symmetry Fd-3m
    C  0.0  0.0  0.0
end
```

```
nwpw
    ewald_rcut 3.0
    ewald_ncut 8
    lmbfgs
    xc pbe96
    dplot
        vectors $NWCHEM_ROOT.movecs
        density total $NWCHEM_ROOT.cube
    end
end
driver
    clear
    maxiter 40
end
set nwpw:cif_filename diamond.opt
set includestress .true.
task pspw optimize ignore
task pspw pspw_dplot
```

The geometry block includes a specification of the lattice parameters and the point group. It is only necessary to specify a single atom position, as all atoms will be carbon. The nwpw block specifies a number of parameters specific to the pseudopotential implementation. Note that there is no definition of basis vectors.

> EXERCISE 8.7. Perform the calculation indicated by the script above. Note that the script is not complete. The header can be obtained from the previous examples. The pspw_dplot task will create a cube file. Plot the electron density in the unit cell.

From the plane wave calculation, we obtain a lattice parameter $a = 3.654$ Å and a C–C distance of 1.582 Å. This is a reasonable result—the bond distances are 0.04 Å longer than the experimental values—not as precise as one might hope but it can be improved somewhat by tinkering with the parameters, although generally at the expense of significantly longer computational times.

We plot a slice of the valence electron density from the pspw_dplot task in figure 8.6. As the core electrons are represented by the pseudopotential, the density does not peak at the nuclear center; this is an artifact of the calculation. As we saw in the molecular calculations, the electron density is strongly peaked around the nuclear centers and this would be the case here as well.

Macroscopic properties of the crystals are determined through statistical mechanics methods that connect the microscopic behavior of the system with macroscopic observables. The partition function $\mathcal{Z}_N$ of a system of N

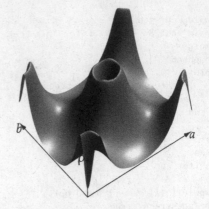

FIGURE 8.6. The valence electron density peaks around the nuclear centers. The total electron density, including the core electrons, would peak at the nuclear centers.

particles defined by the Hamiltonian H is given by the following relation:

$$(8.4) \qquad \mathcal{Z}_N(T) = \frac{1}{h^{3N}N!} \int d^{3N}\mathbf{q}\, d^{3N}\mathbf{p}\, e^{-H(\mathbf{q},\mathbf{p})/k_BT},$$

where $\mathbf{q}$ are the generalized coordinates and $\mathbf{p}$ are the conjugate momenta of the particles. The Helmholtz free energy is then obtained from the partition function directly:

$$(8.5) \qquad F(T,\mathcal{V},N) = -k_BT \ln Z_N(T).$$

Alternatively, one can consider what is termed the microcanonical ensemble and consider the density of states Ω_N defined as follows:

$$(8.6) \qquad \Omega_N(E) = \int d^{3N}\mathbf{q}\, d^{3N}\mathbf{p}\, \delta(E - H).$$

The entropy of a system at a particular energy E is proportional to the density of states:

$$(8.7) \qquad S(E,\mathcal{V},N) = k_B \ln\left[\frac{\Omega_N(E)\Delta}{h^{3N}N!}\right],$$

where Δ represents a small (infinitesimal) difference in energy. It can be considered to be the experimental resolution. The partition function and density of states can be connected in a fundamental way.

If we start with the following identity:

$$1 = \int dE\, \delta(E - H)$$

and multiply both sides by the partition function, we obtain the following:

$$Z_N(T) = \frac{1}{h^{3N}N!} \int dE\, \delta(E-H) \int d^{3N}\mathbf{q}\, d^{3N}\mathbf{p}\, e^{-H/k_B T}$$

$$(8.8) \qquad = \frac{1}{h^{3N}N!} \int dE\, \Omega_N(E)\, e^{-E/k_B T}$$

where in the last step, we utilized equation 8.6 and enforced the action of the delta function. Consequently, the partition function can be seen to be the Laplace transform (with respect to $1/k_B T$) of the density of states.

Thus, a possible means for obtaining the partition function of a complex system would be to utilize the density of states. Often this simply exchanges one difficult problem for another but the **band** modules within NWChem can let us obtain an estimate of the density of states. Mercifully, because thermodynamic properties depend upon the logarithm of the partition function, we can sometimes obtain reasonable results even though our ability to calculate the exact density of states is limited.

Figure 8.7 represents an estimate of the density of states in diamond, taken from a plane wave band calculation. The NWChem script utilized the same **geometry** block as before, invoking the Fd$\bar{3}$m symmetry. Instead of using the **pspw** task, we use the **band** task, with an additional block specific to the density of states calculation:

```
task band optimize ignore
nwpw
  virtual 26
  dos-grid 5 5 5
end
task band dos
nwpw
  virtual 16
  brillouin_zone
    zone_name fccpath
    path fcc 1 gamma x w k gamma
  end
  zone_structure_name fccpath
end
task band structure
```

We note that specifying the number of virtual orbitals and the **dos-grid** parameter causes the running time to jump significantly (about a factor of ten) over that obtained with the default values. One can adjust those values, or omit the block entirely, if computational resources are an issue. The **brillouin_zone** block in the script is used to alter the path from the default, as an example; it is not essential to the calculation of the band structure.

EXERCISE 8.8. Insert the script fragment above in place of the task pspw optimize block in an NWChem script and compute the diamond density of states and band structure. The output files (.dos and .restricted_band) can be easily read through the Import command. The .dos file contains three columns: energy (Hartrees), density and cumulative density. Use the ListLinePlot function to reproduce figure 8.7.

FIGURE 8.7. The calculated density of states (black line) displays a gap in the region of 12.3–16.2 eV in which there are no allowable states. The (scaled) cumulative density (gray line) depicts a nominal square root dependence (light gray line).

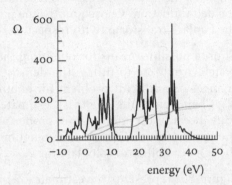

The density of states obtained from the suggested calculation is depicted in figure 8.7. Notably, there is a gap in the density of states for energies between about 12.3 eV and 16.2 eV. For these energies, there are no allowed states. This is known as a *band gap*.

In figure 8.7, we also depict the cumulative (integrated) density of states as the gray curve. Interestingly, a number of approximate models suggest that the density of states should scale like the square root of the energy. We have also included a curve depicting a square-root dependence on energy. While the curves do not coincide, we can infer that the square-root model will capture the bulk of the energy dependence. This is quite comforting, because it is simple to calculate square roots and difficult to compute densities of states. Provided that one is looking for explanations of gross behaviors, it is quite reasonable to use a square-root model as an initial description of a complex system.

It probably has occurred to most students that there is a tremendous amount of information to absorb about calculations in the solid state. Even depicting the charge density becomes problematic because we are essentially dealing with a three-dimensional problem. One tool that solid state physicists developed to cope with the vast amount of information is the band structure plot. Essentially, one plots the band energies as a function of the wavenumber **k** along a path in the reciprocal space that follows the symmetry points we indicated previously. Such a plot for diamond is depicted in figure 8.8.

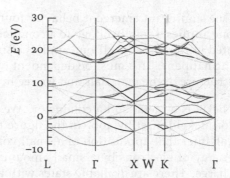

FIGURE 8.8. The energies of different bands in diamond are traced along straight lines between symmetry points in the Brillouin lattice. The zero in energy corresponds to the top of the valence band.

Such "spaghetti plots" are rather formidable to interpret. With practice, one can begin to make sense of the information but it is a lengthy process. We note that in a similar vein one cannot interpret a musical score without training, so it will take some effort to interpret the information in the structure plot. We notice immediately the gap between energy levels noted in the density of states plot (figure 8.7). We also note that near the center of the Brillouin zone (Γ), there is a large energy gap between the filled levels (below 0) and the unfilled levels (above 0). This is characteristic of insulating materials.

EXERCISE 8.9. NWChem is not particularly user friendly but, actually, is one of the more user-friendly computational chemistry codes. Like most codes, the outputs are invariably in units of Hartrees and Bohrs. The **band structure** calculation produces a file that contains the values of k along the specified pathway and the energies of the bands at each k-point. For the fcc Brillouin lattice, we have $K = (3/8, 3/8, 3/4)$, $L = (1/2, 1/2, 1/2)$, $W = (1/2, 1/4, 1/2)$ and $X = (1/2, 0, 1/2)$. Determine the numerical values for the key points along the path depicted in figure 8.8. See if you can reconstruct the figure.

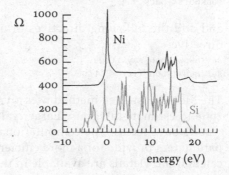

FIGURE 8.9. The density of states of silicon (gray curve) displays a band gap like that observed in diamond. For nickel (black curve, offset by 400.), there is no band gap, characteristic of metals.

One significant emergent behavior found in materials is conductivity. If one applies a small voltage to a conductor, an electronic current is generated in the material. If one applies a small voltage to an insulator, nothing happens. This must be dependent upon the nature of the microscopic structure of the materials and we see in figure 8.9 a possible clue.

Nickel is a metal and a conductor. We observe that the density of states for nickel (offset by 400. in the figure for clarity) is always positive. We can infer that application of a small voltage V to nickel will increase the energy of electrons by a small amount: $\Delta \mathcal{E} = eV$, where e is the electric charge. There are available states with the energy $\mathcal{E} + \Delta \mathcal{E}$, so the electrons are "promoted" to the so-called conduction band. These turn out to be states that are delocalized from the lattice sites, giving the electrons the mobility to move throughout the crystal.

Non-conducting materials are classed as insulators or semiconductors, on a somewhat empirical basis, by the size of the band gap present in their densities of states. Large band gaps are insulators and small band gaps are semiconductors, like silicon.

EXERCISE 8.10. Silicon adopts the same crystal structure as diamond. Compute the silicon density of states. (Repeat the diamond calculations but replace C with Si.) Note that silicon has the same number of valence electrons as carbon, so the computational requirements will be somewhat larger than for carbon.

EXERCISE 8.11. Nickel forms a crystal lattice with Fm$\overline{3}$m (Fm-3m) symmetry. With 28 electrons, calculation times will be significantly longer than for carbon. As a metal, the calculations are also technically more challenging. Add the following to the nwpw block:

```
smear fermi
scf anderson outer_iterations 0 kerker 2.0
monkhorst-pack 3 3 3
```

and add the following directives prior to the task band optimize directive:

```
set nwpw:kbpp_ray .true.
set nwpw:kbpp_filter .true.
```

The Monkhorst-Pack method uses a 3×3×3 sampling in the Brillouin zone to estimate integrals. Larger values than 3 will provide better results but increase the computational time significantly. The kbpp parameters provide for a more efficient, better numerically behaved calculation. Details are available in the NWChem documentation.

8.2. Surface States

For those students with access to adequate computational resources, it is a straightforward exercise to investigate other simple systems like GaAs. One can obtain .cif files of most simple minerals from the crystallography databases, so good starting points are available but one quickly runs into the issue of computational resources. For students with access, it is possible to generate research-grade examples to study with NWChem, which was designed explicitly to scale on large numbers of processors. Although, it should be said, it is also straightforward to ask seemingly simple questions that are not computationally tractable, even on the largest supercomputers. We won't follow that pathway here but there are whole courses devoted to the study of solid state materials and suggest those are more appropriate venues.

Instead, we shall turn our attention briefly to the problem of what happens at surfaces? At a surface, the regular bonding network of the unit cell is disrupted and, as a practical matter, the trick that Bloch exploited to describe the behavior of the bulk materials is no longer applicable. The loss of translational symmetry is a distinct blow to theoretical efforts to describe material properties.

In order to study surface properties, one needs a model that includes enough atoms that a large portion of the interior of the model is equivalent to the bulk. That is, one wants a model in which the surface states do not dominate the behavior. Hence, there is a need to perform relatively large calculations with thousands of atoms or resort to sophisticated means to enforce periodic boundary conditions on one side of a large block and free space on the other. Both approaches are used in research groups today and lie beyond the realm of what we might hope to accomplish.

To obtain some hint as to the behavior of finite systems, let us build a model of a small cluster of carbon atoms. This can be done with the Mercury software by taking the diamond .cif file and selecting packing, which generates a model with 192 carbon atoms. This is more than we can presume to treat given limited computing resources, so by selecting only the carbon atoms within 3.7 Å of the origin, we produce a cluster of 32 carbon atoms, as depicted in figure 8.10. If we now conduct three minimization steps (maxiter 3), in a pspw calculation, the cluster relaxes just a small amount.

Actually obtaining the minimum energy structure for 32 carbon atoms is a daunting problem. There is no particular numerical strategy available to find a global minimum. In practice, there are numerous local minima that can trap the conjugate gradient minimization algorithm into stopping before a global minimum has been determined. A number of strategies to

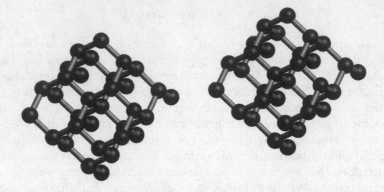

FIGURE 8.10. The initial cluster is indicated on the left and, after three minimization steps, the cluster relaxes to the structure on the right.

explore the parameter space have been developed to overcome this problem but all are computationally intensive. We'll discuss some of them subsequently. This is another case where experiment triumphs over our meager theoretical efforts. Recall that Curl, Kroto and Smalley were investigating small clusters but were able to sort through vast numbers of possible final states by the expedient method of allowing the 10^{16} carbon atoms in the gas stream to conduct their own minimizations, in real time.

For our immediate purposes, we are not seeking a global minimum. What will happen if we take one hundred or one thousand minimization steps is that the 32-atom cluster will deform itself into a wide variety of curious shapes because most (all) of the atoms are part of the surface. In order to prevent those unwanted deformations, one needs to add many more atoms to the simulation and we are back to intractable. In three steps, none of the atoms has moved very far but the bonds have readjusted.

FIGURE 8.11. After three minimization steps, the C–C distances have relaxed somewhat from the typical 1.55 Å distance characteristic of a carbon-carbon single bond.

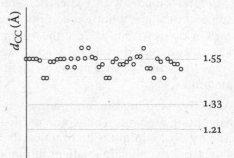

In figure 8.11, we depict the different C–C distances after the minimization steps. For the interior carbon atoms (leftmost in the plot), the distance remains at the 1.55 Å distance characteristic of singly bonded carbon, as we found for ethane. Several of the surface atoms have distances midway between doubly bonded (ethene = 1.33 Å) and singly bonded. None are approaching the triply bonded (ethyne) distance of 1.21 Å. In fact, the distance is quite close to the 1.40 Å C–C distance we found in the benzene ring. As we shall see subsequently, this distance is characteristic of the *aromatic* bonds found in graphene.

> EXERCISE 8.12. Set up the 32 carbon atom simulation as described and optimize the structure for three steps. What bond distances do you obtain?

Fortunately, determination of the actual nature of material surfaces has been made much more tractable by the invention of a new form of spectroscopy by the German physicist Gerd Binnig. Binnig's idea was to measure the electron tunneling current between a probe and crystal surfaces. Tunneling is a quantum phenomenon that arises from the fact that the electronic wave functions decrease exponentially away from a crystal surface. Thus, even though the wave function is small, there is some finite probability that electrons could traverse the barrier, giving rise to a small current. The current will flow only if the tip is sufficiently close to the surface. Binnig and the Swiss physicist Heinrich Rohrer were able to construct such a device by March of 1981.[6]

The pair used three piezoelectric devices to position the tip independently in x, y and z. Measurements were made by scanning the probe in the x-y plane. A feedback loop that sensed the tunneling current controlled the position of the tip above the surface (z-direction). As the tip traversed the sample, the tunneling current was kept constant and the applied voltage was measured. Collecting many such scans allowed the production of images like those in figure 8.12, which represents a nickel surface cut along the (110) plane.

It took Binnig and Rohrer over two years of development to achieve the first scan results that demonstrated that they were actually measuring the tunneling current. Along the way they struggled with a number of design issues like vibration isolation. A device based on magnetic levitation was constructed and abandoned, with the first working apparatus simply

---

[6]Binnig and Rohrer were awarded the Nobel Prize in Physics in 1986 "for their design of the scanning tunneling microscope ." They shared the prize with Ernst Ruska, who was cited "for his fundamental work in electron optics, and for the design of the first electron microscope."

FIGURE 8.12. Three dimen-
sional rendering of the
nickel (110) surface obtained
through STM measurements.
Image originally created by
IBM Corporation.

suspended from a rubber band. More sophisticated schemes were implemented subsequently.

We mentioned above that the device used piezoelectric devices for sample
manipulation but no one at the time had ever determined whether the devices provided a linear distance response over the atomic-scale distances
that Binnig and Rohrer were attempting to sample. This was determined,
self-consistently, from the scanning data. Although the proportionality
constant relating the change in distance Δx to the change in applied voltage ΔV turned out to be smaller than originally thought, a linear voltage
ramp in the piezoelectric device produced a periodic change in the output
signals that matched the anticipated lattice spacing.

A more formidable problem turned out to be the generation and stability
of the probe tips. Originally envisioned as some sort of conical form with
a small radius of curvature at the end, Binnig and Rohrer quickly recognized that the tip ends were rugose, resembling mountain ranges more
than smooth surfaces. A simplistic idealization is depicted in figure 8.13.
Nevertheless, at some point on the end of the tip, there is an island of
atoms that extend beyond others in the area and the tunneling current is
restricted to these atoms.

Indeed, the exponential dependence on distance means that the magnitude of the current from atoms two atomic layers removed from the tip
end is decreased by a factor of a million. Binnig and Rohrer were quick to
recognize that this is an important feature of their device: the tunneling
current is limited to an area on the scale of the atomic size. Moreover, by
changing the value of the tunneling current and the bias voltage of the
tip, they could sample different portions of the local density of states at
atomic scale. The Scanning Tunnelling Microscope (STM) that they developed was a new tool for spectroscopy on surface states.

FIGURE 8.13. The scanning tip (light gray) positioned above a crystal surface (dark gray) can be biased to produce a tunneling current with an exponential dependence upon the distance h between the closest atom in the tip and the surface.

The initial experiments confirmed that the interactions of the tip with the surface remodelled both the surface and the tip. Initially, this was an obstacle to progress, as it was difficult to complete many (any) scans before the tip geometry changed and an offset was introduced into the response curves. Eventually, once the nuances of working with new equipment were understood, it was recognized that the STM could be used in an active rôle in surface studies, not just as a passive recorder of the system properties. The STM could purposefully move atoms adsorbed onto the surface and could investigate the bonding patterns of these adatoms to the underlying surface atoms. Because the STM probed the local density of states, it was possible to also tune those interactions to be sensitive to atoms in the layer below the surface.

The technology provided a massive improvement over the information available from low energy electron diffraction (LEED) experiments. Analogous to the information available from x-ray diffraction experiments, LEED studies are more sensitive to surface properties but beam widths are vastly larger than atomic scale, typically micrometer-scale, that produce average properties over a wide number of surface sites.

Ironically, some of the early criticism of STM methodology was that crystal surfaces looked much like what people had anticipated. Given that most new imaging technologies force significant changes to the prevailing theories, general agreement with current thinking about surface states was surprising. Additionally, the scanning surfaces like those produced in figure 8.12 were thought to have been computer-generated; they were simply too good. Ultimately, physicists found that the information obtained about surface bonding from the LEED measurements was also reproduced through STM measurements, although on an atomic scale. This provided the necessary closure that the STM results were, in fact, real.

Binnig went on to invent the atomic force microscope and built the first working prototype with the American physicist Calvin Quate and the

Swiss physicist Cristoph Gerber in 1986.[7] By placing an STM tip on a cantilever, Binnig and colleagues created an atomic-scale profilometer. Stylus profilometers had been in use for many years, measuring the micrometer-scale features of surfaces. The AFM provides the ability to study surfaces with nanometer resolution. In their original implementation, Binnig, Quate and Gerber used a second STM to control the cantilever position but this has been supplanted by a simpler technique in which a light beam is reflected from the cantilever surface, providing the needed positional sensitivity, as is sketched in figure 8.14.

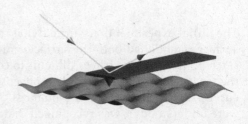

FIGURE 8.14. Placing a diamond tip on the end of a cantilever produces an instrument capable of imaging the surface features of insulators. The tip position is sensed by a laser beam reflected from the cantilever surface.

The significant advantage of the AFM is that the sample need not be a conductor, providing the means to investigate the atomic-scale structure of a host of new materials, including biological materials because the cantilever can be operated immersed in fluid. The cantilever possesses a resonant frequency, typically in the kiloHertz range, that depends upon its dimensions and the material properties of its constituents. As the tip approaches the surface of the sample, it interacts with the sample material and the resonant frequency of the cantilever changes. There are a number of different operational modes for the AFM, categorized by whether or not the tip is allowed to contact the surface. Each has advantages and disadvantages for specific applications and this adaptability underlies the widespread application of AFM methods.

EXERCISE 8.13.   One means to study the behavior of small samples is to use constraints on the atoms. This is accomplished in NWChem via two means. The first is to fix atom positions and the second is to harmonically constrain a two-atom distance. If we return to the 32 carbon simulation, add the following directive to the input script:

```
set geometry:actlist 1:15 17 19:25 29:31
```

[7]Binnig, Quate and Gerber published "Atomic force microscope" in the *Physical Review Letters* in 1986. They were awarded the Kavli Prize in Nanoscience in 2016 ""for the invention and realization of atomic force microscopy, a breakthrough in measurement technology and nanosculpting that continues to have a transformative impact on nanoscience and technology."

This will fix the six atoms at the bottom of the cluster and permit the others to move. What is the result of a 20-step optimization?

8.3. Nanoscale Materials

Chemists studied colloids for many years before the advent of the scanning probe microscopes pioneered by Binnig and colleagues made their debut. Interest in the behavior of nanometer-sized materials expanded rapidly with the ability of researchers to directly probe and even manipulate their samples. Along with the new measurement technology for the field came a new name: nanotechnology. Departments, journals and grant funding mechanisms were rapidly renamed as new discoveries proliferated.

As we have seen previously, the introduction of a new measurement technology can vastly remake the fabric of the scientific endeavor. Such was the case with the scanning probe instruments. Not only could a number of problems be answered that could not be addressed previously but a host of new applications arose as well. For example, if one were to affix a molecule to the AFM tip and place other molecules of interest on the surface, then one could study the forces generated as the two molecular species were brought together. This would provide insights into the details of the chemical binding process that were completely inaccessible previously. This is not to say that previous generations of chemists didn't ponder these sorts of questions but only that they did not have the tools at their disposal to investigate chemical reactions in this fashion.

One problem that we sidestepped earlier in the discussion was the determination of optimal structures of small clusters. This was essentially the problem that Smalley, Curl and Kroto were investigating when they stumbled across the fullerenes. While one might argue that the study of small clusters is not terribly interesting in and of itself, the result is that a wide variety of new areas of research have derived from those early investigations. Beyond the fullerenes, chemists and physicists began investigating carbon nanotubes and now graphene sheets.

Graphene is an allotrope of carbon, in which the carbon atoms form hexagonal arrays in two-dimensional sheets. The sheets are weakly connected, so powdered graphite has long been utilized as a lubricant in mechanical devices. The Russian physicists Andre Geim and his student Konstantin Novoselov were interested in the properties of graphene and found an extraordinarily simple means for obtaining samples: they simply applied adhesive tape to a graphite block and removed the top graphene layer.[8]

---

[8]Geim, Novoselov and colleagues published "Electric field effect in atomically thin carbon films," in *Science* in 2004 and "Two-dimensional gas of massless Dirac fermions in graphene"

This provided the means for studies of this remarkable two-dimensional material.

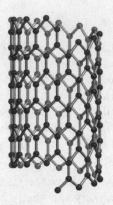

FIGURE 8.15. Carbon nanotube can be envisioned as graphene sheets wrapped into a cylinder (left). It is possible for the tubes to contain a helical pitch (right).

Prior to this work, others had investigated the properties of carbon nanotubes: graphene sheets rolled into a tube, as depicted in figure 8.15. The electrical properties of the nanotubes depend upon both the diameter and helicity. When the graphene sheet is formed into a circle, like poultry fencing, the hexagons of one row can step up or down to the next row (or subsequent rows) when the edges meet, as depicted in the right image in the figure. This helicity can be detected in single-walled nanotubes as the helicity causes the hexagons to deform slightly, breaking the six-fold symmetry.

The nanotubes can be defined in terms of two indices n and m that count the number of hexagons arranged azimuthally around the tube and the vertical offset, respectively. When $|n - m| = 3k$, for k an integer, then the nanotube is a metal. When $|n - m| = 3k \pm 1$, the tube is a semiconductor, otherwise the tube will be semimetallic.

EXERCISE 8.14. The six-membered carbon rings that form the graphene sheet are composed of equivalent carbon atoms. Choose one as the origin and then select the two atoms that are two bond lengths away from the origin. The vectors from the origin to those atoms form the two-dimensional basis vectors $\mathbf{a}_1$ and $\mathbf{a}_2$ for the sheet. A nanotube with indices (n, m) is defined by the rectangle with corners at the origin and $n\mathbf{a}_1 + m\mathbf{a}_2$. What is the structure of a nanotube with indices $(n, m) = (12, 12)$?

in *Nature* in 2005. They were awarded the Nobel Prize in Physics in 2010 "for groundbreaking experiments regarding the two-dimensional material graphene."

Both graphene and carbon nanotubes are the subjects of intense scrutiny. They possess interesting electrical properties that can be altered by chemical modification. Fabrication of devices remains a particular area of interest. While the properties of particular forms of nanotubes are known through experimental and theoretical means, the ability to build, reliably and repeatably (and cheaply), many copies is still a work in progress.

While NWChem is capable of nanotube simulations, it is unlikely that most students will have access to the computing resources necessary to march further along this pathway. Instead, we shall return to a more tractable problem: that of defining the "optimal" structure of small clusters. Numerically, this amounts to finding the structure with the lowest energy but, while easily stated, it is not a particularly simple task. There are many degrees of freedom within even small clusters, so typical strategies that make use of the energy gradients to determine the optimization pathway can be trapped within shallow, local minima and never find the global minimum. Indeed, it is likely that there is an ensemble of minimum energy states that are very close in energy but which may be structurally quite different.

An alternative strategy to minimization of a cost function was developed in analogy to a common practice in material science. Long used by metalworkers, the technique is known as simulated annealing. The process proceeds by first heating the system to circumvent any energy barriers; recall that barrier heights are proportional to $\exp(-\mathcal{E}/k_B T)$. The system is then allowed to cool and, optionally, can be optimized through the usual minimization strategies. The technique is predicated on the ability to conduct dynamics on the sample constituents.

A workable strategy for quantum molecular dynamics was provided by the Italian physicists Roberto Car and Michele Parrinello in 1986.[9] Prior to the Car-Parrinello work, the most common approach to molecular dynamics utilized the Born-Oppenheimer approximation, in which nuclear motions were decoupled from electronic structure. For each time step, the nuclear positions and velocities were updated and then the electronic structure was obtained by optimizing the density with respect to the new nuclear positions. This is a sensible approach, given that nuclei are vastly heavier than electrons, and works well for reasonable temperatures. Unfortunately, the optimization step is quite costly and limited the applicability of quantum methods in solving dynamics problems.

---

[9]Car and Parrinello published "Unified approach for molecular dynamics and density-functional theory" in the *Physical Review Letters*. The pair was awarded the Dirac Medal by the Abdus Salam International Center for Theoretical Physics in 2009.

Car and Parrinello avoided this difficulty by including the electronic structure calculations directly. They proposed a Lagrangian of the following
form:

$$(8.9) \qquad \mathcal{L}_{CP} = \frac{1}{2} \sum_i M_i \left[\frac{d\mathbf{R}_i}{dt} \right]^2 + \frac{\mu}{2} \sum_l \int d^3\mathbf{r} \left| \frac{d\psi_l(\mathbf{r}, t)}{dt} \right|^2 - \mathcal{L}_{KS},$$

where the summation over i is over the nuclei and the summation over
l is over the orbitals. The fictitious electron mass μ is chosen to be large
compared to the nuclear masses, to prevent energy transfer and preserving the electrons in their ground state. The last term in equation 8.9 is the
standard Kohn-Sham Lagrangian for the electronic degrees of freedom, so
all of the tools developed for density functional theory can be utilized.

As a simple example, we can study the case of twelve boron atoms. Initially, the atoms are arranged in an icosahedral geometry. Theoretically,
there are several suggestions as to the optimal geometry that the cluster
should attain. Conducting simulations of the cluster should help provide some insight into the actual behavior. Note that this will likely depend on details of the model, so a single calculation will not magically
reveal the secrets of the universe. The script below will scale the temperature upwards for several cycles before rescaling the temperature back
downward. The times here are in terms of the atomic unit of time $\tau = 2.418884326509(14) \times 10^{-17}$ s. The number of times steps is modest; research efforts would extend the trajectory significantly but this is enough
for our objectives.

```
geometry
  B  1.06848703 -1.06977718 -0.27191204
  B -0.41738004 -1.54046760  0.30786278
  B  1.54044867  0.41755176  0.31097113
  B -1.53464475 -0.41595403 -0.30802728
  B  0.41455946  1.53816406 -0.30497372
  B -1.06506013  1.06638042  0.26996858
  B  0.69163266 -0.69066193  1.27228704
  B -0.26229140 -0.90393650 -1.22051191
  B  0.89747884  0.25958856 -1.21386685
  B -0.69798902  0.69567491 -1.28227908
  B  0.26356093  0.90344215  1.22217740
  B -0.89880225 -0.26000463  1.21830395
end
```

```
nwpw
  simulation_cell
    boundary_conditions aperiodic
    SC 20.0
  end
  cutoff 10.0
  lmbfgs
  Car-Parrinello
    fake_mass 500.0
    time_step 5.0
    loop 10 100
    scaling 1.0 2.0
    emotion_filename b12.00.emotion
    xyz_filename b12.00.xyz
  end
end
task pspw energy
set cpmd:init_velocities_temperature 300.0
task pspw car-parrinello
unset cpmd:init_velocities_temperature
task pspw car-parrinello
task pspw car-parrinello
task pspw energy
task pspw car-parrinello
nwpw
  Car-Parrinello
    scaling 0.99 0.99
    emotion_filename b12.01.emotion
    xyz_filename b12.01.xyz
  end
end
task pspw energy
task pspw car-parrinello
task pspw car-parrinello
task pspw car-parrinello
```

The simulation is a bit lengthy but produces some 400 frames of data. The software is capable of several different output files than enable one to investigate various parameters. The .xyz file contains a sequence of co-ordinates and velocities for each time step. The emotion file contains the output times and various energy information. For example, the kinetic energy associated with the fictitious electron mass is depicted in figure 8.16 for the heating phase of the simulation. Here, the energy can be seen to increase in four steps. This is an indication that the electrons are moving away from their ground state, so the purpose of the task energy directive is to reset the electron density before beginning the cooling process.

Students may have noticed that there is a fair amount of craftsmanship on display in these calculations. This is typical of the level of detail associated with research programs. There are nuances and distinct limitations to all methods, including experimental methods. One must learn how to compensate for shortcomings and obtain consistent results. Like learning to read music, learning to conduct scientific investigations will take time and effort.

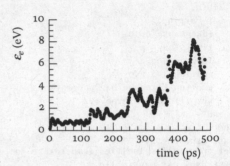

FIGURE 8.16. The energy associated with the fictitious mass from the initial portion of the simulation rises in four steps.

EXERCISE 8.15. Run the Car-Parrinello simulation indicated by the script. (Note: this is an example from the NWChem tutorial series, so output files may be available on the website.) Plot the different energies in the .emotion file. Extract the final frame from the large .xyz file. Does the structure tend towards icosahedral?

EXERCISE 8.16. The .xyz files written by the dynamics routines also include the atom velocity. As a result, the Import function fails. Construct a module to read the .xyz files. Use the ListAnimate function to display all of the structures on the path. Hint: NWChem writes a blank comment line, which the Read function will ignore. Also, at the end of the file, the Read function will return the parameter EndOfFile. The test SameQ[nextline,EndOfFile] will return True if the nextline resulting from the file read has the value EndOf File.

A more involved approach to annealing can be implemented by an exponential decay method. Here the temperature is reduced with an exponential function $\exp(-t/\tau)$, where the time is in units of the atomic time. The value in the script corresponds to $\tau = 1$ ps. The cooling loop runs for 5000 steps, corresponding to about 6 ns of simulation time. The script for this option is indicated below:

```
nwpw
  simulation_cell
    boundary_conditions aperiodic
    SC 20.0
  end
  cutoff 10.0
  lmbfgs
  Car-Parrinello
    Nose-Hoover 250.0 3500.0 250.0 3500.0
    fake_mass 500.0
    time_step 5.0
    loop 10 1000
    scaling 1.0 1.0
    emotion_filename b12.10.emotion
    xyz_filename b12.10.xyz
  end
end
task pspw energy
set cpmd:init_velocities_temperature 3500.0
task pspw car-parrinello
unset cpmd:init_velocities_temperature
nwpw
  Car-Parrinello
    SA_decay 4.134d4 4.134d4
    Nose-Hoover 250.0 3500.0 250.0 3500.0
    loop 10 5000
    emotion_filename b12.11.emotion
    xyz_filename b12.11.xyz
  end
end
task pspw car-parrinello
task pspw optimize ignore
```

EXERCISE 8.17.   From the annealing simulation, plot the final structure. Does it look like the original icosahedral structure?

EXERCISE 8.18.   Another program that can be quite useful is the molecular visualization code VMD, available from the developers at the University of Illinois. VMD can cope with the .xyz files that NWChem writes. Install VMD and use it to examine the Car-Parrinello simulation results. After reading the file, the default Drawing Method is Lines. Change this to VDW and add a second representation where the Drawing Method is Dynamic Bonds. As the simulation evolves, the bonds will be redrawn based on the selected distance.

EXERCISE 8.19.   Repeat the second simulation but for the thirty-two carbon atoms we used previously. Limit the number of steps to 3000 in the second Car-Parrinello step. Describe the resulting structure in terms of single, double, etc., bonds. Is the final structure symmetric?

Perusing the options available in the user documentation for NWChem is a daunting task. There are many methods, not all are compatible. We have examined a few examples but have not exhausted all possibilities. The reason so many approaches are available is that each has advantages and disadvantages. Each has a limited scope of applicability, some of which is determined by the amount of cpu time that can be devoted to the enterprise. Sometimes, brute force is good enough. Often, one must be clever to obtain sensible results in a reasonable time. It would be nice to simply type in the coordinates and tell the model to go calculate the density of states but, in reality, such an undertaking may not be computationally feasible. Perhaps quantum computing may render our discussions irrelevant but that subject will be taken up in a later chapter.

8.4. Superconductivity

One of the most remarkable examples of emergent behavior is superconductivity, first observed by the Dutch physicist Heike Kamerlingh Onnes in 1911.[10] Kamerlingh Onnes set out earlier in his career to complete the liquefaction of all known gases and developed a laboratory to accomplish that goal.

By 1908, he had succeeded in liquefying helium, the element with the lowest boiling point. With no other elements left to liquefy, Kamerlingh Onnes thought, at first, to press on to lower temperatures and form solid helium but was held back by technical details. Along the pathway to liquid helium, for example, Kamerlingh Onnes had developed the requisite thermometry to determine temperature. He had calibrated the resistance of platinum wires against other sources at higher temperatures and also utilized a gaseous helium thermometer, capable of measuring temperatures down to about 1.5 K. To go lower in temperature would have required construction of a new apparatus, so Kamerlingh Onnes paused in his quest for lower temperatures and decided instead to study the behavior of materials at low temperature.

Kamerlingh Onnes pragmatic decision to utilize the existing apparatus quickly proved to be scientifically profitable. Lowering the temperature in platinum had led to a lower resistance, and a low-temperature thermometer. When Kamerlingh Onnes investigated the behavior of mercury at low temperatures, he found that the resistance dropped by a factor of

---

[10]Kamerlingh Onnes published his results in the *Communications of the Physics Laboratories of the University of Leiden*. Most were then reprinted in the *Proceedings of the Koninklijke Nederlandse Akademie van Wetenschappe*. He was awarded the Nobel Prize in Physics in 1913 "for his investigations on the properties of matter at low temperatures which led, inter alia, to the production of liquid helium."

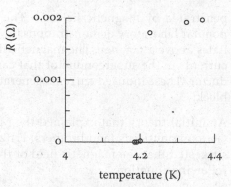

FIGURE 8.17. The resistance of a thin thread of mercury drops to 230 $\mu\Omega$ at 4.2 K and still smaller values at lower temperatures. The transition is abrupt, not gradual.

a million over a very short temperature span, as indicated in figure 8.17. At 4.2 K, the resistance dropped precipitously at lower temperatures and then jumped back at higher temperatures. This is characteristic of critical behavior in many-body systems. The material property (resistance) changes reversibly and abruptly at the transition temperature.

Kamerlingh Onnes quickly established that other metals had this property that he termed supraconductivity. His naming of the phenomenon didn't stick, today we talk about superconductivity. Nonetheless, Kamerlingh Onnes found that lead becomes superconducting at about 6 K and tin at 4 K. This enabled a number of new studies that avoided the many tribulations associated with mercury. For example, one can form wires from lead and tin but mercury has to be maintained within a glass capillary tube and is subject to mechanical fracture.

In 1914, Kamerlingh Onnes constructed a small coil from lead wire and applied a voltage, anticipating that he would be able to achieve very large magnetic field strengths due to the extraordinarily large currents that should flow in the superconductor. Unfortunately, at a rather low field strength the lead transitioned back into the normal resistance state. This was a blow to the intended program of established a high-magnetic field laboratory but for Kamerlingh Onnes did notice that when he removed the voltage source from the circuit that the current persisted. A ring of supercurrent flowing in the lead coil could deflect a compass. Positioning a copper coil adjacent to the compass also generates a magnetic field that can be arranged to cancel the deflection of the superconducting current. For an hour or more, no noticeable change in the compass needle was detected.

Over time, a number of additional superconductors were discovered and additional properties were added to the list of physical phenomena. In particular, the German physicists Fritz Walther Meißner (Meissner) and Robert Ochsenfeld discovered in 1933 that superconductors could not be penetrated by magnetic fields.[11] The Meißner effect forms the basis of popular laboratory demonstrations in which a superconducting block levitates above a magnet. The magnetic field of the magnet induces surface currents in the superconductor that cancel the field within the superconductor. These induced currents generate the lifting force that levitates the block.

An initial theory that explained the Meißner effect was produced by the German physicists (and brothers) Fritz and Heinz London in 1935. They suggested that the magnetic field of the superconductor must satisfy the following equation:

$$(8.10) \qquad \nabla^2 H = \kappa^2 H,$$

now known as the London equation. A solution is available immediately:

$$(8.11) \qquad H = H_o \, e^{-\kappa \cdot r},$$

where κ is the magnitude of the vector $\boldsymbol{\kappa}$ and H_o is a constant.

> EXERCISE 8.20. Demonstrate that equation 8.11 is a solution to equation 8.10.

From equation 8.11, the constant κ has the dimension of an inverse length and $1/\kappa$ is known as the London penetration depth. It marks the characteristic rate of decrease of external magnetic fields into the superconductor. The London equations provide an explanation of the macroscopic behavior of superconductors but do not provide a microscopic justification for the existence of superconductivity.

Despite considerable interest in developing such a theory, none were successful until the 1957 work of John Bardeen, his postdoctoral assistant Leon Cooper and student J. Robert Schrieffer. Bardeen and David Pines had published an article in 1955 in which they concluded that electrons in a crystal could feel a small attractive force, despite the fact that there was significant Coulomb repulsion for free electrons.[12] The motivation for this work was the experimental observation of an isotope dependence on the critical temperature in mercury. If superconductivity was simply due to electronic interactions, then there should be no dependence upon which mercury isotopes formed the lattice. It was plausible, then, that electron-lattice interactions could provide the means for the superconducting state.

---

[11] Meißner and Ochsenfeld published ""Ein neuer Effekt bei Eintritt der Supraleitfähigkeit" in *Naturwissenschaften*.
[12] Bardeen and Pines published "Theory of the Meissner effect in superconductors" in the *Physical Review*. Pines then left Illinois for a tenure-track position at Princeton.

The next pivotal step was provided in a paper by Cooper in 1956, who argued that electrons could form weakly bound pairs.[13] Cooper was convinced that this explained superconductivity but Bardeen was dubious that they had yet completed the entire picture. In the interim came the announcement that Bardeen won the Nobel Prize for the invention of the transistor, along with William Shockley and Walter Brattain.[14] While Bardeen was off drinking champagne, Schrieffer found a means to treat the Cooper pairs coherently and this provided the missing step.

What Schrieffer recognized was that only electrons at the Fermi surface would participate in the pairing interactions. In normal metals, electrons are free to migrate into the conduction band when a small potential is applied. By contrast, in superconductors, there is a small band gap that prohibits promotion to the conduction band and, instead, allows the formation of Cooper pairs. The kinetic energy of the Cooper pairs can be written as follows:

$$(8.12) \qquad \mathcal{E}_K = 4\Omega_0 \int_0^{\hbar\omega} d\varepsilon \, g(\varepsilon),$$

where Ω_0 is the density of states at the Fermi energy, $g(\varepsilon)$ is the probability that a Cooper pair exists with energy ε and $\hbar\omega$ is a small (constant) energy above the Fermi energy.

Interactions between Cooper pairs lead to an interaction energy given by the following expression:

$$(8.13) \qquad \mathcal{E}_I = -4\Omega_0^2 V \int_0^{\hbar\omega} d\varepsilon \int_0^{\hbar\omega} d\varepsilon' \left\{ g(\varepsilon)\left[1 - g(\varepsilon)\right] g(\varepsilon')\left[1 - g(\varepsilon')\right] \right\}^{1/2},$$

where V is the (assumed to be common) interaction strength.

The total energy $\mathcal{E}_K + \mathcal{E}_I$ can be determined by varying $g(\varepsilon)$, with the following result:

$$(8.14) \qquad \mathcal{E}_K + \mathcal{E}_I = -\frac{2\Omega_0(\hbar\omega)^2}{e^{2/\Omega_0 V} - 1}.$$

This is the prediction of the BCS theory.[15] The product $\Omega_0 V$ is independent of isotopic effects, so the entire dependence on isotope is governed by the $(\hbar\omega)^2$ behavior, which is in agreement with experimental observation.

---

[13]Cooper's "Bound electron pairs in a degenerate Fermi gas" was published in the *Physical Review*.

[14]Shockley, Bardeen and Brattain were awarded the Nobel Prize in Physics in 1956 "for their researches on semiconductors and their discovery of the transistor effect."

[15]Bardeen, Cooper and Schrieffer published "Microscopic theory of superconductivity" as a Letter to the *Physical Review* and an expanded "Theory of superconductivity" later in 1957. The three were awarded the Nobel Prize in Physics in 1972 "for their jointly developed theory of superconductivity, usually called the BCS-theory."

Remarkably, the Cooper pairs are not formed by adjacent electrons, the individual electrons that participate in the pairing interaction are widely separated, with a coherence length on the order of 100 nm. Additionally, the behavior of the Cooper pairs in equation 8.14 reflects boson statistics. A common derivation in introductory statistical mechanics courses is the calculation of the average occupation number $\overline{n}_\alpha$ for the state $|\alpha\rangle$. For fermions, the occupation number can only be 0 or 1, reflecting the Pauli exclusion principle. As a result the occupation number has the following form:

$$(8.15) \qquad \overline{n}_\alpha^{\mathrm{F}} = \frac{1}{e^{(\varepsilon-\mu)/k_BT} + 1},$$

where μ is the chemical potential, or the Fermi energy. For bosons there is no restriction on the number of particles that can occupy the same state. This results in the following formula for the occupation number:

$$(8.16) \qquad \overline{n}_\alpha^{\mathrm{B}} = \frac{1}{e^{(\varepsilon-\mu)/k_BT} - 1}.$$

The difference is the sign in the denominator but this has an enormous effect on the occupation number.

> EXERCISE 8.21. Plot the occupation numbers from equations 8.15 and 8.16. Use units where $\mu = 1k_BT$. Plot over the domain $0 \leq \varepsilon \leq 10k_BT$.

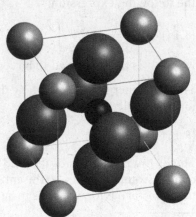

FIGURE 8.18. Cubic perovskites have space group Pm$\overline{3}$m. The faces of the cube are oxygen atoms (gray) and the corners are occupied by one metal (light gray) and the center of the cube is occupied by a second metal (dark gray).

The Cooper pairs are composite bosons, reminiscent of the bosons utilized in the IBM theory discussed earlier. The pairing force, though, arises through interactions of the electrons with the lattice through quanta known as phonons. The resulting pairing force is small and, so, superconductivity was restricted to very low temperatures. This situation changed drastically in 1986, when German physicist J. Georg Bednorz and his Swiss colleague K. Alexander Müller were studying the low temperature properties of metal oxides. Much to their delight, the pair found a perovskite material based on the rare earth elements barium and lanthanum that demonstrated a critical temperature above 10 K.[16] Adding copper to the mix $(Ba_xLa_{5-x}Cu_5O_5(3-y))$ produced a samples that had a transition temperatures in the 30 K range. The nominal perovskite structure is depicted in figure 8.18.

The Bednorz/Müller discovery provoked amazement in the superconductivity community. They were working with insulators not conductors and the observation of superconductivity in an insulating material was astonishing. They provoked a flurry of competition and set off a furious race to drive the critical temperature to unforeseen heights. Without any particular theoretical guidance, experimenters depleted all stocks of rare earth elements, which became unattainable at any price. In very short order, the critical temperature of the copper perovskite superconductors rose above the liquid nitrogen temperature of 77 K. A timeline for the advances in critical temperature is provided in figure 8.19.

These discoveries signalled an extraordinary advance with commercial applications looming in the future. Liquid helium is expensive and in limited supply. Liquid nitrogen is ubiquitous and cheap. While commercialization has not proceeded with the pace initially envisioned, there are some applications. For example, the RF filters used in cell towers use high temperature superconductors to improve the detector sensitivity.

At present, the mechanism for high temperature superconductivity is still in dispute. The perovskite materials are complex and the superconductive state is not homogeneous but reflects the layered nature of the crystal structure. The ceramic materials are brittle and difficult to characterize, owing to the fact samples are generally composed of small grains of unaligned domains. Initially, there was great confusion in the field due to the extraordinarily rapid pace and experimental limitations. Different experimental groups found vastly different critical temperatures for what should have been the same material but was, in fact, different due to fabrication issues. In time, the discourse became somewhat more civilized and two models for high temperature superconductivity have emerged as the most viable description of the phenomenon. The first was put forward by Philip Anderson, who suggested a model of the cuprate superconductors as a quantum spin liquid. The second model was developed by David Pines and collaborators Phillippe Montoux and Alexander Balatsky, who

---

[16]Bednorz and Müller were awarded the Nobel Prize in Physics in 1987 "for their important breakthrough in the discovery of superconductivity in ceramic materials." The award came just a year after they published "Possible high T_c superconductivity in the BaLaCuO system" in the *Zeitschrift für Physik B*.

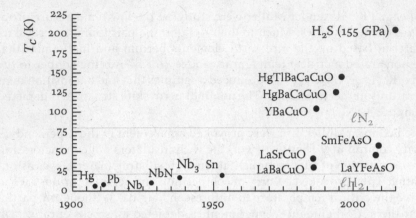

FIGURE 8.19. The highest known critical temperatures of superconducting materials has risen dramatically over time. Key temperatures are the 4.2 K boiling point of helium, the 22 K boiling point of H_2 and the 77 K boiling point of N_2 (light gray lines).

have suggested that the cuprates function as an antiferromagnet, where the spin on each copper atom is antiparallel to the adjacent spins.[17]

There is also a new class of superconducting materials based on iron arsenide compounds that were discovered in 2008. These have a crystal structure that is even more complex than the perovskites but should provide another arena in which the models can be tested. The field of superconductivity remains an active research enterprise. One of the principal difficulties retarding progress in the field is the lack of a definitive experimental outcome that can differentiate the proposed models. This is often the status of fields in their early stages of development.

[17]Anderson's "The resonating valence bond state in La_2CuO_4 and superconductivity" appeared in *Science* in 1987. Monthoux, Balatsky and Pines published "Toward a theory of high-temperature superconductivity in the antiferromagnetically correlated cuprate oxides" in the *Physical Review Letters* in 1991.

Light and Matter

We began in Chapter 2 with a discussion of Maxwell's equations that unified the treatment of electromagnetic phenomena. The coupled, vector differential equations are quite formidable but it is possible to solve them in a number of cases. Additionally, a number of numerical techniques have been developed that permit solving the equations in cases in which analytic solutions are not possible: for asymmetric geometries, for example. Prior to Maxwell's work, the nature of light was a subject of controversy. Newton held that light possessed a corpuscular nature, based upon his own observations that light rays travelled in straight lines until they encountered a surface. At that point, his model broke down but he was unwilling to consider an alternative theory.

Today, we recognize that Newton's ray model of light can be interpreted as a high-frequency approximation to Maxwell's equations. When light from a source strikes an object, we see that the object casts a shadow. Of course, if we look more closely, we can see that light diffracts around the edges and that the shadow does not have as sharp an edge as one might initially assume. As a first approximation, though, one might well describe the behavior of light beams with just a few simple rules. This is the approximation known as geometric optics. As Newton was a master of geometry, we can see how this sort of theory would have held great aesthetic appeal.

9.1. Geometric Optics

The essence of geometric optics is conveyed by figure 1.8. When a light ray (thin beam) strikes an interface where there is a discontinuity in the constitutive properties, part of the beam is reflected and part is transmitted, emerging at a different angle. The relationship between the incoming and outgoing ray angles has been known for a long time. In 984, the Persian scholar Abu Said al-Ala Ibn Sahl produced a manuscript *On burning mirrors and lenses* that is the first known statement of what is today called

© Mark A. Cunningham 2018
M.A. Cunningham, *Beyond Classical Physics*,
Undergraduate Lecture Notes in Physics,
https://doi.org/10.1007/978-3-319-63160-8_9

Snell's Law after the Dutch physicist Willebrord Snell.[1] While the relationship was determined by experiment initially, a success of Maxwell's theory is that the relationship can be derived from the equations.

Even though we know that surfaces at the atomic scale are quite rough, the wavelength of visible light is much greater than the atomic scale and thin beams actually illuminate an area that encompasses tens of thousands of atoms or more. Hence, we can utilize the approximation that there is a step change in electromagnetic properties at a surface, like that of a lens or mirror.

From the first of the Maxwell equations 2.1, we can observe that, near an interface and in a small area like that pictured in figure 9.1, the component of the electric displacement **D** that is normal to the surface will depend upon any surface charge σ. We can write the following expression:

$$(9.1) \qquad \mathbf{D}_2 \cdot \mathbf{n}(\mathbf{r}_2 + \delta)\Delta A - \mathbf{D}_1 \cdot \mathbf{n}(\mathbf{r}_2 - \delta)\Delta A = \sigma \, \Delta A,$$

where we have made use of the fact that the normal component of **D** is orthogonal to the sides of the tube. Taking the limit where $\delta \to 0$, equation 9.1 provides a boundary condition on the electric displacement. Because the differential area ΔA is arbitrary, it can be eliminated from the equation. If we utilize the constitutive relations for the displacement, we have a relationship for the electric fields at a boundary:

$$(9.2) \qquad \epsilon_2 \mathbf{E}_2(\mathbf{r}_2) \cdot \mathbf{n}(\mathbf{r}_2) - \epsilon_1 \mathbf{E}_1(\mathbf{r}_2) \cdot \mathbf{n}(\mathbf{r}_2) = \sigma.$$

Here, ϵ_i is the dielectric permittivity in medium i.

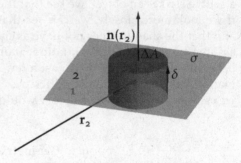

FIGURE 9.1. In the vicinity of the point $\mathbf{r}_2$ on the interface between volumes 1 and 2, we can construct a small tube with cross-sectional area ΔA that extends just above and below the surface.

Similarly, Faraday's equation 2.3 can be used to derive a boundary condition on the component of the electric field parallel to the surface. At a microscopic level, as depicted in figure 9.2, the path integral over the

---

[1] The attribution to Snell is another historical curiosity. Snell discovered the relationship in 1621 but never published his findings. After his death, the outline of a treatise on optics was found amongst his papers. His work only became known when it was cited in 1703 by Christiaan Huygens in his *Dioptrica*.

small loop in the neighborhood of the point r_1 can be seen to produce the following result:

$$(9.3) \qquad E \cdot dl \rightarrow E_2 \cdot dl - E_2 \cdot \delta - E_1 \cdot \delta - E_1 \cdot dl + E_1 \cdot \delta + E_2 \cdot \delta,$$

where we have assumed that the normal vector is constant over the dimension of the loop. Note that all of the terms involving δ vanish, provided that dl is suitably small. So, we are left with the following result:

$$E_2 \cdot dl - E_1 \cdot dl = -\frac{d}{dt}\Big[B_2 \cdot n(r_1)\delta \, dl + B_1 \cdot n(r_1)\delta \, dl\Big].$$

Now as we take the limit where δ and dl vanish, the right-hand side vanishes and we are left with the following result, where here $n(r_2)$ is the normal to the surface:

$$(9.4) \qquad E_2 - [E_2 \cdot n(r_2)]n(r_2) = E_1 - [E_1 \cdot n(r_2)]n(r_2).$$

That is, the components of the electric field parallel to the surface are equal.

EXERCISE 9.1. Justify the terms (especially signs) in equation 9.3.

FIGURE 9.2. We can construct a small loop in the vicinity of r_1 that runs parallel to the surface at a height δ, pierces the surface and then returns in the opposite direction. Here the normal is defined with respect to the loop, not the interface between media.

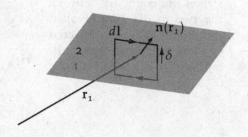

Using the same strategy for the magnetic induction and magnetic field, we can derive the following boundary conditions:

$$(9.5) \qquad \mu_2 H_2 \cdot n(r_2) = \mu_1 H_1 \cdot n(r_2) \quad \text{and}$$

$$(9.6) \qquad H_2 - [H_2 \cdot n(r_2)]n(r_2) + n(r_2) \times K = H_1 - [H_1 \cdot n(r_2)]n(r_2),$$

where K is a current sheet embedded in the interface. The component of the magnetic field normal to the surface is continuous at a boundary, as is the component of the magnetic induction parallel to the surface, in the absence of surface currents.

EXERCISE 9.2. Following the derivations of the boundary conditions on the electric field, justify the boundary conditions on the magnetic field expressed in equations 9.5 and 9.6.

If we work in the Fourier transform domain, then the fields in the medium where the incident ray begins, call it medium 1, are the sum of incident and reflected rays:

$$(9.7) \qquad \tilde{E}_1 \, e^{i k_1 \cdot r} = \tilde{E}_I \, e^{i k_1 n_I \cdot r} + \tilde{E}_R \, e^{i k_1 n_R \cdot r}.$$

An illustration of the ray geometry is provided in figure 9.3. Note here that both incident and reflected rays propagate with the same wavenumber k_1 but with different directions n_I and n_R, respectively. In medium 2, there is only the refracted ray:

$$(9.8) \qquad \tilde{E}_2 \, e^{i k_2 \cdot r} = \tilde{E}_T \, e^{i k_2 n_T \cdot r}.$$

We can recover Snell's law by enforcing the boundary conditions on the fields at the interface.

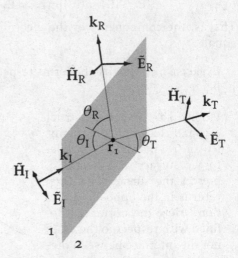

FIGURE 9.3. The incident ray strikes the boundary between the two media (1 and 2) at the point r_1. A reflected ray and refracted (transmitted) ray are generated.

We can simplify the discussion a bit by assuming that the interface is planar, or at least has a small radius of curvature at the point r_1. With this assumption, the general problem of fields can be decomposed into two components: one with $\tilde{E}$ parallel to the surface and a second with $\tilde{H}$ parallel to the surface. In either case, we will have a boundary condition that has the following form:

$$(9.9) \qquad \tilde{F}_I \, e^{i k_1 n_I \cdot r_1} + \tilde{F}_R \, e^{i k_1 n_R \cdot r_1} = \tilde{F}_T \, e^{i k_2 n_T \cdot r_1},$$

where F is one of the fields. Note that in equation 9.9, the only spatial dependence is in the exponential factors, the transformed fields are functions of the wave vectors k_i. Consequently, for equation 9.9 to be correct for all points on the surface r_1, the exponential factors must all be equal:

$$(9.10) \qquad k_1 n_I \cdot r_1 = k_1 n_R \cdot r_1 = k_2 n_T \cdot r_1.$$

We can, for the moment, choose the surface to have a normal in the z-direction and the surface to lie at the $z = 0$ plane. (This isn't necessary but simplifies the following discussion.) Then, equation 9.10 can be expanded into the following:

$$(9.11) \quad k_1(\mathbf{n_I})_x \, x + k_1(\mathbf{n_I})_y \, y = k_1(\mathbf{n_R})_x \, x + k_1(\mathbf{n_R})_y \, y = k_2(\mathbf{n_T})_x \, x + k_2(\mathbf{n_T})_y \, y$$

Again, for this equation to hold, the terms in x and y must hold individually. Thus, we must have the following:

$$(9.12) \quad k_1(\mathbf{n_I})_x = k_1(\mathbf{n_R})_x = k_2(\mathbf{n_T})_x \quad \text{and} \quad k_1(\mathbf{n_I})_y = k_1(\mathbf{n_R})_y = k_2(\mathbf{n_T})_y$$

We note that, in a polar coordinate system centered on $\mathbf{r_1}$, we can write $\mathbf{n_I} \cdot \mathbf{n}(\mathbf{r_1}) = \cos\theta_I$. The remaining components of $\mathbf{n_I}$ would be given by $(\mathbf{n_I})_x = \sin\theta_I \cos\varphi$ and $(\mathbf{n_I})_y = \sin\theta_I \sin\varphi$. Thus, for an arbitrary azimuthal angle φ, we must have that the following relations hold:

$$(9.13) \qquad\qquad k_1 \sin\theta_I = k_1 \sin\theta_R = k_2 \sin\theta_T.$$

The first equality can be simplified to state that the angle of incidence is equal to the angle of reflection. The other equality provides Snell's law: the sines of the angle of refraction and angle of incidence are related by the wave vectors. Recall that $k^2 = \omega^2 \mu\epsilon + i\omega\mu\sigma$, where σ is the bulk conductivity of the medium. Hence, for conductive media, there is an imaginary component to the wave vector and waves in the conductor become evanescent. The field falls off exponentially within the media with a characteristic length, the skin depth, $\delta = [2/\omega\mu\sigma]^{1/2}$.

So, we can turn now to the problem of determining the reflected and refracted fields given an incident field. We can accomplish this by enforcing the boundary conditions on the fields at the interface between the media. As we mentioned, the problem can be decomposed into two separate cases: either the incident electric (TE) or magnetic (TM) field is parallel to the surface. The direction of the other field is governed by Maxwell's equations 2.22–2.25. Consider first the case of TE incidence. For simplicity, let us continue to utilize a coordinate system in which the normal to the surface is in the z-direction and let us, without loss of generality, choose the polarization of the incident electric field to be in the x-direction: $\tilde{\mathbf{E}}_I = \tilde{E}_I \hat{\mathbf{x}}$. From the orthogonality conditions on the fields and the wave vector, this means that the wave vector can only have y and z components: $\mathbf{k_I} = k_1(0, \sin\theta_I, \cos\theta_I)$. From the results in equation 9.13, we also can show that the other wave vectors are of the form: $\mathbf{k_R} = k_1(0, \sin\theta_I, -\cos\theta_I)$ and $\mathbf{k_T} = k_2(0, \sin\theta_T, \cos\theta_T)$. The incident magnetic field is obtained from Faraday's law, equation 2.24, from whence we obtain the following:

$$(9.14) \qquad\qquad \tilde{\mathbf{H}}_I = \frac{1}{\omega\mu_1} \mathbf{k_I} \times \tilde{\mathbf{E}}_I = \frac{\tilde{E}_I k_1}{\omega\mu_1} (0, \cos\theta_I, -\sin\theta_I).$$

We now need to determine the field strengths of the reflected and refracted waves: $\tilde{E}_R$ and $\tilde{E}_T$. From the boundary condition on the electric fields tangential to the surface, we obtain the following:

(9.15) $$\tilde{E}_I + \tilde{E}_R = \tilde{E}_T.$$

From the boundary conditions on the tangential components of the magnetic field, we obtain the following:

(9.16) $$\frac{k_1}{\omega \mu_1}\left[\tilde{E}_I \cos\theta_I - \tilde{E}_R \cos\theta_I\right] = \frac{k_2}{\omega \mu_2}\left[\tilde{E}_T \cos\theta_T\right].$$

After a suitably tedious amount of algebra, we are led, at last, to the Fresnel equations, named for the French physicist Augustin-Jean Fresnel[2]:

(9.17)
$$\frac{\tilde{E}_R}{\tilde{E}_I} = -\frac{k_1 \mu_2 \cos\theta_I - k_2 \mu_1 \cos\theta_T}{k_1 \mu_2 \cos\theta_I + k_2 \mu_1 \cos\theta_T} \quad \text{and} \quad \frac{\tilde{E}_T}{\tilde{E}_I} = \frac{2 k_1 \mu_2 \cos\theta_I}{k_1 \mu_2 \cos\theta_I + k_2 \mu_1 \cos\theta_T}.$$

As we can see in figure 9.4, the coefficients are smoothly varying functions of the incident angle. This is actually a bit misleading. From Snell's law, equation 9.13, we can deduce that the angle of refraction is given by the following expression:

(9.18) $$\theta_T = \sin^{-1}\left[\frac{k_1}{k_2}\sin\theta_I\right].$$

Obviously, this equation is well defined when $k_1 < k_2$ but what happens when $k_1 > k_2$? In this case, there will be angles θ_I where the argument in the square brackets is greater than one and for these values, the $\sin^{-1}$ function becomes complex and the fields evanescent.

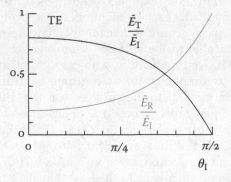

FIGURE 9.4. The normalized reflection and refraction coefficients are plotted as a function of incident angle. Ratios of $k_2/k_1 = 1.5$ and $\mu_2/\mu_1 = 1$. were used in this example.

EXERCISE 9.3. Plot the real and imaginary parts of $\sin^{-1} x$ for the domain $0 \leq x \leq 5$.

---

[2]Fresnel published his treatise *Mémoire sur la diffraction de la lumière* in 1818, for which he won the prize offered by the Académie des Sciences.

EXERCISE 9.4. Consider a beam defined by the incident wave vector $\mathbf{k}_I = k_1(0, \sin\theta_I, \cos\theta_I)$ and that possesses an incident electric field $\tilde{\mathbf{E}}_I = \tilde{E}_I(\cos\alpha, \sin\alpha\cos\theta_I, -\sin\alpha\sin\theta_I)$. Show that the incident field can be decomposed into TE and TM components.

EXERCISE 9.5. Plot the scaled reflection and refraction coefficients for $k_2/k_1 = 1.5$ and $\mu_2 = \mu_1 = 1$. What happens when the ratio of wavenumbers is reversed?

The solutions for transverse magnetic fields can be obtained similarly. We note that the magnitude of the magnetic field is proportional to the electric field:

$$H = \frac{k}{\omega\mu}E.$$

Using this, we can again solve for the Fresnel equations for TM incidence:
(9.19)
$$\frac{\tilde{E}_R}{\tilde{E}_I} = -\frac{k_2\mu_1\cos\theta_I - k_1\mu_2\cos\theta_T}{k_2\mu_1\cos\theta_I + k_1\mu_2\cos\theta_T} \quad \text{and} \quad \frac{\tilde{E}_T}{\tilde{E}_I} = \frac{2k_1\mu_2\cos\theta_I}{k_2\mu_1\cos\theta_I + k_1\mu_2\cos\theta_T}.$$

At first glance, the Fresnel coefficients appear similar but, as we note in figure 9.5, there are significant differences.

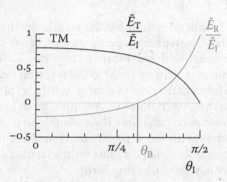

FIGURE 9.5. For the same parameters used in generating figure 9.4, the Fresnel coefficients for TM incidence display different behavior. The reflected wave vanishes at the Brewster angle θ_B.

For example, using the same parameters as before, the reflected wave coefficient changes sign at the angle θ_B, known as the Brewster angle after the Scottish cleric turned physicist David Brewster.[3] Brewster noted that unpolarized light (TE+TM) incident on a crystal became polarized (TE) upon reflection when the sum of incident and refracted angles was $\pi/2$.

EXERCISE 9.6. Show that, when $\tilde{E}_R = 0$, the sum of the incident and refracted angles becomes $\theta_I + \theta_T = \pi/2$, provided that the materials are not magnetic ($\mu_i = 1$). Hint: Use the $\sin^{-1}$ addition formula:

$$\sin^{-1}x + \sin^{-1}y = \sin^{-1}\left[x(1-y^2)^{1/2} + y(1-x^2)^{1/2}\right].$$

---

[3]Brewster published "On the laws which regulate the polarization of light by reflection from transparent bodies" in the *Philosophical Transactions of the Royal Society of London* in 1815.

EXERCISE 9.7.   Plot the TM Fresnel coefficients for $k_2/k_1 = 1.5$ and $\mu_i = 1$. What happens if the ratio k_2/k_1 is less than one?

The Fresnel equations provide the means for determining the behavior of optical systems. Draw an incident ray and propagate it through the system by drawing straight lines until it strikes and interface. Compute the local normal and then generate reflected and refracted rays. Propagate those until they strike interfaces and continue. While such a prescription is quite tedious, it is just the sort of thing at which computers excel.

As a practical matter, there are subtleties associated with the fact that the dielectric permittivity of glasses, for example, is not independent of frequency. This frequency dependence is known as dispersion. The name arises from the observation that prisms spread (disperse) incident light into its spectral components. Dispersion is a useful feature when trying to determine spectra but not a useful feature when trying to produce images. In lens applications, this frequency dependence leads to chromatic aberration: the focal length depends upon the color. Much of the design of lens systems for cameras and microscopes is predicated on minimizing the effects of dispersion.

The phenomenon has its origins in the microscopic behavior of materials. Electromagnetic energy incident on a material will predominantly interact with the local electron population. In conductors, this will drive currents proportional to the electron density. In insulators, where there is a band gap, there is a more subtle response as the electron density adapts to the external fields. A simple model for the dielectric response involves resonance behavior. If we assume that electrons behave as harmonic oscillators with some particular resonance frequency ω_0 and that the electromagnetic fields oscillate with a frequency ω, then the dielectric permittivity has the following form:

$$(9.20) \qquad \epsilon = \epsilon_0 \left[1 + \frac{ne^2}{m_e} \sum_j \frac{n_j}{\omega_j^2 - \omega^2 - i\alpha_j\omega} \right],$$

where n is the number of electrons in a unit volume, n_j are the fraction with resonant frequency ω_j and α_j is a (small) damping factor. Instead of a single resonant frequency, there are many frequencies ω_j corresponding to states within the matrix.

EXERCISE 9.8.   Plot the real and imaginary parts of the function $f(\omega, \alpha) = [1 - \omega^2 - i\alpha\omega]^{-1}$ over the domain $0 \leq \omega \leq 3$, for $\alpha = 0.03$ and 0.1. What is the behavior above and below the resonance frequency $\omega = 1$? What is the effect of alpha?

Equation 9.20 provides a simple explanation for the general behavior of dielectric media, in which the "dielectric constant" is not actually a constant, but a slowly varying function of frequency. Beyond the issues with dispersion, a problem that has plagued astronomers and opticians from Galileo to Newton to modern days is the very limited range of dielectric constants. All glasses have a dielectric constant in the neighborhood of 3–5, resulting in an index of refraction of about 1.5. As a result, the amount of focussing through lenses is quite restricted.

EXERCISE 9.9.   Use values of $k_2/k_1 = 1.5$ and $\mu_1 = \mu_2 = 1$ and compute the refracted paths for incident rays that are parallel to the centerline of the lens. Consider cases where the offset from the centerline has the values $0.05R...0.5R$ in steps of $0.05R$. For simplicity, consider the case where $R = 1$ and the center of the lens is at $(2,0)$.

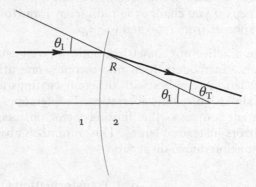

FIGURE 9.6. Refraction at an interface with a curved surface leads to converging rays.

In figure 9.6, we illustrate one incident ray striking a curved surface. In this case, the surface is spherical, with a constant radius of curvature. Such lenses are relatively easy to construct but making them larger does not lead to better optical properties. The refracted rays do not converge to a point but instead form a caustic surface. As a result, lens makers are restricted to using lenses with radii of curvature that are large compared to the lateral dimension of the lens.

If we place two curved surfaces back to back, as illustrated in figure 9.7, then the initial beam of parallel rays converges but not to a point. Rays far from the centerline are bent across the centerline at points short of the radius of curvature. For rays near the axis, the effect is much less pronounced. The caustic surface is the envelope formed from the rays on the right of the lens. In an ideal lens, all of the rays would cross the axis at a point. For real lenses, this cannot happen with spherical surfaces.

EXERCISE 9.10.   Construct a *Mathematica* function that can trace a ray through two spherical surfaces. Each sphere has a center at some

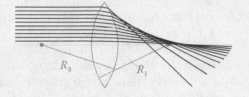

Figure 9.7. For a thick lens with two curved surfaces, rays from the periphery of the lens cross the centerline (light gray) before those from near the axis.

distance c_i along the x-axis and a radius of curvature R_i. Incident rays run parallel to the x-axis at a height $y = h$. Use the `Circle` and `InfiniteLine` functions to identify the intersections of the incident ray with sphere 1 and then the refracted ray with sphere 2. Hint: use the `NSolve` function.

For a series of rays, see if you can reproduce figure 9.7. What happens if you change the radii from 1 and/or move the centers of the spheres from (2,0) and (0.35,0)?

The practice of geometrical optics is well established, as are the limitations. In modern light microscopes, one utilizes not individual lenses but lens assemblies with different coatings and indices of refraction to minimize chromatic aberration. In telescopes, where light intensity from distant sources is the limiting factor on image production, one utilizes mirrors instead of lenses. This minimizes absorption of faint signals and distortion due to dispersion.

9.2. Transformation Optics

Many investigators from Galileo and Newton and onwards have experimented with different glass formulas to see if some magical addition might make a material with a refractive index of 10 or 100. None have as yet succeeded. In 1964, the Russian physicist Victor Veselago wrote an interesting paper on the consequences of having negative values of the permittivity and permeability.[4] One might first imagine that such circumstances would have no effect whatsoever. Recalling that $k^2 = \omega^2 \mu \epsilon$, it seems that if both μ and ϵ were to change sign, then the signs would simply cancel.

In fact, an inspection of our discussion of the boundary conditions on the electric and magnetic fields displayed in equations 9.2 and 9.5 reveals a remarkable result. The fields parallel to the surface are unaffected by the

[4]Vesalago published "The electrodynamics of substances with simultaneously negative values of ϵ and μ" in *Uspekhi Fizicheskikh Nauk* in 1964. It appears in English translation in *Physics Uspekhi* in 1968

change in sign of ϵ and μ but the normal components of the fields must change signs. Indeed, from equations 2.24 and 2.25, we had originally concluded that the vectors $\tilde{\mathbf{E}}$, $\tilde{\mathbf{H}}$ and $\mathbf{k}$ formed a right-handed set of orthogonal vectors. If ϵ and μ are negative, then the vectors will remain orthogonal but now have a left-handed sense.

This result is rather perplexing. We know that the Poynting vector is defined as $\mathbf{S} = \mathbf{E} \times \mathbf{H}$ and that this defines the flow of energy in the electromagnetic field. With negative values for ϵ and μ, we find now that the wave vector $\mathbf{k}$ points in the opposite direction from the Poynting vector. This situation is illustrated in figure 9.8, where medium 2 is presumed to have negative values for the electromagnetic properties. The figure depicts TE incidence, where the electric field is parallel to the surface.

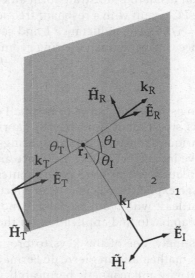

FIGURE 9.8. A TE wave is incident on the surface between two media. For a region (2) where μ and ϵ are negative, the refracted wave vector points in the opposite direction from the Poynting vector. Refracted rays are on the *same* side of the normal as the incident ray.

EXERCISE 9.11. Construct a figure that represents a TM wave incident on a material with negative values of μ and ϵ. Where is the refracted ray?

Remarkably, in figure 9.8, the refracted ray is on the same side of the normal as the incident ray. This result is not a consequence of TE incidence; it also holds for TM incidence. As Veselago discovered, the change in refractive properties of what he called left-handed media bring about significant changes to the optics of any systems that could utilize them. At the time, there were no known media that possessed such properties, so Veselago's ideas remained dormant.

In the late 1990s, the British physicist John Pendry and his collaborators began experimenting with periodic metallic arrays, demonstrating that a number of unusual effects could be obtained from an array of wires, for example. In 1999, Pendry and team began utilizing split-ring resonators, demonstrating that the small, metal rings had effective electromagnetic properties that could be tuned to a wide variety of values, at least over a modest bandwidth in the microwave region.[5] A sketch of the split-ring geometry is provided in figure 9.9. Shortly thereafter, metamaterials with negative values of μ and ϵ were fabricated and their behaviors were largely what Veselago had predicted.

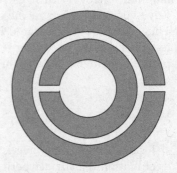

FIGURE 9.9. Two metal rings deposited on an insulating substrate form an effective LC circuit with a resonant frequency $\omega_0^2 = 1/LC$. The inductance L and capacitance C depend upon the ring geometry.

The ring sizes were chosen to be small compared to the microwave wavelengths studied but were readily fabricated with PC board technology. The various diameters and gap spacings provided the ability to choose the effective inductance and capacitance and provide a nearly arbitrary resonant frequency. This gave experimenters the opportunity to utilize equation 9.20 explicitly to generate an effective dielectric constant of arbitrary sign, at least within a reasonable bandwidth. There followed something of a scramble to find applications of the new technology.

Interestingly, one of the keys to progress was initially set forth by the French mathematician Pierre de Fermat in 1662. If we allow the dielectric permittivity and magnetic permeability to become functions of space and not just stepwise changing, then we can define the optical path of a ray as follows:

(9.21) $$s = \int_A^B d\mathbf{l} \cdot \mathbf{k}.$$

Fermat's observation was that the actual path taken by a light ray was the one that minimized the path:

(9.22) $$\delta s = 0.$$

[5]Pendry and coworkers published "Magnetism from conductors and enhanced nonlinear phenomena" in the *IEEE Transactions on Microwave Theory and Techniques* in 1999. Pendry, Thomas Ebbesen and Stefan Hell shared the Kavli Prize in Nanoscience in 2014 "for their transformative contributions to the field of nano-optics that have broken long-held beliefs about the limitations of the resolution limits of optical microscopy and imaging."

Allowing the electromagnetic parameters to become arbitrary functions of space makes equations 9.21 and 9.22 quite general.

Indeed, one can use Fermat's principle to trace the paths of rays through materials that possess a gradient in ϵ, for example. Such curved pathways explain the shimmering visible in the distance over hot roadways. The permittivity depends upon temperature, which is highest on the road surface and decreases in the vertical direction. Rays of light from the sky that occur at shallow enough angles will be bent back away from the road surface.

One might then ask the following question: "how can we obtain a particular ray curvature?" That is, what distribution of permittivity and permeability will lead to a particular ray path? The answer to these sorts of questions can be found through the language of differential geometry. We have not yet spent much time discussing the geometry of curved spaces but finding such a mathematical language occupied Einstein in his search for a general theory of relativity.

We introduced tensor notation earlier to demonstrate that Maxwell's equations could be made manifestly Lorentz invariant. That is, Maxwell's equations lead to a relativistically sensible theory. We did not, though, explore the implications of such a treatment. What we find now is that the minimal curve that Fermat described in 1662 is precisely the geodesic in a curved geometry. The definition of the geodesic is the curve with the shortest path between two points in space(time) that preserves the tangent vector to the curve. In Cartesian coordinates, this is simply a straight line. On the surface of a sphere, the geodesic is an arc of a great circle.

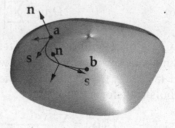

FIGURE 9.10. If we consider a curve from point **a** to **b** on a curved surface, a natural coordinate system can be defined as the direction along the curve **s**, the normal to the surface **n** and the cross product **s** × **n**, that points to the right of the curve.

What Einstein found in his studies of differential geometry is that, along a path, one can define, locally, a Cartesian coordinate system. As we have indicated in figure 9.10, at each point on the path from **a** to **b**, we can define the tangent along the direction of the path **s**, and the normal to the surface **n**. From these two, we can define a direction perpendicular, as in the figure. As we move along the pathway, the direction vectors change, which brings us to the issue of how one might be able to define derivatives. If we are at point **a**, for example, and take a small step in the **s**-direction, this will take us off of the curve.

Mercifully, mathematicians have worked out the details of defining how to move about the path. Because we are interested in the distance along the path, we naturally will choose to work in a universe where a metric can be defined. We introduced the metric tensor g in chapter 2, where the differential element ds can be defined as follows:

$$(9.23) \qquad ds^2 = \sum_{j,k} g_{ij} dx^j dx^k,$$

where the infinitesimal elements dx^j are displacements.

In order to compensate for moving off of the path, as might happen if we naïvely stepped off in the **s**-direction, we need what the mathematicians call a connection. This is provided by the Christoffel symbols:

$$(9.24) \qquad \Gamma^m_{jk} = \frac{1}{2} \sum_l g^{lm} \left[\frac{\partial g_{jl}}{\partial x^k} + \frac{\partial g_{kl}}{\partial x^j} - \frac{\partial g_{jk}}{\partial x^l} \right].$$

The covariant derivative of a basis vector $\mathbf{e}_k$ is then defined as follows:

$$(9.25) \qquad D_j \mathbf{e}_k = \sum_m \Gamma^m_{jk} \mathbf{e}_m.$$

This expression characterizes the fact that the intrinsic coordinate system that we are using varies along the path.

For a vector field $\mathbf{u}$, the covariant derivative becomes the following:

$$(9.26) \qquad D_j \mathbf{u} = \sum_{km} \left[\frac{\partial u^m}{\partial x^j} + \Gamma^m_{jk} u^k \right] \mathbf{e}_m.$$

The Christoffel symbols are not tensors, even though we have used covariant and contravariant indices. If we consider changing coordinate systems from $(x_1, \ldots, x_n)$ to $(y_1, \ldots, y_n)$, then it can be shown that the Christoffel symbols transform as follows:

$$(9.27) \qquad \Gamma^m_{jk}(\mathbf{y}) = \sum_{pqr} \frac{\partial^2 x^p}{\partial y^j \partial y^k} \frac{\partial y^m}{\partial x^p} + \Gamma^r_{pq}(\mathbf{x}) \frac{\partial x^p}{\partial y^j} \frac{\partial x^q}{\partial y^k} \frac{\partial y^m}{\partial x^r}.$$

Note that the first term on the right-hand side of equation 9.27 contains second derivatives of the coordinates with respect to the new coordinates. These all vanish in the theory of special relativity, where it was assumed that all coordinate transformations were linear. The second term on the right-hand side is precisely what we would obtain for a tensor with two covariant indices and one contravariant index. The result of the covariant derivative on a vector is to generate a second-rank tensor and the Christoffel symbols are the essential piece that ensures that happens.

EXERCISE 9.12. As a simple example, compute the Christoffel symbols for spherical coordinates:

$$(x, y, z) = (r \sin \theta \cos \varphi, r \sin \theta \sin \varphi, r \cos \theta).$$

Hint: In Cartesian coordinates, all of the Γ^m_{jk} vanish, so equation 9.27 can be used directly.

We can now explain how all of this complexity is warranted in the current discussion on optics. If we permit the constitutive equations to take their tensor form, we have now the following forms for the components of electric displacement and magnetic induction:

$$(9.28) \qquad D^j = \epsilon_0 \sum_k \epsilon_{jk} E^k \quad \text{and} \quad B^j = \mu_0 \sum_k \mu_{jk} H^k.$$

Here, we follow the usual notation and extract the vacuum factors of ϵ_0 and μ_0 explicitly. The divergence equations can now be written as follows:

$$(9.29) \qquad \epsilon_0 \sum_{jk} \frac{\partial \epsilon_{jk} E^k}{\partial x^j} = \rho \quad \text{and}$$

$$(9.30) \qquad \mu_0 \sum_{jk} \frac{\partial \mu_{jk} H^k}{\partial x^j} = 0.$$

The curl equations have the following form:

$$(9.31) \qquad \sum_{jk} \varepsilon^{ijk} \frac{\partial}{\partial x^j} E_k = -\sum_k \frac{\partial \mu^{ik} H_k}{\partial t} \quad \text{and}$$

$$(9.32) \qquad \sum_{jk} \varepsilon^{ijk} \frac{\partial}{\partial x^j} \epsilon_{jk} H_k = J^i + \sum_k \frac{\partial \epsilon^{ik} E_k}{\partial t}$$

Here ε^{ijk} is the antisymmetric Levi-Civita tensor and ϵ^{jk} is the dielectric permittivity tensor.

What we can recognize from equations 9.29–9.32 is that the electromagnetic permeability and permittivity tensors can be treated as an effective metric tensor. If we can tune ϵ and μ into any sort of spatial dependence we desire, then we can effectively create a curved geometry in which electromagnetic waves will propagate. In particular, they will propagate along geodesics of the curved space.

As an example, consider the transform of the radial coordinate in the following expression:

(9.33) $$r' = \begin{cases} a - (3a - 2b)\left[\dfrac{r}{b}\right]^2 + (2a - b)\left[\dfrac{r}{b}\right]^3 & r < b \\ r & r \geq b. \end{cases}$$

As illustrated in figure 9.11, this transformation maps the radial coordinate in the domain $0 \leq r \leq b$ into the range $a \leq r' \leq b$ and leaves the coordinate unchanged elsewhere.

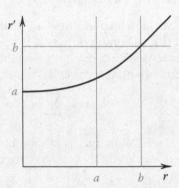

FIGURE 9.11. The transform in equation 9.33 maps the radial coordinate r into the radial coordinate r', leaving a hole of radius a.

Figure 9.12 depicts the geodesics that arise in the transformed coordinate system. Away from the center of the coordinate system, where $r > b$ (gray circle), the geodesics are straight lines, as we would find in a homogeneous medium. Inside the radius b, the geodesics curve, approaching the inner radius a only tangentially. Light rays will follow these curved paths, avoiding the central area.

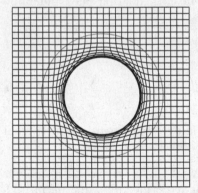

FIGURE 9.12. The transformation in equation 9.33 results in a mapping of the usual coordinate space into one with a hole of radius a. The grid lines here are geodesics and depict the paths of light rays.

The concept of transformation optics arises from the observation that, if the electromagnetic properties possess the spatial dependence defined in equation 9.33, then Maxwell's equations in real space will possess solutions that have the properties of solutions in the transformed space. That is, if metamaterials can be fashioned to have the appropriate values of μ and ϵ, then one can direct electromagnetic energy in ways that were previously considered impossible.

> EXERCISE 9.13. Plot the transformation defined in equation 9.33.
> Use $a = 2.5$ and $b = 4$. How does this change if you modify a and b?
> Now construct a function to plot the two-dimensional mapping of
> the point $(x, y) \rightarrow (x', y')$ using the transform. Use the ListLinePlot
> function to draw a series of lines over the range $-5 \leq x, y \leq 5$ using
> steps of 0.201. (The line through the origin is tricky; we'll avoid it.)
> You should be able to reproduce the geodesics in the figure.

A proof-of-principle experiment was conducted by the American physicist David Smith and his colleagues in 2006.[6] Smith and his students constructed several concentric layers of split-ring resonators, each layer tuned to produce effective ϵ and μ values that performed a transformation like that defined in equation 9.33. In figure 9.13, some of the results of the experiment can be observed. In the leftmost panel, the fields were measured within an empty scattering chamber, where the nominal wave direction is toward the bottom of the page. The wave field is approximately what we would expect of a plane wave propagating through the chamber.

In the center and right panels, a copper tube with the diameter indicated by the inner circle was placed in the chamber surrounded by metamaterial designed to have the appropriate μ and ϵ values at 10 GHz; this material extended to the radius of the outer circle. The fields depicted in the center panel are the scattered fields measured at 9 GHz, away from the design frequency. This image has the incident field subtracted. There are interference effects visible in the backward direction and diffraction effects visible in the shadow of the tube. For an observer in the far field of the tube, the scattered radiation will be detectable and the presence of the tube can be inferred.

In the right panel, the frequency has been raised to 10 GHz, where the metamaterial was designed to provide the correct μ and ϵ behavior. The scattering is greatly reduced and is in good agreement with numerical results obtained by Smith and his students that simulated the actual parameter profile they implemented. This gives rise to the hope that more sophisticated metamaterials can be devised to more closely match the ideal transformation specified in equation 9.33.

In the popular press, this demonstration has been called an invisibility cloak, which is a rather large overstatement. In principle, it is true that, if

---

[6]Smith, Pendry and coworkers published "Metamaterial electromagnetic cloak at microwave frequencies" in *Science* in 2006 and "Scattering cross-section of a transformation optics-based metamaterial cloak" in the *New Journal of Physics* in 2010.

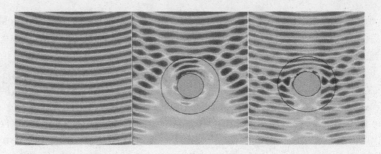

FIGURE 9.13. Microwave field intensity inside a scattering chamber (left panel) depicts nominal plane wave propagation (downward). A copper tube (inner black circle) surrounded by metamaterial within the outer black circle induces scattering and diffraction effects. At 9 GHz, away from the design frequency, scattering effects are large (center panel). These are greatly reduced at the nominal 10 GHz frequency (right panel). Image courtesy of David R. Smith, Duke University.

there are no far-field scattering effects, anything within the radius a would be unobservable. In practice, Smith and his students have found that the scattering from the metamaterial is vastly greater outside the relatively narrow frequency band in which the effective permeability and permittivity have the desired values. At present, the bandwidth over which parameter values can be tuned is rather small. Extending that bandwidth is the focus of numerous research groups.

EXERCISE 9.14. Conformal mapping is a strategy utilized in finding solutions to the Laplace equation in curious geometries. It has fallen from favor with the advent of numerical computation but also illustrates the use of coordinate transformations. Show that the following transformation:

$$w = \frac{z-i}{z+i}$$

maps the upper half plane into the unit circle. Here, $z = x + iy$ and $w = x' + iy'$.

Indeed, the advent of additive manufacturing processes has opened a new pathway for constructing novel three-dimensional materials. Smith and students utilized PC board technology to make split-ring resonators on flat surfaces and then curled them into cylinders. With a three-dimensional printer, one can create vastly more complex forms. It is likely that the modest results depicted in figure 9.13 can be greatly improved, at least at microwave frequencies. Constructing such devices that would be operational at optical frequencies is still speculative.

9.3. Quantum Optics

The existence of a conformal symmetry within Maxwell's equations was unexpected and has given rise to numerous research programs devoted to exploiting the phenomena. In addition to the electromagnetic cloaking that we have discussed, subwavelength imaging has made significant advances over the previous state of the art. That such fundamental new behaviors of Maxwell's equations have just been discovered begs the question about what yet remains to be discovered.

What we have been discussing still lies within the province of classical field theory and does not invoke any particular quantum mechanical description of matter. The electron cloud that exists within the material has been reduced, on average, to net polarization and magnetization effects that give rise to the bulk permittivity and permeability. A key element to the success of the classical theory is the long wavelength of visible light with respect to the lattice spacings. As electromagnetic energy impacts matter, numerous lattice sites are illuminated and quantum effects are difficult to disentangle. Presently, we do not have the capacity to compute the electron charge density with enough precision to make a tidy connection to Maxwell's equations. Nevertheless, there are places where quantum phenomena are visible.

Astronomers, for example, have observed correlations in photon counting experiments that can be attributed to the quantum nature of light itself. In 1954, British astronomers Robert Hanbury Brown and Richard Twiss proposed a new interferometric technique and two years later reported on the first implementation of their idea.[7] Essentially, Hanbury Brown and Twiss proposed to use two spatially separated detectors to simultaneously sample the incoming light from distant stars, amplify the output and then correlate the outputs. This process would provide them with the ability to create an observing platform with the effective aperture of the separation distance and not limited by the size of the separate receiving antennas. They noted that the electromagnetic intensity appeared to be correlated in their results.

EXERCISE 9.15.   Correlation is a powerful technique in time series analysis. Construct input signals from the following function:

$$f(t) = \sin[\pi(t + \tau)/3.4]\sin[\pi(t + \tau)/9.1].$$

Create a (discrete) input by taking values from $0 \le t \le 100$ in steps of 0.1. Use values of $\tau = 0$ and 0.3. Now add two noise vectors to the raw data by using the RandomReal function, and setting the width of

---

[7]Hanbury Brown and Twiss published "Correlation between photons in two coherent beams of light" and "A test of a new type of stellar interferometer on Sirius" in *Nature* in 1956.

the distribution to be around 0.8. Plot the (noisy) data. Can you see a correlation? Now use the ListConvolve function to correlate the data. How does the result compare to the original (noiseless) data? What happens if the noise level is increased? How does the result depend on the time offset?

The phenomenon was given firmer theoretical support by American physicist Roy Glauber in 1963.[8] Glauber was able to demonstrate that the correlation of the two signals was twice that obtained from a classical calculation by utilizing the quantum nature of photons. The expectation value of the product of the two photon intensities can be written in terms of creation and annihilation operators:

$$\langle I(x)I(y)\rangle = \langle a^\dagger(x)a^\dagger(y)a(x)a(y)\rangle$$

$$(9.34) \qquad = \langle a^\dagger(x)a(x)\rangle\langle a^\dagger(y)a(y)\rangle + \langle a^\dagger(x)a(y)\rangle\langle a^\dagger(y)a(x)\rangle.$$

When the optical path lengths are equal $x = y$, one obtains a factor of two, just as Hanbury Brown and Twiss had observed.

As Glauber found, the quantum nature of the photon/electron has to be taken into account in order to provide a successful description of low-intensity measurements. In particular, the *coherence* of electromagnetic waves is intimately connected to the underlying quantum state of the system. As detectors became more sensitive and applications now attempted to measure single-photon interactions, the quantum nature of light became manifest. This remains an active area of current research. Light interacting with small samples can excite various resonance modes within the material, in a fashion that is not particularly well described by a bulk dielectric constant.

Developments in technology like those epitomized in the work of John Hall and Theodor Hänsch provided extraordinarily precise tools for the study of light/matter interactions. Hall and Hänsch utilized lasers capable of short bursts of light to create what is now termed a frequency comb. Recall that, in the Fourier domain, short time-domain signals extend across a wide frequency band. The most extreme example is the delta function, which requires all frequencies. Placing the laser source material into a cavity of length L causes general reinforcement of the wave fields within the cavity that are matched to the cavity length. These fields will

---

[8]Glauber published "Photon correlations" in the *Physical Review Letters* and an expanded "The quantum theory of optical coherence" in the *Physical Review*. Glauber was awarded half of the 2005 Nobel Prize in Physics "'for his contribution to the quantum theory of optical coherence." The other half was split between John Hall and Theodor Hänsch "for their contributions to the development of laser-based precision spectroscopy, including the optical frequency comb technique."

be separated in frequency by an amount $\Delta v = c/2L$. There are several techniques available, like acoustic-optical modulation of the cavity mirror, to reinforce the cancellation at frequencies other than those desired.

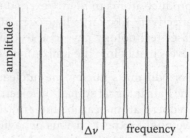

FIGURE 9.14. The frequency content of the output of a mode-locked laser consists of a series of narrow (Gaussian) components separated by a constant Δv.

A sketch of the structure of a mode-locked laser is depicted in figure 9.14. Here, we illustrate an overall envelope of the intensity around the nominal frequency of operation. Within the bandwidth of the laser, only selected frequencies, separated by Δv, are found. In the simplest implementation, the bandwidth ranges over a factor of two in frequency (an octave). In this instance, the laser pulse can be directed through a crystal in which two-photon interactions provide a component with twice the frequency of the input laser ($2v$). The comb also has a component $v + N\Delta v = 2v$, for some value of N. Adjusting the optics of the lasing system can ensure that those two values are identically the same, to within the width of the Gaussian components. Overall, pulse widths on the order of a few femtoseconds are routinely available and efforts are underway to extend the technology into the attosecond domain.

EXERCISE 9.16. Assume that the envelope of the laser pulse in the time domain has a Gaussian profile with a width Γ and that the center frequency of the pulse is v_0. What is the Fourier transform? How does that change as a function of Γ?

As a practical matter, laser technology has advanced to the point that the SI system of units has been modified to utilize the new technology to define the base units. The velocity of light is defined as 299 792 458 m·s^{-1} and the meter as the distance light travels in 1 s. The physical artifact defining the length of the meter has been retired, making laboratory standards more accessible. The second is now defined in terms of the frequency of a particular hyperfine transition in ^{133}Cs: 9 192 631 770 such transitions make up 1 s.

Such precision has enabled a number of detailed investigations into the nature of the quantum world. For example, we have seen that the world is composed of particles that possess a spin quantum number and that the behavior of fermions with half-integral spin and bosons with integral spin are quite different at low temperatures. Additionally, it is possible for fermions to couple into composite bosons, as we have seen in nuclear physics and superconductivity. In the late 1980s, physicists were able to produce atoms in highly excited states, particularly alkali atoms that look effectively like large hydrogen atoms with a single valence electron. The laser technology gave rise to the ability to selectively populate Rydberg levels just below the ionization energy.

For odd-mass nuclei, the nuclear spin is half-integral and, coupled with the half-integral spin of the electron, the total angular momentum of these Rydberg atoms can be integral. They are effectively bosons. As a result, we might expect curious behavior at low temperatures and a number of groups worldwide began the hunt for what is called the Bose-Einstein condensate. If these alkali atoms are bosons, then at low temperatures they should occupy the same quantum state, a possibility forbidden to fermions.

The first technical hurdle is to gather a (large) number of Rydberg atoms into a small volume. This can be accomplished through optical trapping.[9] Notionally, a laser beam with an energy just below a transition in the atom of interest will not produce transitions unless the atom motion causes the light to be blue-shifted enough for the transition to take place. After absorbing the photon, the atom will subsequently re-emit the photon in a random direction. Over many such collisions, the atom's momentum will dissipate. In the hunt for Bose-Einstein condensation, researchers also added a quadrupole magnetic field to the trap, creating a magneto-optical trap.

The optical trapping process permits researchers to reach temperatures in the μK range in the trapped atoms. To cool the atoms even further, one utilizes evaporative cooling. In essence, the atoms in the cloud are illuminated with radio frequency (rf) waves ($\approx$4 MHz), that can cause a spin flip in the prepared atoms. The atoms with flipped spins are no longer kept in the trap. By incrementally decreasing the frequency of the rf field, the warmest of the remaining atoms can be systematically removed from the trap. The result is a decrease in temperature into the 100 nK range. This level of cooling was obtained in 1995 by Carl Wieman and Eric Cornell at NIST, in an assembly of rubidium atoms and shortly thereafter by Wolfgang Ketterle at MIT using sodium.[10] The achievement of a Bose-Einstein

---

[9]The Nobel Prize in Physics 1997 was awarded jointly to Steven Chu, Claude Cohen-Tannoudji and William D. Phillips "for development of methods to cool and trap atoms with laser light."

[10]Wieman and Cornell and coworkers published "Observation of Bose-Einstein condensation in a dilute atomic vapor" in *Science*. Ketterle and coworkers published "Bose-Einstein condensation in a gas of sodium atoms" in the *Physical Review Letters*. The three were

condensate was a decade-long trek through a host of technical obstacles.

> EXERCISE 9.17. The magnetic field from a current loop involves el-
> liptic integrals and a representation is available in the *Mathematica*
> documentation for the EllipticE function. Define two functions
> Br[r_,z_] and Bz[r_,z_] by cutting and pasting the results in the
> example. Use the StreamPlot function to examine the magnetic
> field lines produced by a coil of radius $R = 1$, located a distance 1.1
> from the origin. A quadrupole magnet is composed of four coils
> in opposition. Construct a quadrupole field from the sum of four
> current loops and use the StreamPlot function to visualize the field
> lines. Compute the magnitude of the field and plot it with the Plot3D
> function. Hint: two of the coils are oriented orthogonally to the first
> pair, exchanging the sense of r and z.

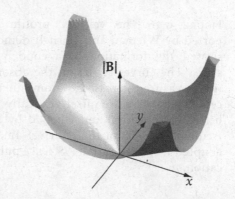

FIGURE 9.15. The magnitude
of the magnetic field vanishes
in the center of an array of
current loops.

In figure 9.15, we illustrate the magnitude of the magnetic field of a quadrupole magnet. The field is created by two pairs of magnets, one pair with
north poles opposing and another pair, at 90° from the first pair, with
south poles opposing. This creates a bowl-shaped region in which the
magnetic field vanishes in the center. Unfortunately, with no magnetic
field, there is also no confining force. As they had expected, when Wieman
and Cornell managed to finally capture enough rubidium atoms and cool
them sufficiently, the atoms dribbled out the bottom of the trap.

This situation was not unexpected but solving this particular issue had
not been the highest priority. Fortunately, Cornell proposed using a rotating magnetic field applied to the quadrupole magnets. The time orbiting
potential (TOP) field fluctuated faster than the cooled atom cloud could

---

awarded the 2001 Nobel Prize in Physics for "for the achievement of Bose-Einstein condensation in dilute gases of alkali atoms, and for early fundamental studies of the properties
of the condensates."

follow, enabling the attainment of the Bose-Einstein condensate. A significant question, of course, is how might one determine that a condensate had been achieved? Like superfluid ^{3}He, there are no sparks that fly from the chamber or intense vibration of the laboratory.

> EXERCISE 9.18.   Repeat the quadrupole magnet calculations, this time with one of the magnets having a 20% larger field. What is the shape of the resulting trap? Where is the minimum?

Initially, Wieman and Cornell simply illuminated the chamber and measured the absorption by whatever was in the middle. Production of the condensed state should lead to narrower features. This is what they observed, as illustrated in figure 9.16.

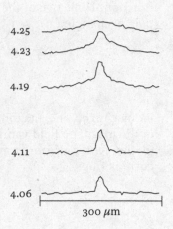

FIGURE 9.16. The velocity profile reported by Wieman and Cornell demonstrated the formation of a condensed state. The curves represent the absorption measured across the atoms in the cell. The curves are offset vertically for better visualization and labelled by the frequency (MHz) of the rf drive. Image adapted from the authors' *Science* publication.

With his apparatus, Ketterle was able to produce over one hundred times the density observed by Wieman and Cornell and was able to study the condensed state in more detail. Notably, Ketterle and his students observed interference fringes in the optical density when two condensates were allowed to mix. Interference is a property of wave phenomena, from which we can infer that the collection of atoms in the condensate act like waves, as we found earlier for electrons. For all of the limitations inherent in quantum theory, there is no other ready explanation for the observation of interference of the atoms than the atoms behave like waves and can be described by wavefunctions.

9.4. Quantum Computing

Beyond just the creation of a relatively arcane new form of matter, the technology developed along the pathway to Bose-Einstein condensation provides exquisite control over systems at extraordinarily low temperatures. Despite the rather oxymoronic implications of the term laser cooling, the technology has prompted an array of high-precision measurements of the fundamental properties of matter. For example, in 2010, Chin Wen Chou and collaborators at NIST used the technology to build a clock based on $^{27}Al^+$ ions and achieved a precision of a part in 10^{17}. With this level of precision, Chou and his colleagues could observe relativistic effects at laboratory scale.[11] Special relativity requires corrections proportional to $[1 - v^2/c^2]^{-1/2}$, which is generally considered to be negligible at terrestrial velocities. Nevertheless, Chou's team were able to observe the effects in ions with velocities on the order of 10 m/s and were able to detect the general relativistic corrections associated with moving the clock a meter higher in the laboratory.

Precise control over assemblies of atoms has launched research into the development of quantum computers. At the heart of computing, as envisioned by Alan Turing, is the concept of the state of the machine.[12] The state can be represented by symbols or, in modern computers, by arrays of bits that represent the symbols. Computation is the process of manipulating the symbols by means of logical operations in a stepwise fashion.

The allure of quantum computing is that quantum states exist as superpositions of many states. For example, two electrons must form an antisymmetric pair, represented by the following wavefunction:

$$(9.35) \qquad \psi = \frac{1}{\sqrt{2}}\Big[|\alpha\rangle|\beta\rangle - |\beta\rangle|\alpha\rangle\Big],$$

where we can take $|\alpha\rangle$ to be the spin-up state and $|\beta\rangle$ to be the spin-down state. That is each electron is simultaneously in both spin-up and spin-down states. Only the process of measuring the spin of one of the electrons determines the spin of the other.

One can envision encoding a particular problem into a (large) number of quantum bits, which contain the superposition of many states. Operations on the bits through unitary transformations affects the phase of the state but not the amplitude. Selecting a particular desired result by measuring the values of some subset of the bits collapses the quantum state into the desired solution. Assuming that preparation of the state can be accomplished with a time dependence that is linear in the number of bits,

---

[11]Chou published "Optical clocks and relativity" in *Science*.

[12]Turing published "On computable numbers, with an application to the Entscheidungsproblem" in the *Proceedings of the London Mathematical Society* in 1936. The "decision problem" proposed in 1928 by the German mathematician David Hilbert asks if a machine could verify the truth of a mathematical statement solely through the use of the defining axioms.

then the quantum algorithm vastly outperforms other implementations on a host of applications.

We have already seen in multi-body problems that there are emergent phenomena, such as aromatic chemical bonds, that are difficult to obtain from present day calculations. Quantum computing, particularly applied to the solution of quantum phenomena, is quite attractive. As we have seen in some measure, larger basis sets can help with calculations of electron density but one must always truncate the basis somewhere. With a quantum computer, we might be able to sidestep the sum over intermediate states because the coherent superposition of the bit elements is already performing that summation implicitly.

Developing a practical quantum computer is a subject of ongoing research. Pieces, such as stable quantum bits, have been demonstrated, although maintaining the stability of the states is an ongoing subject of study. Quantum computing derives its advantage from the coherence of the quantum state. Decoherence will obviate that advantage.

One particularly interesting aspect to equation 9.35 is that each electron is simultaneously in both spin up and spin down states. One might imagine separating the electrons somehow to large distances, where the four-vector describing the electrons is spacelike. Einstein and his collaborators Boris Podolsky and Nathan Rosen entertained just such a thought in 1935 and came to the conclusion that quantum mechanics must somehow be incomplete.[13] Their conclusion was motivated, in part, by an inconsistency in how physical properties can be determined. In quantum mechanics, eigenvalues of non-commuting operators cannot be simultaneously determined. Yet, Einstein and his colleagues pointed out that, for objects separated in space, the determination of the spin of one electron must instantaneously determine the spin of the other. Einstein referred to this as "spukhafte Fernwirkung," which is routinely translated as spooky action at a distance, although spukhafte could also be interpreted to mean mystical or ghostly. In any case, this is inconsistent with information propagating at the velocity of light.

In fact, Einstein was incorrect. Two recent experiments conducted at NIST and in Vienna demonstrate that physical systems display just such counterintuitive behavior.[14] Both groups worked with entangled photon states, in which nonlinear effects in a crystal generate two photons from

---

[13]Einstein, Podolsky and Rosen published "Can quantum-mechanical description of physical reality be considered complete?" in the *Physical Review*.

[14]Shalm *et al.* published "Strong loophole-free test of local realism" and Guistina *et al.* published "Significant-loophole-free test of Bell's theorem with entangled photons" in the *Physical Review Letters* in 2015.

an initial incident photon. The output photons exist in a superposition of polarization states. Measuring the polarization of one photon determines the polarization of the other. When the polarization measurements are conducted at spacelike distances, the outcome of each (distant) experiment agrees with the other.

Entangled photons are also envisioned as the basis of secure communications. If two distant individuals exchange pairs of entangled photons, each will be able to use the sequence to deduce the contents. In a simplistic fashion, the two polarization states can be considered as 0 or 1 and a string of 0s and 1s can be used to encode a message. Of course, one recipient will receive the inverted message "0101" instead of "1010" but they can readily deduce which interpretation to use. If another individual intercepts one of the signals, the act of measuring the polarization to deduce the signal will cause the original quantum state to collapse. As a result, the interception can be detected.

We are presently not close to a world populated by quantum computers or secure electronic messaging but there is progress on many fronts. Researchers can trap individual atoms in linear and two-dimensional arrays and prepare them into coherent states. These states can be manipulated using a number of techniques that produce transformations akin to Fourier transforms, for example.

One of the pedestrian motivations for quantum computing, particularly favoring optical techniques, is reducing the power consumption of modern computers. The most expensive part of high-performance supercomputers is the air-handling system. The chips are small but are placed in comparatively large boxes simply to provide room for capacious air flow for cooling.

This thermal energy arises from the flow of currents through the myriad of interconnects, whose resistance scales like the inverse of the cross-sectional area. Additionally, flipping a bit from 1 to 0 requires work, if bits are realized as small pools of electrons held in place by electromagnetic fields. This is truly surprising but Ohm's law works well over extraordinary length and time scales, from the nanoscopic to the national power grid.

One can formally prove that the ultimate computer will require $kT \ln 2$ of energy to flip a bit between one of two different states but modern computers that rely on electronic currents consume far more than the ideal minimum. As a result, a host of different methods to define a bit have been investigated. It is possible to use either electron spins or nuclear spins as embodiments of the states 0 and 1. We have significant experience with experimental methodologies for coaxing spin states into either up or down configurations. The difficulty is that bit density in present-day electronic memory is quite high. Replicating that density with splin lattices is quite challenging.

Modern memory systems also incorporate error detection/correction codes. As feature sizes continue to decrease, the ability to make perfect arrays of memory also decreases. Rather than rejecting chips with manufacturing defects, it is possible to encode eight bits of information in more than eight bits of physical storage. The coding schemes permit multiple-bit errors to be detected and corrected. As a result, modern memory chips do not have to be perfect to still function perfectly. Repeating this capability in spin-lattices or multi-photon systems is difficult.

In fifty years, of course, this section will undoubtedly have to be revised. Early automobiles were not reliable enough to supplant horse-driven wagons and there weren't enough paved roads to entice most citizens to embrace modern technology. There are numerous groups exploring pathways to replace the current state-of-the-art technology with more capable, cost-effective solutions.

X

Biological Systems

Advances in our understanding of optics have had great impact on the study of biological systems. Beginning with the Dutch draper-turned-optician Antonie van Leeuwenhoek whose microscopes made visible the cellular structure of organisms in the late 1600s, the geometrical optics that we have discussed played a fundamental rôle in the developing field of biology. The wavelength limitations of visible light can be overcome with electron microscope, providing considerably more resolving power.[1] Further advances in cryo-electron microscopy have now brought the resolution down to the nanometer scale and beyond, in fortuitous cases. This is verging on the scale of atomic resolution within molecules that are composed of millions of atoms.

Together with the advent of synchrotron sources of x-rays and powerful computer systems, we now have the capacity to visualize the biological machinery with unprecedented clarity. From these observations, we know that cells are highly structured and that the chemical reactions that take place within the cells are high regulated. These observations provide clues into cellular function, including a possible explanation for one of the largest questions in biology: how can life exist, given that entropy must continually increase?

10.1. Diffusion

At the microscopic level, the condensed phase is governed by diffusive processes. These processes can be characterized by considering individual particle motion to be described by random walks in three dimensions. As a result, the concentration of a drop of dye molecules in a glass of water eventually moves to a state in which dye molecules are equally distributed throughout the volume. There is no additional force of nature

---

[1] As mentioned previously, the German physicist Ernst Ruska was awarded half of the Nobel Prize in Physics in 1986 "for his fundamental work in electron optics, and for the design of the first electron microscope." Ruska shared the award with Binnig and Rohrer for their development of STM.

© Mark A. Cunningham 2018
M.A. Cunningham, *Beyond Classical Physics*,
Undergraduate Lecture Notes in Physics,
https://doi.org/10.1007/978-3-319-63160-8_10

that directs this dispersion; systems simply proceed to the most probable state and the most probable state is the one in which dye molecules are evenly distributed.

In order to study this process in more detail, we shall now undertake the use of a molecular dynamics code NAMD that uses a classical representation of molecular forces. The principal reason for using a specialty code instead of implementing a dynamics code within *Mathematica* scripts is that there is a tremendous amount of technical sophistication incorporated in modern molecular dynamics codes. In essence, the code simply integrates Newton's equations of motion to find the new positions of all the atoms at the next time step. In practice, this is not simple. We shall simply sidestep many technical issues and utilize the insights of our forebears. Ultimately, we will need to understand the inner workings but can put that aside for the time being.

In the 1970s, computers were not capable of solving quantum mechanical problems of any sensible size, so an alternative strategy was developed. The quantum interactions were parameterized and the discipline of molecular mechanics was created. Quantum calculations of the electron density or molecular orbitals were replaced by a vastly simplified Hamiltonian that represents a chemically static picture of the condensed phase. Unlike quantum calculations like those that we performed with NWChem, molecules within the molecular mechanics framework do not change over time. Covalent bonds are not formed or broken, although hydrogen bonds are. Molecular mechanics provides a means for studying the structural behavior of large systems; today, simulations of millions of atoms are possible. Molecular mechanics does not offer the ability to study the chemical transformations at work in biological systems but a hybrid methodology in which a small portion of the atoms are treated quantum mechanically and the remainder classically has also been developed.[2] This so-called QM/MM strategy does provide the ability to study chemical processes but, at present, remains a technically difficult enterprise.

For the moment, we will stick to the classical treatment of molecular systems. There is no unique classical representation of a quantum system. As we have seen, the electron density around any particular nuclear center is not an integral multiple of the fundamental electron charge. Indeed, as chemists have long known, different elements display different electron affinities. As a result, there are several different strategies for determining the parameterization of the quantum state. In the vernacular,

---

[2]Martin Karplus, Michael Levitt and Arieh Warshel were awarded the Nobel Prize in Chemistry in 2013 "for the development of multiscale models for complex chemical systems."

these parameterizations are known as force fields, which is not a particularly descriptive, or good, choice but we have encountered odd choices of nomenclature previously. As a technical note, it is not possible to simply move from one force field to another. Each has its own philosophy and enforces internal consistency but is generally incompatible with other force fields.

The choice we shall make here is to utilize the CHARMM force field developed originally by Martin Karplus and his students at Harvard. Among the more popular choices, Arieh Warshel and Michael Levitt developed the CFF force field and Peter Kollman and students developed the AMBER force field. All are roughly equivalent but one has to make a choice. The CHARMM force field was developed purposefully for use in biological macromolecules like proteins and nucleic acids, so we shall utilize that work here. We note that NWChem also has a molecular mechanics implementation that uses the AMBER force field. Students could also utilize these codes but the NAMD/VMD combination is likely to be found more forgiving and friendly to the uninitiated.

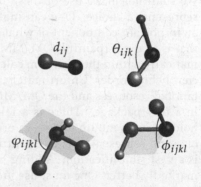

FIGURE 10.1. The Hamiltonian contains terms including up to four-body terms like those indicated at right. Two-body interactions enforce distances d_{ij}. Three body terms enforce angles θ_{ijk}. There are two types of four body interactions that enforce dihedral angles ϕ_{ijkl} and so-called improper dihedrals φ_{ijkl}.

All of the force fields have terms up to four body interactions, like those pictured in figure 10.1. They correspond to the following Hamiltonian terms:

$$\mathcal{H}_b = \frac{1}{2} \sum \kappa_b (|\mathbf{x}_i - \mathbf{x}_j| - d_{ij})^2 + \frac{1}{2} \sum \kappa_\theta (\theta - \theta_{ijk})^2$$

$$(10.1) \qquad + \sum_n \kappa_\phi [1 - \cos n(\phi - \phi_{ijkl})] + \frac{1}{2} \sum \kappa_\varphi (\varphi - \varphi_{ijkl})^2.$$

Here, the summations run over all atoms that are identified as bound to one another. In the molecular mechanics methodology, one has numerous atom types for each element but with a general preference for keeping the number of types (relatively) small. This strategy recognizes that doubly bound carbon atoms behave differently than singly bound carbons and that there may need to be subtypes within doubly or singly bound. Thus, while all force fields begin with the bonded Hamiltonian terms specified in equation 10.1, all of the force fields diverge thereafter.

In addition to the bonded interactions, all of the force fields contain Coulombic interactions that arise from assigning each atom a static partial charge that reflects their electron affinity. Hydrogen bonding is accounted for by a Lennard-Jones potential as in the following expression:

$$(10.2) \quad \mathcal{H}_{nb} = \sum \varepsilon_{ij}\left[\left(\frac{R_{ij}}{|\mathbf{x}_i - \mathbf{x}_j|}\right)^{12} - \left(\frac{R_{ij}}{|\mathbf{x}_i - \mathbf{x}_j|}\right)^6\right] + \sum \kappa_e \frac{q_i q_j}{|\mathbf{x}_i - \mathbf{x}_j|},$$

where here the summation runs only over atoms that are not bound to one another. The last term in equation 10.2 is, of course, the Coulomb interaction. There are other terms possible, including five- and six-body terms and, recently, polarization effects have begun to be included in molecular force fields, replacing the simple static charge model.

We shall not make use of these most recent enhancements. Again, this represents a choice. One can make the argument that, if polarization is an important component of what you intend to study, the best way to do that is to use a (painful) QM/MM strategy. Polarization is handled automatically within the quantum calculations, provided that they are of sufficiently high order. Unfortunately, this requires significantly more computational resources and the QM/MM methodology does not exist currently in a tidy package. There is a fair amount of standing on one foot and holding one's mouth just so, in order to get the calculations to work. The classical representation of quantum phenomena through the force fields is a vast simplification. Adding more terms will not necessarily make it markedly better. One must use judgement.

Many man-years of effort have gone into the construction of the force field parameters. Many quantum calculations of small molecules, including structural optimization and computation of vibrational frequencies were conducted. Structural information from crystallographic sources were also included in fitting the parameters of the classical Hamiltonian. Hydrogen bonding parameters and partial charges were obtained through simulations like that depicted in figure 7.9, in which a water molecule is allowed to interact with the target molecule. There are still ambiguities remaining, so the CHARMM force field developers made choices, like the charge on methyl groups, that are maintained consistently.

Interestingly, the earliest force fields lumped methyl groups into a single, methyl atom. This was done because the computational power available was not sufficient to handle all atoms independently. As more computational power became available, a significant effort was made then to revise the force fields and incorporate all atoms independently, as was first done for the CHARMM19 parameterization. There is significant work today to build so-called coarse-grained representations in which groups of atoms are again consolidated into effective atoms, to handle very long simulations of very big macromolecular assemblies, like the entire ribosome. For our purposes, we will utilize the CHARMM32 force field release, which include all atoms and support simulations with proteins and nucleic acids. These are accessible from the CHARMM force field development web site.

Performing a molecular mechanics simulation is vastly more complex than the quantum simulations we conducted earlier.[3] First one must provide a list of molecules, their constituent atoms and the requisite parameters from equations 10.1 and 10.2. If the parameters do not exist, they must be either taken by analogy from some molecule already parameterized or developed from scratch. For our initial simulation, we shall consider the outcome of a small droplet of methanol in a water bath. The files prefixed with meoh provide the requisite NAMD configuration files to conduct a short simulation.

NAMD requires the following files as input:

(1) a structure file that defines the molecules and their connectivities,
(2) a parameter file that defines the force field parameters,
(3) a coordinate file that defines initial positions and
(4) a configuration file that defines what tasks NAMD is required to perform.

NAMD may also require other files, depending upon the tasks at hand. In particular, NAMD is designed to be run in multiple steps. The basic output is the list of atom positions at designated time intervals. NAMD also writes intermediate files that can be used to restart from wherever the previous simulation finished.

The structure file is obtained from the auxiliary program psfgen supplied with NAMD/VMD. It utilizes a topology file provided by the CHARMM developers that defines the molecular connectivities. This is complicated by the fact that proteins and nucleic acids are polymers, with each amino acid in a protein, for example, connected to an adjacent amino acid. The coordinates are provided by .pdb files that are the lingua franca of the Protein Database. The .pdb files contain the list of amino acids and their constituents, the positions and other information about the structure.

---

[3]Scripts for all exercises are provided as supplemental material.

EXERCISE 10.1. Install NAMD, download the CHARMM32 files and conduct the simulations specified in the configuration files meoh_03, meoh_04 and meoh_05. The last 200 ps of the simulation represent a constant temperature and pressure simulation of the diffusion of methanol in water. Note that the numerical progression defines the order. Files can be named anything but numerically increasing at least encodes the sequence of events.

In the meoh example provided, a dozen methanol molecules are solvated in a box of water about 40 Å in size. Initially, the methanol molecules are harmonically restrained to maintain their positions and the entire box of water undergoes 1000 steps of conjugate gradient minimization to clear bad contacts. Solvation is performed by superimposing a sufficient number of copies of a large box of equilibrated water over the desired volume. Water molecules outside the target volume are deleted, as are those within some prescribed radius from anything (methanol in this case) inside the box. This solvation process can occasionally result in water molecules sitting atop whatever was inside the box, leading to numerical crashes. The initial minimization avoids that particular difficulty.

The minimization is followed by a short dynamics run (in meoh_03) where the volume is held constant and the harmonic restraints on the methanols are maintained. The next simulation extends the trajectory again with restraints on the methanols but here the pressure is held constant. Finally, the so-called production run changes the time step to 2 fs and again holds constant pressure and, at last, the methanol atoms are permitted to roam freely. The harmonic restraint values are obtained from the occupancy column of the .harm files, which are simply copies of the original .pdb file with modified values of the occupancy column.

EXERCISE 10.2. The NAMD code utilizes the tcl language to parse the configuration files. Some portions of the scripts are simply tcl base language and other portions are specific keywords for NAMD. The set command is a tcl primitive that sets a variable to a particular value. See if you can decipher how the configuration files conduct the sequence of events stated in the text.

The primary output of the NAMD runs are the trajectory (.dcd) files. These contain the atom positions at designated time intervals (1 ps). The reason for using molecular dynamics is to obtain estimates of the ensemble properties. An alternative to molecular dynamics is Monte Carlo sampling, where atoms are randomly moved at each trial and an average is built up by weighting each sample by the Boltzmann factor $\exp[-\mathcal{E}/k_B T]$. In principle, a suitably long time average is comparable to the ensemble average—this is known as the ergodic principle. Because samples from adjacent time intervals are highly correlated, the trajectory files are sampled much less frequently.

EXERCISE 10.3. Use VMD to view the simulation results. First, define a New Molecule by loading the meoh_o2.psf file. Then add the original coordinates from the meoh_o2.pdb file to the molecule and subsequent trajectory (.dcd) files. The final 200 steps contain the production data. Add a new Representation, selecting one of the methanol molecules and use the CPK drawing mode. Turn off the display of the remaining atoms. Describe the trajectory of the methanol.

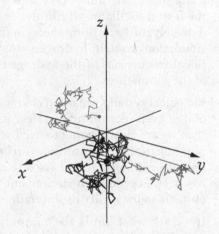

FIGURE 10.2. The oxygen atom position for two methanol molecules is plotted as a function of time. One of the molecules (gray) exited the simulation domain; its periodic image re-entered on the opposite side.

We can utilize the VMD code to visualize the trajectories and perform analysis. One objective of the methanol simulation is to study the diffusion process. In figure 10.2, we plot the trajectories of two methanol molecules (specifically the oxygen position). What we observe is that by the end of the simulation, the two molecules are quite distant, despite having started only Ångstroms apart. We also observe the effect of periodic boundary conditions in the simulation. One of the methanol molecules exited the simulation volume; its periodic image then entered from the opposite side. This is denoted by the line stretching across the plot, which can be seen to be an artifact of the boundary conditions not an error in the coding or some other mysterious feature.

EXERCISE 10.4. The track_meoh.tcl script can be used with VMD to extract the oxygen coordinates from the simulation data. In the VMD window or the Tk Console, enter the following commands:

```
source track_meoh.tcl
track_meoh 3
```

The source command causes VMD to read the file that defines a new procedure. Invoking the procedure causes VMD to read the coordinates associated with the methanol molecule with RESID 3 and write them to the file track_meoh_03.dat. Plot the data for two of the methanol molecules using Import to read the file and ListPoint-Plot3D to view the trajectories.

If we now look into some of the details of the calculations, we notice that the time steps during the production phase of the simulation were 2 fs. This is an exceedingly short time but is one that is forced onto us by the constraints of numerical stability. In order for the integration to be numerically stable, we cannot take time steps that are longer than the period of the fastest oscillator within the ensemble. This condition is comparable to the Nyquist sampling theorem found in signal processing. Because our simulations contain hydrogen atoms, the time steps are set by the vibrational frequencies of the hydrogens. This is a major constraint on utility of the simulations.

Biological systems are often characterized by processes that have millisecond or even second, time scales. If our fundamental time step is 10^{-15} s, then we are clearly going to need 10^{12} to 10^{15} steps, or more, to reach into physiologically relevant times. (If a processor can take 10^6 steps per second, we will need 10^6 to 10^9 s to conduct that many steps.) Much effort has been expended to circumvent this issue and it remains a significant obstacle to the utility of simulation.

If we ask what limits the computational speed, then we find that computation of the Coulomb interaction dominates. The bonded interactions actually take very little time to compute, largely because they are local. The non-bonded interactions, in principle, require a double sum over all atoms in the model. In practice, one utilizes a refinement of a method devised by the German physicist Paul Peter Ewald for computing the potential in periodic systems.[4] Ewald's strategy provides an efficient solution in the Fourier domain, so part of the NAMD configuration file defines the parameters for the discrete Fourier transforms that perform the Ewald summation.

Additionally, cutoffs on electromagnetic interaction are enforced, in which atoms beyond a particular radius are not included in the summations. As the biomolecules are largely electrically neutral, this strategy accounts for the fact that long-range electrostatic interactions are screened. For biological systems, the cutoff distance is typically 13 Å and sets a weak constraint on the size of the simulation model. Generally, one would like to

---

[4]Ewald published "Die Berechnung optischer und elektrostatischer Gitterpotentiale" in the *Annalen der Physik* in 1921.

avoid direct interactions between the target molecule and its periodic images. Including a water buffer around the molecule of interest of sufficient thickness can ensure that this does not occur. This is termed a "dilute" approximation. Unfortunately, this can require ≈10 Å of water around the protein, adding tens of thousands of water molecules, which then can dominate the calculational effort.

An alternative is to utilize implicit water models, in which the bulk water outside the molecule of interest is simply treated as an effective dielectric constant. Liquid water possesses a large dielectric constant due to the fact that the molecule possesses a static dipole moment in addition to its intrinsic polarizability. Application of an electric field to water can flip a large fraction of the water into alignment with the field. A complication arises, though, when water molecules are hydrogen-bonded to the molecule of interest. These water molecules are not free to follow external fields and their effective dielectric constant is greatly reduced. Managing the transition between oriented water molecules adjacent to the molecule of interest and bulk water molecules that can be represented by a dielectric constant is an ongoing research project. This balancing act also impacts the ability to conduct simulations at constant pressure.

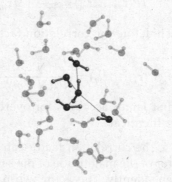

FIGURE 10.3. The water molecules within 5 Å of the central water molecule taken from a step in the simulation display structure. Four molecules within the selection are within hydrogen bonding distance (3.2 Å from oxygen to oxygen).

The TIP3 parameterization of water utilized in the CHARMM force field provides a reasonable representation of water, which is an extraordinarily difficult substance to model properly. Even very high level quantum calculations do not currently provide exact models of water behavior but the TIP3 model captures much of the physical properties. Figure 10.3 represents one step from the simulation that demonstrates the fact that there are networks of hydrogen bonding within the bulk water. In real water, there are also proton transfer reactions taking place, which is not contained within the static model of water used within CHARMM. Here, the same two hydrogen atoms are permanently attached to each oxygen.

EXERCISE 10.5.   In VMD, select a water molecule from the simulation. Create a representation of all atoms within 4 Å of the molecule.
How many adjacent water molecules are in a conformation that suggests they may be hydrogen bonded? How does that number change
at the next time step?

Examining individual water molecules within the simulation, one can
readily see that the water molecules generally have four other water molecules within nominal hydrogen bonding distance. The participants change
as the simulation progresses; the hydrogen bonding network is not static.
It is possible to quantify this observation. VMD possesses several analysis tools. One of them computes the radial distribution function $g(r)$. In
general, the probability of finding N particles distributed with values in
the ranges r_1 to $r_1 + dr_1$ through r_N to $r_N + dr_N$ can be obtained from the
partition function:

$$(10.3) \qquad P(r_1,\ldots,r_N)\,dr_1\cdots dr_N = \frac{e^{-\mathcal{E}/k_BT}}{\mathcal{Z}_N}\,dr_1\cdots dr_N.$$

One can also define the probability of finding some subset m of those N
particles by integrating over the remainder:

$$(10.4) \qquad P(r_1,\ldots,r_m) = \frac{1}{\mathcal{Z}_N}\int dr_{m+1}\cdots dr_N\, e^{-\mathcal{E}/k_BT}.$$

One can then define the correlation factor $g^m(r_1,\ldots,r_m)$ of m particles as
follows:

$$(10.5) \qquad g^m(r_1,\ldots,r_m) = \frac{N!}{N^m(N-m)!}\frac{V^m}{\mathcal{Z}_N}\int dr_{m+1}\cdots dr_N\, e^{-\mathcal{E}/k_BT}.$$

Here, the first factor accounts for overcounting due to the fact that the
particles are identical.

When $m = 2$, the correlation factor can generally be demonstrated to be
solely a factor of the magnitude of the separation $|r_1 - r_2|$ and not the sepa-
rations independently. This is the two-particle correlation that we call the
radial distribution function. The radial distribution functions computed
from several simulations are illustrated in figure 10.4. At long distances,
$g(r) \to 1$, depicting that there is no long range correlation. The reason
that $g(r)$ is interesting is that crystallographers measure a structure factor
$S(q)$ that is essentially the Fourier transform of $g(r)$. Consequently, one
can compare computational results with experiment, although the comparison is not exactly straightforward.

EXERCISE 10.6.   Select the Radial Distribution Function analysis tool
from the Extensions tab in VMD. Choose both selections to be name
OH2 and set the First frame to be 163 (starting with the production

FIGURE 10.4. The radial distribution functions for water
oxygen atoms interacting
with other water molecules
(OH2-OH2, black), and
methanol (OH2-OG, gray
and OH2-CB, light gray)
were computed with VMD.
The hydrogen bond from
the hydroxyl (OH) group of
methanol is comparable to
that of water. The methyl
group (CH_3) does not form
hydrogen bonds and is, on
average, more distant.

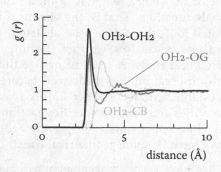

data). Compute $g(r)$. Repeat for second selections of name OG and
name CB. Plot the resulting distribution functions.

We note that there is a marked difference in the radial distribution factors
between the methyl carbons (CB) and the water oxygen (OH2) atoms. The
methyl group does not participate in hydrogen bonding and, as a result,
the average distance between the heavy (O-C) atoms is nearly 4 Å. We
illustrate the difference in figure 10.5, where a snapshot of the molecules
surrounding one of the methanol molecules is taken from the production
portion of the simulation.

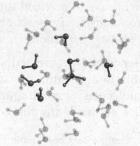

FIGURE 10.5. Water molecules
adjacent to the methyl group
of methanol are not aligned in
hydrogen bonding conformations. This is the essence of
hydrophobicity.

Here, we can observe that the hydroxyl group (OH) has water molecules
in the vicinity that are within hydrogen bonding distance but the methyl
group (CH_3) does not. This is the essential difference between hydrophilic
and hydrophobic interactions. These differences are more apparent in
color in VMD than they are on the printed page.

10.2. Molecular Structure and Recognition

One of the central dogmas of molecular biology is, succinctly, structure equals function. That is, the chemical function of a molecule is directly related to its three-dimensional structure. Remarkably, the origins of this idea date at least to 1894 when the noted German chemist Emil Fischer declared that enzymes must recognize their substrates like a lock and key.[5] This observation was made well before the atomic model of matter was an established fact. Fischer simply reasoned that enzymes select specific molecules and conduct specific reactions. Altering the substrate in almost any chemical manner will greatly reduce the reaction efficiency. Ergo, there must be some geometrical constraints on potential substrates.

As we have seen, the geometry associated with atoms bound to carbon differs depending on the bond type: single, double or triple. As we shall see from crystal structures of enzymes and their substrates, the substrate often fits within a small niche, tucked into place by a series of hydrogen bonds with the enzyme. Indeed, at an atomic scale, Fischer's suggestion has been repeatedly validated.

Our knowledge of the structure of biological molecules comes, in large measure, from x-ray diffraction experiments. As William Lawrence Bragg found, if the wave vector of an incident x-ray is $\mathbf{k}_o$ and it emerges with a wave vector $\mathbf{k}$, maxima in the diffracted beam will occur at points where $\mathbf{k} - \mathbf{k}_o = h\bar{\mathbf{a}} + k\bar{\mathbf{b}} + l\bar{\mathbf{c}}$, where $\bar{\mathbf{a}}$, $\bar{\mathbf{b}}$ and $\bar{\mathbf{c}}$ are the reciprocal vectors of the lattice. Formally, one can demonstrate that the diffraction pattern is the magnitude of the Fourier transform of the charge distribution. Thus crystallographers continue to fend with what is known as the "phase problem." If one knew the phase of the Fourier transform in addition to the magnitude, then one could perform the inverse Fourier transform and recover the electron density. In practice, this is not a simple undertaking but much progress has been made.

In 1946, the science had progressed sufficiently that Dorothy Crowfoot Hodgkin was able to determine the structure of penicillin.[6] In dealing with larger biomolecules, crystallographers also had to deal with the fact that biomolecules are made up almost exclusively of hydrogen, carbon,

---

[5]Fischer published "Einfluss der Configuration auf die Wirkung der Enzyme" in the *European Journal of Inorganic Chemistry*. Fischer was awarded the 1902 Nobel Prize in Chemistry "in recognition of the extraordinary services he has rendered by his work on sugar and purine syntheses."

[6]Crowfoot and colleagues published "X-ray crystallographic investigation of the structure of penicillin" as a chapter in *Chemistry of Penicillin*, Princeton University Press in 1949. Crowfoot was awarded the Nobel Prize in Chemistry in 1964 "for her determinations by X-ray techniques of the structures of important biochemical substances."

nitrogen and oxygen. These are all low-mass atoms with relatively few electrons.

The first protein structures to be solved were those of hemoglobin by Max Perutz and myglobin by John Kendrew.[7] Perutz and Kendrew utilized a technique in which diffraction data were acquired on a crystal and then the crystal was soaked in a solution containing a heavy metal, like mercury, and the measurements repeated. The metals can coordinate to specific locations within the protein and stand out like beacons amidst the carbons and other light elements. Obtaining diffraction data from crystals soaked with two different heavy metals, Perutz and Kendrew were able to obtain guesses for their unknown phases and, ultimately, reconstruct the molecules.

EXERCISE 10.7.   Conduct an NWChem optimization for the small molecule N-methylacetamide pictured at right. Describe the geometry of the nitrogen atom. What is the distance between the carbon and oxygen?

Proteins are composed of some twenty amino acids, polymerized by what are termed peptide bonds, like the one depicted in the exercise. From the drawing, we anticipate that the carbon atom will be doubly bonded to the oxygen, with a shorter bond length than that found in methanol, for example. The nitrogen, as drawn, should have a pyramidal form. Instead, the electron density is delocalized and the carbon-nitrogen bond becomes aromatic. As can be seen in figure 10.6, the four central atoms (HNCO) become planar, with the result that protein conformations are greatly limited. In particular, due to the aromatic nature of the C-N bond, rotations around this bond are strongly suppressed. The amino acids terminate on one end (N terminal) with an amino group (NH_3^+ in solution) and on the other with a carboxyl (COO^- in solution). Polymerization proceeds with a condensation reaction that frees a water molecule.

As became apparent from the crystal structures, proteins possess structure but the initial structures did not have the spatial resolution to resolve precisely how the protein function arose. A significant improvement in crystallographic technique was provided by Herbert Hauptman and Jerome Karle who developed direct methods for determining the unknown phase.[8] A variety of additional improvements are ongoing. Today,

---

[7]Perutz and Kendrew were awarded the Nobel Prize in Chemistry in 1962 "for their studies of the structures of globular proteins."

[8]Hauptman and Karle were awarded the Nobel Prize in Chemistry in 1985 "for their outstanding achievements in the development of direct methods for the determination of crystal structures."

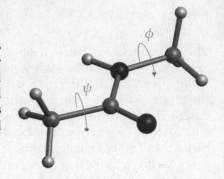

Figure 10.6. The peptide bond is an aromatic bond in which the four inner atoms are planar. The bond between the nitrogen (black) and the outer carbon is rotatable (ϕ), as is the bond between the C terminal carbons (ψ).

one typically will utilize a selenomethionine substitute for any sulfur-bearing methionine residues in the protein. This avoids the problem of soaking the crystals with heavy metal solutions and hoping for the metal ions to be adsorbed. Indeed, the modern HKL3000 software currently deployed at synchrotron laboratories can resolve structures nearly automatically. Moreover, the software incorporates "best practices" and performs a number of quality control tests on the resulting structures, enabling even inexperienced graduate students to produce structures of quality comparable to those produced by seasoned crystallographers.

Exercise 10.8. Download the 1TEM pdb file from the Protein Database (rcsb.org). This can be performed within VMD through the Extensions...Data tab or from the website itself. This is a 1.95 Å structure of the TEM-1 β-lactamase from *E. coli* with a sacrificial substrate bound in the active site. Use the NewRibbons Drawing Method to identify the secondary structures in the protein backbone. Use the CPK Drawing Method to render the substrate (resname ALP). Identify all residues within 5 Å of the substrate. Find the hydrogen bonds that orient the substrate within the binding pocket. (Hint: Type 2 while the mouse is in the graphics window. This will change the mouse mode to measure distances. Heavy atom distances of about 3 Å suggest hydrogen bonding. Create a graphics representation using SURF Drawing Method to identify the binding pocket. Change the substrate Drawing Method to VDW. Does the substrate fit the binding pocket?

A typical difficulty encountered by crystallographers is that enzymes are very efficient catalysts. As a result, the rate-limiting step in the chemical process that converts reactants into products is the diffusion of the reactants and, subsequently, products in and out of the active site. The crystal structures obtained by crystallographers provide, essentially, the ensemble averaged structure. So, even if the protein crystals are soaked with the reactant molecules, most of the time they are not coordinated within the active site and do not appear in the structure.

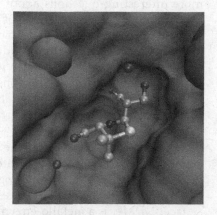

FIGURE 10.7. The TEM-1 β-lactamase hydrolyzes penicillin and is a major source of antibiotic resistance. Here, a substrate analog (ball-and-stick) has been covalently bound within the active site which is represented as the niche within the protein surface. Two water molecules (spheres) are also observed.

Without actually capturing an image of the substrate in the active site, the mechanisms associated with binding and subsequent chemistry remain elusive. There are a number of strategies available to crystallographers that can sidestep this problem, such as working at cryogenic temperatures to reduce the chemical reaction rates. As another example, in figure 10.7, chemist Shariar Mobashery and his students used a molecule similar to that of the native substrate but which did not permit all of the requisite chemistry to be conducted.[9] As a result, a fragment of the analog remains covalently attached to the protein, providing confirmation of the active site and support for the chemical mechanisms at work.

A detailed investigation of the structure in this example reveals that all of the atoms within the substrate that can form hydrogen bonds are hydrogen bonded either directly to the protein or indirectly through water molecules. This is most often what we mean by molecular recognition. A hydrogen-bonding network forms that orients the target molecule within the active site.

Not all cases of molecular recognition are quite so specific. Indeed, one of the most challenging issues in antibiotic resistance is the emergence of new enzymes that are capable of hydrolzying a large number of different antibiotic agents. Such promiscuous behavior indicates that the binding of substrates is significantly less specific. A recent example is the discovery of a new enzyme NDM-1, named after the initial patient from New

_____

[9]Mobashery and his students published "Crystal structure of 6α-hydroxymethyl-penicillanate complexed to the TEM-1 β-lactamase from *Escherichia coli*: Evidence on the mechanism of action of a novel inhibitor designed by a computer-aided process" in the *Journal of the American Chemical Society* in 1996.

Delhi, that efficiently hydrolyzes a broad range of β-lactam antibiotics. We won't discuss molecular evolution in any great detail but one of the more interesting questions about enzymes involves gain of function or loss of function mutations. One of the early crystal structures of a TEM-1 enzyme in complex with an antibiotic substrate was obtained from a mutant copy of the enzyme that rendered it no longer able to hydrolyze the substrate. Instead, the substrate was covalently bound to the substrate; the missing amino acid was responsible for the chemical step that released the hydrolyzed substrate. Being able to predict such changes in function is a subject of great interest in current biomolecular research.

> EXERCISE 10.9.   Download the 4HL2 structure of NDM-1 in complex with a hydrolyzed substrate. Describe the secondary structures of the protein. Identify hydrogen bonds between the protein and substrate. Use the SURF representation to visualize the binding surface. NDM-1 is a metallo-enzyme that utilizes zinc ions to effect the catalytic reaction.

The Protein Database now contains well over 100,000 structures, so one might ask if it would be possible to find small molecules to fit within the binding sites of any particular enzyme? A simple approach is to search through comparable databases of small molecules and try to fit each into the known structure. While the approach may seem simple, it is actually a rather complex undertaking but has been applied with some success in developing leads for new drugs. The basic approach is to simplify the geometry of the substrate molecule into an array of partial charges. The electrostatic energy is then computed with the substrate oriented in a range of angles with respect to the protein surface. A variety of scoring functions have been devised to cull through the results to find any molecules fit well enough for further study. The method is relatively rapid to apply but does not have good accuracy. Often the best scoring compounds do not possess the best predicted binding affinities.

Part of the reason for the limited success of the molecular docking approach is due to the fact that proteins are not rigid bodies. As the biochemist Daniel Koshland pointed out in 1958, enzyme/substrate interactions will invariably result in what he called an induced fit, where each molecule shifts its structure to accommodate the other.[10] Indeed, there are many examples of substrates that appear to be completely enveloped by protein, without any clear picture as to how they might have arrived at their destination.

---

[10]Koshland published "Application of a theory of enzyme specificity to protein synthesis" in the *Proceedings of the National Academy of Sciences USA*.

In most cases, this situation arises from what is called allosteric changes in the protein structure, where loop sections of the backbone permit hinge-like motion. In these systems, the protein opens to admit the substrate and then closes to conduct the chemistry. This motion is often enabled or disabled by binding of cofactors elsewhere on the protein and provides a mechanism for regulating the amount of product that the enzyme is permitted to produce.

> EXERCISE 10.10.   Download the 1CTS and 2CTS structures. These two examples represent the enzyme citrate synthase in its open and closed states, respectively. In order to see the difference more clearly, use the RMSD Calculator tool under the Extensions/Analysis tab to align the proteins. The substrate CIT is visible in both. How does the structure change between open and closed states?

For the simple case of a binding pocket on the surface of an enzyme, we can quantify what we mean by molecular recognition. If we consider an enzyme E and its substrate S, the first step in the reaction is the formation of what is termed the Michaelis complex:[11]

$$(10.6) \qquad E + S \underset{k_2}{\overset{k_1}{\rightleftharpoons}} ES \xrightarrow{k_{cat}} E + P.$$

Here the first term in equation 10.6 represents the enzyme and substrate separately in solution. Each of these is characterized by a free energy, typically the Gibbs free energy for situations of constant temperature and pressure. The intermediate state, in which the substrate is aligned within the active site of the enzyme by hydrogen bonds but has not been chemically modified, also is characterized by a free energy. The probability of forming that state is proportional to the Boltzmann factor:

$$(10.7) \qquad P \propto e^{-[\mathcal{G}(ES) - \mathcal{G}(E+S)]/k_B T},$$

where here $\mathcal{G}$ is the Gibbs free energy: $\mathcal{G} = \mathcal{G}(p, \mathcal{V}) = \mathcal{E} + p\mathcal{V} - TS$. Notionally, from the Michaelis complex, the chemical reaction proceeds through a transition state in which chemical bonds are reformed to the product state and, ultimately, to the release of products back into solution as indicated in the final step of equation 10.6. Here, details of the chemical pathway are simply lumped into an overall rate of the catalyzed reaction k_{cat}.

This simplified version of the catalytic process incorporates the concepts that the rate of formation of the Michaelis complex k_1 competes with the unbinding of the substrate and its release back into solution k_2 but that the forward reaction into products is unidirectional. This is certainly not

---

[11]The German chemist Leonor Michaelis and Canadian physician Maud Menten published a mathematical treatment of enzyme kinetics in "Die Kinetik der Invertinwirkung" in the *Biochemische Zeitung* in 1913.

the case as enzymes, depending upon concentration of the reactants and products can work effectively in either direction. What we have assumed is that the free energy associated with the product state is so much lower in energy than the reactant state, that the reverse reaction can be excluded.

EXERCISE 10.11.   Compute the probabilities of occupying an intermediate state where the energy difference is $k_B T$, $3k_B T$, $5k_B T$, $10k_B T$ and $30k_B T$.

Thus, in order to compute the binding affinity, we really need to compute the free energy. This is precisely why we have introduced students to the use of NAMD. We can estimate the ensemble averages through the use of the molecular dynamics simulations. Unfortunately, conducting a simulation to estimate the free energies associated with equation 10.6 is quite difficult. There is a theoretical trick that can be employed to obtain relative free energies.

Consider the binding of an alternative substrate S' by the enzyme. Following equation 10.6, we can produce a complete thermodynamic cycle:

$$(10.8) \qquad \begin{array}{ccc} E + S & \rightarrow & ES \\ \downarrow & & \downarrow \\ E + S' & \rightarrow & ES' \end{array}.$$

Experimentally, we can conduct the two horizontal reactions: take two separate substrates and measure the binding affinities separately. Theoretically, we can conduct the two vertical reactions: mutate the original substrate into the other substrate. This is known as an alchemical transformation, which is only accessible in the computer world.

Following the free energies around the loop, we find that

$$\mathcal{G}(ES) - \mathcal{G}(ES') = \mathcal{G}(E + S) - \mathcal{G}(E + S').$$

The mechanism provided within NAMD to estimate these free energy differences is known as a dual topology method. That is both substrates are included in the simulations simultaneously but without interacting with each other. Their interactions with the enzyme are controlled by a visibility parameter λ, where the effective substrate is given by $S_{\text{eff}} = \lambda S + (1 - \lambda)S'$. Conducting a series of simulations at different values of λ permits an estimate of the free energy difference between initial and final states to be computed. This alchemical free energy perturbation technique provides the best estimates of the relative binding affinities that can be produced presently, because the problem of induced fit is handled intrinsically. As a matter of practice, it is quite difficult to obtain precise values of binding affinities, in part because finite dynamics runs do not guarantee adequate sampling of conformational space. Consequently, the technique, despite its advantages over molecular docking, has not yet seen widespread use.

> EXERCISE 10.12.   The methodology of alchemical free energy perturbation calculations is quite involved. Mercifully, several example cases are provided on the NAMD web site as tutorials. Download the Alchemical Free Energy Perturbation tutorial from the NAMD tutorial collection, and conduct the ethane to ethane example.

Our emphasis to this point on small molecules and hydrogen bonds can be explained by the observation that the binding energy of one hydrogen bond has a nominal value in the range of 60–300 meV (2-12 $k_B T$). In the middle of that range, we might naïvely expect a 5 $k_B T$ advantage for every hydrogen bond made by the substrate with the enzyme. A single hydrogen bond difference can often explain significant differences in binding affinity (see Exercise 10.11).

The picture becomes somewhat less clear when we move beyond small molecule binding to proteins and begin to consider some of the macro-molecular assemblies that arise in biological organisms. These large structures are composed of dozens (hundreds) of subunits that self-assemble into the machinery that conducts some of the essential chemical transformations required by biological systems. We shall take as one example the ribosome, which is an organelle responsible for most protein production within the cell.

The central dogma of molecular biology is that genetic information is encoded within DNA. That primary information is transcribed into messenger RNA (mRNA) which, in turn, is translated into protein in the ribosome. That translation process utilizes the code listed in Table 10.1, which is nearly universal across all species.[12] Amino acids are first attached to the ends of transfer RNAs (tRNAs), one for each amino acid by a group of enzymes known as tRNA-synthetases. The tRNAs all end in the nucleotide sequence CCA, with their respective amino acids attached to either the 2′ or 3′ hydroxyl of the terminal adenosine. Each tRNA has a loop that contains the three-base anticoding sequence. The ribosome captures the mRNA and, in a multistep process, tRNAs are admitted to processing center, their anticodon loops are compared to the coding segment on the mRNA and, if a match is detected, the amino acid is chemically attached to the nascent peptide chain.[13] The structure of the ribosome

---

[12]Francis Crick and coworkers published "General nature of the genetic code for proteins" in *Nature* in 1961. Crick, James Watson and Maurice Wilkins were awarded the Nobel Prize in Physiology or Medicine in 1962 "for their discoveries concerning the molecular structure of nucleic acids and its significance for information transfer in living material."

[13]The Nobel Prize in Chemistry in 2009 was awarded to Venkatraman Ramakrishnan, Thomas Steitz and Ada Yonath "for studies of the structure and function of the ribosome."

TABLE 10.1. The four base pairs adenosine (A), cytosine (C), guanine (G) and uracil (U) of RNA encode for the twenty amino acids, along with punctuation. All proteins start with methionine, so AUG is the start code.

GCU GCC GCA GCG } alanine	GGU GGC GGA GGG } glycine	UCU UCC UCA UCG } serine
AGU AGG CGU CGC CGA CGG } arginine	CAU CAC } histidine	AGU AGC } serine
	AUU AUC AUA } isoleucine	ACU ACC ACA ACG } threonine
AAU AAC } asparagine	UUA UUG CUU CUC CUA CUG } leucine	UGG } tryptophan
GAU GAC } aspartic acid		UAU UAC } tyrosine
UGU UGC } cysteine	AAA AAG } lysine	GUU GUC GUA GUG } valine
CAA CAG } glutamine	AUG } methionine UUU UUC } phenylalanine	UAA UAG UGA } STOP
GAA GAG } glutamic acid	CCU CCC CCA CCG } proline	

is known due principally to the herculean efforts of Vekatraman Ramakrishnan, Thomas Steitz and Ada Yonath, whose x-ray diffraction studies over a number of years provided increasingly precise descriptions of this remarkable machinery.

Charging of the tRNA with its cognate amino acid is the responsibility of the associated tRNA synthetase. A particularly intriguing example is illustrated in figure 10.8, in which we display the structure of the valyl-tRNA synthetase of *Thermus thermophilus* (RCSB id 1IVS). We have thus far focussed on hydrogen bonds as the means for molecular recognition but, in 1958, Linus Pauling noted that there was a significant problem associated with the fidelity of translation for small, hydrophobic residues

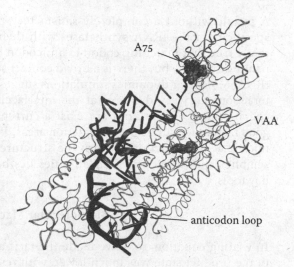

FIGURE 10.8. The tRNA (black) ends in the sequence CCA, with the anticodon displayed on a loop. The tRNA synthetase (gray) recognizes the tRNA through its coding sequence and catalyzes the attachment of the cognate amino acid to the terminal adenosine (A75). A valyl-adenalate analog (VAA) also occupies an active site.

like valine and isoleucine, that differ by a single methyl group.[14] Pauling reasoned that a methyl group provides a solvation energy of about 100–120 meV, or about 5 $k_B T$. As a result, the wrong amino acid should be attached about once in every 100–200 trials. Experiment indicated that errors occurred an order of magnitude less frequently.

EXERCISE 10.13.   Download the 1IVS structure from the Protein Database and open it in VMD. This structure contains two copies (dimer) of the protein and tRNA. Additionally, a substrate analog occupies a potential active site. Identify the secondary structures in the protein. What is the extent of interactions between protein and tRNA? The anticodon loop consists of the bases C34-A35-C36. What residues in the protein interact with these bases?

The solution to Pauling's dilemma was provided by the discovery of a second, editing site in some of the synthetases, located in what is called the head group at the top of the figure. In the 1IVS structure, the acceptor stem of the tRNA occupies what is presumed to be the editing site and a valine analog occupies what is presumed to be the active site responsible for attaching valine to the tRNA. The editing step buys another factor of 30 or so in specificity, bringing the thermodynamic calculations into general agreement with experimental measurements that point to errors in translation of about one in 3000.

---

[14]Pauling's remarks on "The probability of errors in the process of synthesis of protein molecules" were published in the *Festschrift Arthur Stoll*, Birkhäuser Verlag, 1958.

A puzzle without a completely satisfactory resolution at present is the specificity of the tRNA synthetases with their cognate tRNAs. It is certainly true that matching codon to anticodon loops provides a thermodynamic advantage but there is a broad contact surface between protein and tRNA. Molecular dynamics simulations suggest that the tRNA/protein interactions are very dynamic at the interface, except in the codon/anticodon region. There does not exist a current technology that provides good insight for such large interaction areas. Protein-protein and protein-nucleic acid interactions govern the structures of large biomolecular assemblies and predicting their affinities lies beyond the reach of present day tools.

10.3. Biomolecular Machines

In writing equation 10.6, we made the tacit assumption that the energy of the product state was much lower, with respect to the transition state, than the reactant state. In this case, the probability of the reverse reaction in which product is converted back into reactants is considered to be negligible. The attentive student has probably wondered if there are not some desirable chemical reactions that are energetically unfavorable, where the reaction must proceed uphill. In fact, there are many instances of such unfavorable energetics. The charging of tRNA with its cognate amino acid is one such example.

In order to overcome this difficulty, organisms universally utilize adenosine triphosphate (ATP). By incorporating ATP hydrolysis as part of the total reaction, the energetics can be shifted by 0.5 to 0.75 eV (20–30 $k_B T$). As a result, even energetically unfavorable reactions can be converted into a form where equation 10.6 is applicable. Depending on activity, the average human utilizes about their own weight in ATP daily, nearly 1000 kg in times of extreme exertion. The bulk of the ATP is not synthesized *de novo* but regenerated from ADP and phosphate in a large molecular machine known as the ATP synthase, pictured in the cryo-em image in figure 10.9.[15]

The F_o motor segment can be found in the PDB file 1C17 and is composed of two concentric rings of alpha helices formed by 12 copies of a single protein. The helical motif is commonly found in membrane-spanning proteins and is composed on the exterior surface largely of hydrophobic residues that can be solvated in the highly hydrophobic membrane. The

---

[15]The Nobel Prize in Chemistry in 1997 was awarded to Paul Boyer and John Walker "for their elucidation of the enzymatic mechanism underlying the synthesis of adenosine triphosphate (ATP)." The award was shared with Jens Skou "for the first discovery of an ion-transporting enzyme, Na$^+$, K$^+$ -ATPase."

FIGURE 10.9. Cryo-electron microscopy permits the reconstruction of the large-scale structure of ATP synthase. The lower section is embedded in the cell membrane and called the F_o component. The upper section, known as F_1, is connected to the F_o segment through the stator segment at the right of the image. Image courtesy of John L. Rubinstein, The Hospital for Sick Children, Toronto.

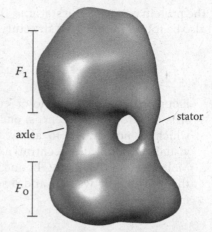

F_o motor is driven by a proton current. The cells use ion pumps to maintain ionic gradients that give rise to a potential difference across the cell membrane. Free protons diffuse through the M chain and are coordinated by glutamic acid residues. The protons are then passed to aspartic acid residues about midway down the rotating core. Once the aspartic acid is protonated, it is much less hydrophobic and the F_o motor turns one twelfth of a revolution.

EXERCISE 10.14. Download the 1C17 structure and open it in VMD. What are the secondary structures of the protein? Create a representation in which aspartic (ASP) and glutamic (GLU) acids are visible. The ends of the helical domains are in solution in water. Does the positioning of these hydrophilic residues make sense given that the sides of the helices are embedded in hydrophobic lipids? Add a membrane segment to the VMD plot using the Membrane Builder utility under the Extensions/Modelling tab. Choose a 100×100 membrane of either type. Fix either the membrane or the protein and rotate/translate the other so that they are roughly aligned. Do the protein helix lengths span the membrane?

Membrane proteins are quite difficult to study, as they do not crystallize readily like globular proteins. This structure was solved through solution NMR spectroscopy. The membrane-spanning helices generally require an hydrophobic environment to stabilize their structure. Indeed, portions of the stator that connect the F_o and F_1 motors are still not resolved at the atomic level. An additional portion of the stator is found in the 2A7U structure, that was also solved using NMR spectroscopy. This portion of the protein connects the stator to the F_o motor. A length of the stator can also be found in the 1L2P structure.

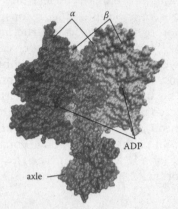

FIGURE 10.10. The F_1 motor consists of three $\alpha\beta$ sequences, one of which is not drawn to allow the visualization of the central axle. Molecules of ADP are bound at the interface between α and β segments.

The F_1 motor segment depicted in figure 10.10 can be found in the PDB file 1E79 and is composed of alternating α-β domains, chains A B C and D E F, respectively. The axle is formed by chains G H and I. As a motor, the F_1 segment is powered by ATP hydrolysis into ADP. There are three ADP molecules bound into active sites in the structure, in the A, B and C chains. Binding of ATP would cause the adjacent chain to shift to encompass the molecule. This causes a rotation of the central axis and also brings the catalytic power of the enzyme to bear, with the result that ATP is hydrolyzed and the axle has revolved to the next interface.

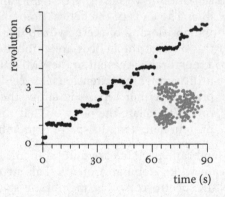

FIGURE 10.11. The F_1 motor motion was observed through videotape. The motor proceeds in a stepwise fashion (black dots), moving one third of a revolution. The angular distribution (gray dots) separates into three separate lobes. Image adapted from Yasuda et al., Cell 1998.

Ryohei Yasuda and colleagues were able to visualize the motion of the F_1 motor in 1998.[16] They created constructs of the α and β proteins that contained several histidine residues at their ends. The histidines bind to

[16]Yasuda and coworkers published "F_1-ATPase is a highly efficient molecular motor that rotates with discrete 120° steps" in Cell.

nickel, so placing a solution of protein on a nickel-coated slide bound the motor in place. They also modified the crown of the F_1 motor to include a segment of the protein streptavidin and added the protein actin to the solution. Actin self-assembles into long tendrils and binds to streptavidin.

In figure 10.11, we illustrate some of Yasuda's results. When a weak solution of ATP was added to the microscope slide, the actin filaments would revolve around the fixed motor locations. The filaments were roughly 1 μm in length, over a thousand times the size of the motor, and could be imaged through a microscope. The researchers found that the end of the filaments moved between three separate locations, 120° apart.

> EXERCISE 10.15. Download the 1E79 structure and open it in VMD. Create ribbon representations of the chains and CPK representations of the ADP molecules. Are all three of the $\alpha\beta$ sectors of the motor equivalent? Where is the axle located? What residues are responsible for coordinating the ADP?

As an ATP synthase, the F_1 motor works in the opposite direction from that observed by Yasuda and coworkers, building ATP from ADP and phosphate. It is powered by the rotation of the F_0 motor. Details of the reactions remain an active research topic, despite our general understanding of the gross features.

10.4. Molecular Toolkit

Many physics tools have been utilized in the study of the ATP synthetase. X-ray crystallography and NMR spectroscopy have been utilized to provide atomic resolution. Cryo-electron microscopy has filled missing gaps in the structure at the nanometer scale and, recently, AFM microscopy has been utilized to study the motors.

One of the most widely used and versatile tools in biophysics is optical tweezers. First reported by Arthur Ashkin from Bell Laboratories in 1970, the device was first realized by Ashkin and coworkers in 1986.[17] The basic principle of optical tweezers for micrometer-sized objects is illustrated in figure 10.12, where we have assumed the object to be spherical.

As light traverses the sphere ($\mathbf{k}_o \rightarrow \mathbf{k}_t \rightarrow \mathbf{k}_f$), there is an overall momentum change $\hbar\Delta\mathbf{k}_2$ on the light. This must, of course, be compensated by a

---

[17]Ashkin published "Acceleration and trapping of particles by radiation pressure" in the *Physical Review Letters* in 1970 and realized his ideas with help from Steven Chu, among others, in "Observation of a single-beam gradient force optical trap for dielectric particles" published in *Optics Letters* in 1986.

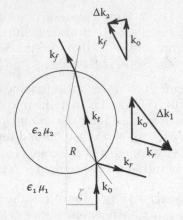

FIGURE 10.12. Light with wave vector $\mathbf{k}_o$ impacting a sphere at a distance ζ from the centerline will refract and reflect. The light exiting the sphere $\mathbf{k}_f$ has a net momentum shift $\Delta p = \hbar \Delta \mathbf{k}_2$.

momentum change in the sphere $-\hbar \Delta \mathbf{k}_2$. If the illumination of the sphere is uniform, symmetry demands that the net momentum transfers (in the lateral directions, at least) will sum to zero, so the sphere will be unaffected by the light. If instead, there is a variation with respect to ζ of the intensity, then there will be a net momentum imparted to the sphere. This momentum tends to push the sphere into the region of greatest intensity. It is possible to arrange the laser intensity to be approximately Gaussian in the ζ direction, resulting in the sphere nominally hovering around the midpoint of the beam width.

Attentive students have probably noticed that the momentum transfer $\hbar \Delta \mathbf{k}_1$ from the reflected beam in figure 10.12 is larger than that of the refracted beam and has a significant component in the direction of the beam. Thus, even if the force on the sphere in the lateral direction is null, there will be a significant force along the beam axis. There are two means for compensating for this issue. In the first, one can use counterpropagating beams. In the second, one can bring the beam to a sharp focus. This has the effect of providing an intensity gradient in the beam direction that also serves to confine the sphere.

EXERCISE 10.16. Redraw figure 10.12 and label all of the angles where the beams intersect the spherical surface. We neglected reflection at the second interface in our discussion (as well as higher order reflections). Draw the reflected wave at the exit point. Considering the Fresnel equations, what are the relative amplitudes of reflected and transmitted rays?

A simplified view of an optical tweezers apparatus is illustrated in figure 10.13. The tweezers are created by light from the laser. The laser output is expanded and then steered towards a dichroic mirror (M1) that reflects the light upward through the objective lens, which brings the beam

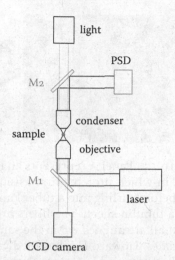

FIGURE 10.13. The optical tweezers are created by the laser field that is passed through the objective lens and focussed on the sample. The laser field is expanded through the condenser lens before being routed to the position sensitive detector (PSD). The sample can be illuminated by another light source and imaged on the CCD camera.

to a tight focus at the sample location. The diverging beam is captured by the condenser lens and reflected by a second dichroic mirror (M2) to a position sensitive detector that is used to determine beam position.

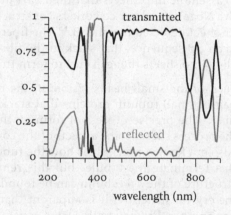

FIGURE 10.14. For incidence at 45° on the mirror surface, this material has a transmission coefficient above 0.85 throughout the domain 450–750 nm. Conversely, the reflection coefficient is above 0.90 throughout the domain 350–420 nm.

Dichroic mirrors have wavelength-dependent reflection and transmission properties. Typical behavior of a dichroic mirror is illustrated in figure 10.14. By using laser illumination for the tweezers outside the signal band, the two frequencies can be kept separate. Thus, in figure 10.13, the laser beam used for the tweezers is entirely reflected by the mirrors, where the alternate light source is passed unhindered.

Optical tweezers have been utilized in a host of different applications. One of the most intriguing biological problems involves the molecular motors kinesin and dynein that traverse microtubules. Microtubules are long, fibrous filaments that are found in eukaryotic cells. In figure 10.15,

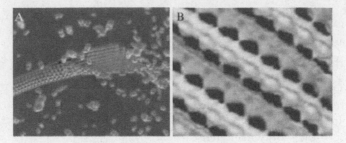

FIGURE 10.15. Panel A represents an idealization of microtubule formation, in which proteins first assemble into small units and then sheets, before curling into a tube. Panel B depicts the cryo-em structure of a tubulin sheet. The fibers are rotated slightly from fiber to fiber but all are aligned $\alpha\beta$ in the same direction. Image courtesy of Eva Nogales, Howard Hughes Medical Institute, University of California, Berkeley and Lawrence Berkeley National Laboratory.

we illustrate the putative construction of a microtubule, that is presumed to assemble into sheets that then curl into tubes, as depicted in panel A. Eva Nogales and her coworkers constructed the cryo-em image of a tubulin sheet depicted in panel B.[18] The fibers are constructed of two proteins in an $\alpha\beta$ sequence, that stack in the fiber direction. Multiple fibers then align into sheets that can curl to form tubes.

Each of the small beads visible in the cryo-em image is composed of a pair of small tubulin proteins that stack in an $\alpha\beta$ sequence, forming the fiber. The precise process of forming the microtubule is not known but the process shares some aspects of the conceptual drawing in figure 10.15. Also not known precisely is how the tubules end. There are cap segments that terminate the tubules but this remains an active research area. A structure of the $\alpha\beta$ tubulin can be found in the PDB structure 1JFF. From the crystal structure, it is apparent that the proteins conform tightly to one another. Predicting this from separate structures

EXERCISE 10.17.   Download the 1JFF structure and open it in VMD. The α and β segments are in chains A and B, respectively. The protein association is enhanced by the binding of GTP at the interface. Create a VDW representation of the GTP and identify its binding location. What are the secondary structures of the proteins? What are the interactions (hydrogen bonds) at the interface?

---

[18]Kenneth Downing and Nogales published "Cryoelectron microscopy applications in the study of tubulin structure, microtubule architecture, dynamics and assemblies, and interaction of microtubules with motors" in *Methods in Enzymology* in 2010.

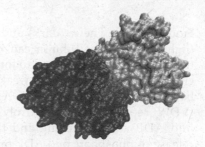

FIGURE 10.16. The α (dark) and β (light) proteins stack into a tightly bound state through hydrogen bonding across the protein-protein interface.

The molecular motors kinesin and dynein traverse the microtubules in opposite directions. Their function is to transport materials across the cell. Precisely how they accomplish this remains an area of active research. Kinesin is composed of two components, each of which is formed by a head group, a long α helix and a tail segment that attaches to the cargo molecule. An example can be found in the PDB structure 3KIN.

EXERCISE 10.18. Download the 3KIN structure and open it in VMD. The long α helices of the protein segments are truncated to a small fraction of their length in order for the protein to crystallize. It is known that the coils form a coiled coil, like two strands of rope wrapped around one another. What is the secondary structure of the protein? How do the head segments attach to the coil? Where does ATP bind?

Each of the head segments is powered by ATP hydrolysis and the pair swing along the microtubule chain, hand over hand, as it were. The complete cycle is not fully understood but is does appear that the head groups can form two different attachments to the microtubule: a weak attachment in which both head groups are bound and a strong attachment in which only a single head group can stay attached. The differences between weak and strong are due to conformational changes in the protein induced by the presence or absence of ATP, ADP and phosphate in the active site. There is an additional coordination of the C-terminal portion of the kinesin protein with the tubulin protein that helps to stabilize the structure but this is not always resolved in the crystal structures as it tends to be disordered (or not ordered in the same way throughout the crystal).

EXERCISE 10.19. Download the 1BG2 structure of kinesin which depicts the conformational shift associated with ATP hydrolysis. Use the RMSD tool to align the 3KIN and 1BG2 structures. What is the difference in the structures? It will help to use the NewRibbons representation.

FIGURE 10.17. The α (dark) and β (light) proteins of tubulin can interact with kinesin (ribbon) to form either a tightly bound or weakly bound complex. Binding of ATP (VDW representation), hydrolysis into ADP and phosphate and the release of phosphate and ADP provoke changes in the kinesin structure.

The combined cryo-em/x-ray diffraction picture of the process is contained in the PDB structure 2PN4, depicted in figure 10.17. Here, the 1JFF and 1BG2 structures were fitted into the cryo-em map of the complex. Note that the C-terminal loop of kinesin also interacts with the α tubulin.

EXERCISE 10.20. Download the 2P4N structure that depicts the interaction between a kinesin head group and the tubulin proteins. This structure was obtained by fitting the 1JFF and 1BG2 structures into a 9 Å cryo-em map of the structure. Where does the kinesin bind?

The picture of this complex machinery that has been developed has utilized several imaging technologies: x-ray and electron diffraction, cryo-electron microscopy and NMR spectroscopy to define structures with resolutions that range from nearly atomic resolution (3 Å) to 10 nm. The high resolution provides details about the specific interactions that are involved in protein-protein binding and the lower resolution images provide large scale structure.

Optical tweezers have been utilized to study the dynamics of the system and to obtain further clues as to the nature of the mechanism. Steven Block and coworkers conducted a series of investigations using an assay in which kinesin tails were chemically attached to silica beads via long linking chains.[19] With low kinesin concentrations, the average number of kinesin molecules per bead was less than one. Microtubules were fixed to a microscope slide mounted on a piezo stage and the silica beads were captured one at a time by the optical tweezers. The captured beads were then moved to a nearby microtubule. If the bead began moving on its own, it was clear that the kinesin head had contacted the microtubule and

---

[19]Brown and colleagues published "Direct observation of kinesin stepping by optical trapping interferometry" in *Nature* in 1993.

was translating via ATP hydrolysis. Drawing a picture of the experiment is quite difficult: the silica beads have a diameter of about 500–1000 nm and the kinesin motors are only a few tens of nanometers in length. The image of an ant dragging a large boulder springs to mind. Nonetheless, Brown found that the motors were capable of exerting more than 5 pN of force.

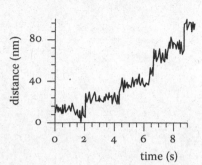

FIGURE 10.18. Under high loads, kinesin pauses at each step, revealing an 8 nm step length.

Brown and coworkers conducted a number of experiments to characterize the kinesin motion, measuring the velocity, the length of travel before the motor released from the filament and the force exerted by the motor. By holding the beads fixed in the optical trap and translating the piezo stage, Brown and his colleagues could apply a load to resist the kinesin motion. As depicted in figure 10.18, individual beads made steps of approximately 8 nm in length. Their results are consistent with the microscopic picture of the head groups of kinesin walking hand over hand along the length of the microtubule, spanning one $\alpha\beta$ dimer per step.

This work represents a remarkable achievement: studying the properties of a single molecule and not deducing the properties of a single molecule from the ensemble average of a host of molecules. This provides significant insight into the mechanical functioning of biological systems that cannot be obtained through other means.

Other physics tools have also been applied to the study of biological systems. Atomic force microscopes can be used to image biological samples and can provide images comparable in resolution to the cryo-em images that we have discussed. AFMs have also been repurposed utilize the cantilever as a means of applying a force to rupture chemical bonds. With a molecule tethered to a surface and the other end tethered to the AFM tip, vertical tip motion can be utilized to exert a force to study, among other topics, protein unfolding. Protein structure is key to its function and how proteins achieve their folded state is a subject of current research. Among the more popular theories, it has been proposed that folding proceeds through a series of checkpoints from unfolded to folded states. AFM studies have indicated that this idea is broadly true. There remain ensembles of states around the checkpoints that effectively broaden the pathway.

On Plasmas

In the preceding chapters, we have discussed matter at increasing sizes and decreasing symmetry. In biological systems, physics tools have been utilized to help identify the structure and mechanisms employed in the machinery of living organisms. All of the discussion though, was centered on materials that are relatively commonplace terrestrially and have omitted what may be the most common form of matter in the universe: plasma. Because matter densities on the planet are high and the temperatures are not, most matter on the planet exists in the form of electrically neutral gases, liquids and solids. There are, of course, ions intermixed but typically at low densities and they are invariably interacting with solvent molecules.

There are large charge imbalances in the upper atmosphere, partially powered by the flux of high-energy particles from the solar wind. A complete understanding of the processes that lead to charge formation and separation is not yet at hand but the fact that it exists is evidenced by spectacular lightning strikes. Views of the planet from space suggest that lightning is a global phenomena, so perhaps the transient nature is not as firm a principle as one might hope. Nevertheless, in most daily activities, we do not often encounter plasmas.

Leaving the planet's surface, we find that plasmas are quite common. Just above the earth's atmosphere are found circulating plasma sheaths known as the Van Allen radiation belts.[1] American physicist James van Allen and his coworkers attempted to launch three Geiger counters into orbit on Explorer rockets. The first and third were successful in reaching orbit. The first payload simply telemetered data continuously back to earth but was only detectable when the satellite passed over a ground receiving station. The third version contained a magnetic tape recorder that recorded an entire orbit's worth of data that was replayed upon command from a ground station.

---

[1] James Van Allen and coworkers published "Observation of high intensity radiation by satellites 1952 Alpha and Gamma" in *Jet Propulsion* in 1958.

© Mark A. Cunningham 2018
M.A. Cunningham, *Beyond Classical Physics*,
Undergraduate Lecture Notes in Physics,
https://doi.org/10.1007/978-3-319-63160-8_11

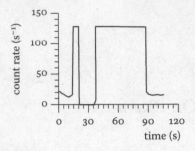

FIGURE 11.1. The Geiger counter rate for the γ mission plateaus at 128 s^{-1}, the maximum rate and, surprisingly, falls to zero for extended durations.

Van Allen and his colleagues were perplexed by the data from the first mission, in which the measured count rate was frequently zero, but the absence of a continuous record of count rate made the data quite difficult to analyze. With the addition of the tape recorder on the third mission, Van Allen was able to obtain records like those depicted in figure 11.1. The explanation is that the zeroes are the manifestation of detector saturation.

When a charged particle strikes the Geiger-Müller tube, it causes a cascade of charged particles that result in a sizeable voltage pulse. This voltage pulse is further amplified through electronic circuits (tubes in 1958) and directed to a counting circuit that only counts pulses above a threshold voltage. When the charged particle rate becomes very high, the Geiger-Müller tube cannot respond with pulses of the same amplitude; the charged-particle cascades are effectively short-circuited by the arrival of the next charged particle. Consequently, no pulses exceed the threshold of the counting circuit. Ironically, zero counts in the detector has to be interpreted as an extremely intense radiation field.

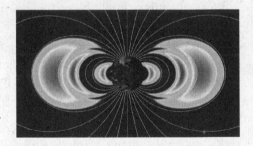

FIGURE 11.2. The plasma density in the Van Allen belts is nominally separated into two lobes. The NASA Van Allen probes have detected that, on occasion, there is additional structure. Image courtesy of NASA.

Since these early experiments, the Van Allen belts have been explored repeatedly. After van Allen's initial discovery, it was found that there are, nominally, two main lobes: an inner and an outer belt, as depicted in figure 11.2. The inner belt is known to be composed primarily of protons that are believed to result from the decay of neutrons generated in cosmic ray interactions with the atoms in the upper atmosphere. The outer belt has a large component of energetic electrons but is understood to be more dynamic. Notably, the Van Allen probes launched in 2012 have provided a much more dynamic picture of the belts than had previously been assumed. The Van Allen probes are two satellites that orbit in reasonably close proximity but at a large enough distance that researchers can gain insight into the physical extent of their observations. That is, over what distances do the events they detect occur?

Electrons with energies up to 2 GeV have been detected in the outer belt, which has been observed to bifurcate into two separate segments and later rejoin. The mechanisms at work that can accelerate electrons to that energy appear to be local to the belt. In addition, the belts are not as symmetrical as depicted in figure 11.2, reacting to events in the solar wind. As we shall see, plasmas are often more complex than we might first assume.

11.1. Fourth State of Matter

An operative definition of a plasma is a collection of charged particles that are not bound. This states arises when electrons have enough kinetic energy that they exceed the binding energy of whichever nuclei are present or the density of particles is so low that collisions that would lead to electron capture are infrequent. Because we are discussing charged particles and unbound electrons, it will generally suffice to treat plasmas classically through Maxwell's equations. This offers great simplifications but we are most concerned about collective effects that are most difficult to describe from a single-particle perspective.

We might initially hope that we can be able to describe plasmas simply through their bulk dielectric permittivity and magnetic permeability. Such an approach was highly successful in the treatment of dielectric materials like glasses. We were able to construct a reasonably complete theory of ray tracing through lens systems. Indeed, there are some cases in which we will be able to treat plasmas through a bulk dielectric; plasmas are generally not magnetically permeable materials.

One such case is the earth's ionosphere, where decreasing density at the top of the atmosphere and the incident solar radiation create the necessary conditions for a plasma to exist. In past years, when analog radio signals were the norm, it was commonplace at night to receive broadcasts from very distant sources. Radio tuners utilized relatively low-fidelity LCR resonant circuits as the band-pass filter in the receiver and these admitted signals from a broad range of sources. Hence, the AM broadcast band seemed filled to overcapacity after dark. It was recognized at the time that this was due in large measure to the reflection of radio waves from an ionized layer at the top of the atmosphere. In recent years, there has been an international effort to characterize the properties of the ionosphere. Known as the International Reference Ionosphere (IRI) project, the members have made public the data and models developed.

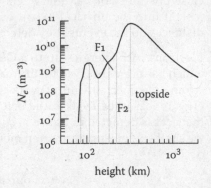

FIGURE 11.3. The International Reference Ionosphere is a collaborative model of the electron and ionic content of the ionosphere. The layers F2 and F1 are particularly important due to their impact on radio transmission.

An example of the electron density as a function of height is depicted in figure 11.3. Note that this is a logarithmic plot. The electron density peaks here at an altitude of about 320 km above the earth's surface and it is this density that is responsible for the reflection of radio waves.

EXERCISE 11.1. The IRI website (irimodel.org) contains links to on-line executable that can evaluate the electron and ion densities. Plot the electron and ion densities as functions of height, latitude and longitude. Choose values that represent the latitude and longitude of your own location.

An electric field applied to a plasma will cause charge separation: the electrons will migrate in the direction opposite to the positively charged ions. This migration will generate an electric field internal to the plasma, $E = \sigma/\epsilon_0$. Where σ is the charge density. If there are n electrons per unit volume, then in an infinitesimal slab of thickness δx, we have $\sigma = -en\,\delta x$. The force acting on that slab will have the following form:

$$(11.1) \qquad m_e \frac{d^2 \delta x}{dt^2} = -eE = -\frac{e^2 n}{\epsilon_0} \delta x.$$

If the external field is oscillatory, then equation 11.1 can be seen to be the harmonic oscillator equation with frequency ω_p given by the following:

$$(11.2) \qquad \omega_p = \frac{e^2 n}{\epsilon_0 m_e}.$$

This is known as the plasma frequency.

The dielectric permittivity can be thought of as arising from the polariz-
abilities of the different states associated with different species of atoms.
A simple model that captures this behavior can be written as follows:

$$(11.3) \qquad \epsilon(\omega) = 1 + \frac{e^2 n}{\epsilon_0 m_e} \sum_j \frac{f_j}{\omega_j^2 - \omega^2 - i\gamma_j \omega}.$$

Here, the ω_j are the frequencies of the states in the atoms or molecules
and γ_j is a dissipative term that is generally small. The f_j can be thought
of as the occupation probabilities for the different states.

For plasmas, this expression simplifies to the following:

$$(11.4) \qquad \epsilon(\omega) \approx 1 - \frac{\omega_p^2}{\omega^2}.$$

The magnitude of the wave vector is obtained, as usual:

$$(11.5) \qquad k^2 = \omega^2 \epsilon \mu = \epsilon_0 \mu_0 (\omega^2 - \omega_p^2).$$

What we see from equation 11.5, is the magnitude of the wave vector becomes purely imaginary for frequencies below the plasma frequency. Recall that plane wave propagation carries the factor $\exp[ikx]$ for propagation in the x-direction, so waves no longer propagate into the plasma and
are instead attenuated.

> EXERCISE 11.2.   From the Fresnel equations for TE incidence on
> a planar layer, what happens when the propagation constant k_2 is
> imaginary? The reflection coefficient R is the square of $\tilde{E}_R / \tilde{E}_I$. What
> is R for k_2 imaginary?

In this era of digital information transmission, one is not often inconve-
nienced by interference and sporadic signals interrupting evening broadcasts but radio reflection from plasma interfaces does still present some
operational difficulties. Satellite transmissions are often victims of so-
called space weather. In particular, returning manned spacecraft are unable to communicate during some of the most hazardous portions of reen-
try. The kinetic energy of the vehicle is reduced from orbital velocity to a
few hundred kilometers per second by compressing and heating the atmosphere surrounding it, thereby creating a plasma sheath that blocks radio
communications. Although this is now a distant memory, in 1969 Apollo
astronauts returned from the surface of the moon. After having ventured
from this planet and leaving footprints on another celestial body, their
return was widely televised. There was an uncomfortably long time that
broadcasters kept repeating that there was a communications blackout, it
was expected and, any time now, we should be hearing from the astronauts. This pause was far longer than the networks would have preferred.

It is one thing to build suspense but good television requires that the storyline keep advancing. Fortunately, all systems worked as advertised and the astronauts reported back that they were swinging under parachutes; people across the world breathed sighs of relief.

11.2. Magnetohydrodynamics

The present discipline of plasma physics owes much of its structure to the Swedish physicist Hannes Alfvén, who began working on a coherent theory of the auroras in the late 1930s.[2] Alfvén treated the plasma as a charged fluid, bringing together ideas from both electromagnetic theory and fluid dynamics. Today, the theory is known as magnetohydrodynamics and represents a solid first step towards understanding the behavior of an extraordinarily complex phenomenon.

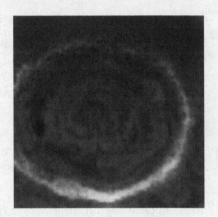

FIGURE 11.4. The north pole of Saturn is marked by an extraordinary hexagonal feature, first imaged by Voyager in 1980, here by Cassini in 2006. The bright feature is the aurora. Image courtesy of NASA/JPL/University of Arizona.

As can be seen in figure 11.4, the north pole of Saturn contains a remarkable hexagonal feature. Surrounding that feature in this night image from the Cassini spacecraft is the bright aurora created as cosmic radiation impacts the upper atmosphere. Auroras are common features around planets with atmospheres and magnetic fields but even a glance at figure 11.4 will indicate that describing the features of the auroras will be a complex undertaking. There is a great deal of structure within the auroras: filaments and swirls that suggest there are interactions that occur over a wide range of length scales.

---

[2]Alfvén was awarded the Nobel Prize in Physics in 1970 "for fundamental work and discoveries in magnetohydro-dynamics with fruitful applications in different parts of plasma physics." He shared the award with Louis Néel, who was cited "or fundamental work and discoveries concerning antiferromagnetism and ferrimagnetism which have led to important applications in solid state physics."

We can, formally, begin with a description of all the different species that make up the plasma in terms of distribution functions in phase space. That is, there is some function $\Phi_j(\mathbf{x},\mathbf{v},t)$ that describes the distribution of particles of type j in space $\mathbf{x}$ with velocity $\mathbf{v}$ at a time t. We can obtain the macroscopic charge density ρ_c and current $\mathbf{j}$ by integrating over velocity and summing over species:

$$(11.6) \qquad \rho_c(\mathbf{x},t) = \sum_j q_j \int d^3\mathbf{v}\,\Phi(\mathbf{x},\mathbf{v},t) = \sum_j q_j n_j(\mathbf{x},t) \quad \text{and}$$

$$(11.7) \qquad \mathbf{j}(\mathbf{x},t) = \sum_j q_j \int d^3\mathbf{v}\,\mathbf{v}\Phi(\mathbf{x},\mathbf{v},t),$$

where q_j is the charge of species j and n_j is the particle density in coordinate space. Note that the mass densities can be similarly defined as $\rho_j = m_j n_j$. These terms are the sources of the electromagnetic fields, so one might hope that invoking conservation in phase space would provide us with an equation for the time evolution of the distribution functions.

We know that phase space is conserved, by which we mean that the following relations hold:

$$(11.8) \qquad \frac{d}{dt}\Phi_j(\mathbf{x},\mathbf{v},t) = 0 = \left[\frac{\partial}{\partial t} + \frac{\partial x}{\partial t}\frac{\partial}{\partial x} + \cdots + \frac{\partial v_x}{\partial t}\frac{\partial}{\partial v_x} + \cdots\right]\Phi_j(\mathbf{x},\mathbf{v},t).$$

The accelerations in equation 11.8 arise from the Lorentz force, so we can incorporate Maxwell's equations through the following:

$$(11.9) \qquad \left[\frac{\partial}{\partial t} + \mathbf{x}\cdot\nabla_\mathbf{x} + \frac{q_j}{m_j}\left(\mathbf{E}+\mathbf{v}\times\mathbf{B}\right)\cdot\nabla_\mathbf{v}\right]\Phi_j(\mathbf{x},\mathbf{v},t) = 0.$$

Here, the gradient operators $\nabla_\mathbf{x}$ and $\nabla_\mathbf{v}$ affect the spatial and velocity spaces, respectively.

Despite the rather appealing simplicity of equation 11.9, we need to admit that it is completely intractable as a starting point for calculations. This should serve as another cautionary note on the need to understand what the equations really mean, not just that they are mathematically sound. At a microscopic level, the distribution functions are actually sums over all of the individual particles, so the functions that appear in the equation must be some sort of ensemble average properties. The fields $\mathbf{E}$ and $\mathbf{B}$ that arise from the microscopic distributions must also be averaged over the ensemble, as computing them explicitly from the particle positions is intractable. Thus we need to interpret the last term in equation 11.9 as the dot product of the ensemble-averaged fields with the gradient of the ensemble-averaged distribution.

Unfortunately, that statement is not true. We have, explicitly,

$$\left\langle \left(\mathbf{E} + \mathbf{v} \times \mathbf{B}\right) \cdot \nabla_{\mathbf{v}} \Phi(\mathbf{x}, \mathbf{v}, t) \right\rangle \neq \left\langle \mathbf{E} + \mathbf{v} \times \mathbf{B} \right\rangle \cdot \left\langle \nabla_{\mathbf{v}} \Phi(\mathbf{x}, \mathbf{v}, t) \right\rangle$$

where the angle brackets mean ensemble average. The dot products at the microscopic scale involve signs that are lost when averaged. The problem is akin to the one we encountered calculating the electron charge density earlier. What are termed *correlations* are lost in the averaging process. As a result, there is no straightforward derivation of equations that describe the time evolution of a plasma. Every approach is an approximation, often a gross oversimplification, that can be justified on some premise or another. Each approach has to be understood for what it contains and what it does not.

The physical properties that define the plasma, at least within the approximations of magnetohydrodynamics, are the particle densities n and the pressure p. The use of a scalar pressure again points to the fact that we are restricting ourselves to a macroscopic view of the plasma: the pressure is an emergent, macroscopic property that arises out of an ensemble average. The basic equations can be stated as follows:

$$(11.10) \qquad\qquad \frac{\partial n}{\partial t} + \nabla \cdot n\mathbf{u} = 0,$$

$$(11.11) \qquad\qquad mn\frac{\partial \mathbf{u}}{\partial t} + \nabla p - \mathbf{j} \times \mathbf{B} = 0,$$

$$(11.12) \qquad\qquad \mathbf{E} + \mathbf{u} \times \mathbf{B} = \sigma\mathbf{j}$$

$$(11.13) \qquad\qquad \mu_0\mathbf{j} = \nabla \times \mathbf{B} \quad \text{and}$$

$$(11.14) \qquad\qquad \frac{d}{dt}\left(\frac{p}{n^\gamma}\right) = 0.$$

Here, we have suppressed indices for species, so the total density $n = n_e + n_i$ is the sum of electronic and ionic components, for example. There is an overall fluid velocity $\mathbf{u}$ where it is implicitly assumed that the fluid velocities of ions and electrons are comparable, on average. This condition arises, in part, from the assumption that the plasma is essentially neutral: $n_i \approx n_e$.

The first equation 11.10 represents the conservation of mass: the time rate of change of fluid within some volume is balanced by the flux of fluid entering and leaving. The second equation 11.11 represents momentum conservation. The next two equations arise from a low-frequency approximation to Maxwell's equations and the final equation 11.14 is a statement that processes within the plasma do not exchange thermal energy with anything exterior to the plasma, i.e., the processes are adiabatic. The superscript $\gamma = c_p/c_V$ is the ratio of specific heats, generally taken to be

5/3. These equations are valid for low frequencies and large scales but are missing a lot of physics. There is no energy flow in or out of the plasma, for example, so the model cannot support radiation or heating of the plasma. Nevertheless, the model does provide a starting point for incorporating such processes.

With a suitable rearrangement of terms, it is possible to demonstrate that the energy density within the plasma is given by the following:

$$E = \frac{p}{\gamma - 1} + \frac{nu^2}{2} + \frac{B^2}{2\mu_0}. \tag{11.15}$$

A parameter that can be used to characterize plasmas can be obtained from the ratio of thermal to magnetic terms:

$$\beta = \frac{2\mu_0 p}{B^2}. \tag{11.16}$$

Small values of β indicate that magnetic effects dominate the plasma physics, where large values indicate that fluid dynamics dominates.

One might now surmise that there is an easy approach to solving the MHD equations. We have, after all, made enormous simplifying assumptions but there is more simplifying required. While it is possible to toss all of the equations onto a computer and see what happens, Alfvén began by considering small perturbations to a steady state solution. Consider then that the magnetic field can be represented as two terms: $\mathbf{B} = \mathbf{B_0} + \mathbf{B_1}$, where $B_1/B_0 \ll 1$. For bookkeeping purposes, one could also use a multiplicative factor $\mathbf{B} = \mathbf{B_0} + \varepsilon \mathbf{B_1}$, where $\varepsilon \ll 1$. We also use the same strategy for the velocity $\mathbf{u} = 0 + \mathbf{u_1}$ and density $n = n_0 + n_1$ and then substitute back into the MHD equations. If we keep only the small terms (linear in ε), we are left with a linearized version of the MHD equations:

$$\frac{\partial n_1}{\partial t} + n_0 \nabla \cdot \mathbf{u_1} = 0$$

$$mn_0 \frac{d\mathbf{u_1}}{dt} + v_s^2 \nabla n_1 + \mathbf{B_0} \times (\nabla \times \mathbf{B_1}) = 0 \tag{11.17}$$

$$\frac{\partial \mathbf{B_1}}{\partial t} - \nabla \times (\mathbf{u_1} \times \mathbf{B_0}) = 0.$$

Here, $v_s = \sqrt{\gamma p_0/n_0}$ is the speed of sound in the plasma. Equations 11.17 can be combined to provide a single equation in $\mathbf{u_1}$. We obtain the following expression:

$$\frac{\partial^2 \mathbf{u_1}}{\partial t^2} - v_s^2 \nabla(\nabla \cdot \mathbf{u_1}) + \mathbf{v}_A \times \nabla \times [\nabla \times (\mathbf{u_1} \times \mathbf{v}_A)] = 0, \tag{11.18}$$

where we have defined the Alfvén velocity as $\mathbf{v}_A = \mathbf{B_0}/\sqrt{\mu_0 m n_0}$.

Exercise 11.3. Substitute the perturbation expansions into equation 11.10. (It might prove advantageous to utilize ε as a bookkeeping aid.) What is the zeroth order equation? What is the first order equation?

Equation 11.18 represents a complicated form of a wave equation for the perturbed velocity $\mathbf{u}_1$. As students should now expect, we will utilize the Fourier transform:

$$(11.19) \qquad \mathbf{u}_1(\mathbf{x},t) = \frac{1}{4\pi^2} \int d^3k \int d\omega\, \tilde{\mathbf{u}}_1(\mathbf{k},\omega)\, e^{i(\mathbf{k}\cdot\mathbf{x}-\omega t)}.$$

Now we can convert the difficult differential equation into a difficult algebraic equation:

$$(11.20) \quad -\omega^2 \tilde{\mathbf{u}}_1 + (v_s^2+v_A^2)(\mathbf{k}\cdot\tilde{\mathbf{u}}_1)\mathbf{k} + \mathbf{v}_A\cdot\mathbf{k}\big[(\mathbf{v}_A\cdot\mathbf{k})\tilde{\mathbf{u}}_1 - (\mathbf{v}_A\cdot\tilde{\mathbf{u}}_1)\mathbf{k} - (\mathbf{k}\cdot\tilde{\mathbf{u}}_1)\mathbf{v}_A\big] = 0.$$

We can interpret the results of our progress best by considering two limiting cases for equation 11.20.

First, consider the case where the direction of propagation $\mathbf{k}$ is perpendicular to the magnetic field direction $\mathbf{B}_0$. Then, from the definition of $\mathbf{v}_A$, we have that $\mathbf{v}_A \cdot \mathbf{k} = 0$. Then, we are left with the following:

$$(11.21) \qquad -\omega^2 \tilde{\mathbf{u}}_1 + (v_s^2+v_A^2)(\mathbf{k}\cdot\tilde{\mathbf{u}}_1)\mathbf{k} = 0.$$

Taking the dot product with $\mathbf{k}$, we find the following dispersion relation:

$$(11.22) \qquad \omega^2 = (v_s^2+v_A^2)k^2.$$

The result is known as a magnetosonic wave that propagates with a velocity proportional to $\sqrt{v_s^2+v_A^2}$.

If we now consider the case where the direction of propagation is parallel to the magnetic field direction, then equation 11.20 can be again simplified. We note that $\mathbf{k} = (k/v_A)\mathbf{v}_A$ and can simplify the equation to the following:

$$(11.23) \qquad (v_A^2 k^2 - \omega^2)\tilde{\mathbf{u}}_1 + \Big[(v_s^2+v_A^2)\frac{k^2}{v_A^2} - 2k^2\Big](\mathbf{v}_A\cdot\tilde{\mathbf{u}}_1)\mathbf{v}_A = 0.$$

Here, we find that there are two kinds of waves. If $\tilde{\mathbf{u}}_1$ is parallel to $\mathbf{v}_A$, then we obtain a dispersion relation as follows:

$$(11.24) \qquad \omega^2 = v_s^2 k^2,$$

which corresponds to a wave propagating with the velocity of sound in the direction of the magnetic field.

The second type of wave, known as an Alfvén wave, occurs when $\tilde{\mathbf{u}}_1$ is perpendicular to the magnetic field. Then $\tilde{\mathbf{u}}_1 \cdot \mathbf{v}_A = 0$ and we find a wave with a dispersion relation as follows:

$$(11.25) \qquad \omega^2 = v_A^2 k^2.$$

This corresponds to a disturbance propagating at the Alfvén velocity v_A and is a magnetohydrodynamic phenomenon.

EXERCISE 11.4. From the last of equations 11.17, what are the magnetic fields B_1 associated with each of the three kinds of waves?

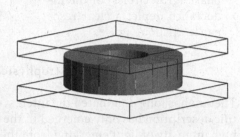

FIGURE 11.5. Jameson constructed a toroidal box from thin-walled stainless steel to hold the sodium. A series-connected set of copper sector elements (gray) generate the azimuthal B_1 field. The B_0 field was generated by a large electromagnet; the pole pieces were large blocks (black) containing the torus.

One of the early experiments designed to study Alfvén waves was conducted by British physicist Antony Jameson in 1964.[3] Jameson built a torus from thin-walled stainless steel and filled it with sodium under an argon atmosphere. Copper was wound circumferentially around the outer diameter of the torus. This coil served both to heat the sodium to liquefy it and as part of the mutual inductance measurement. A set of copper sectors, gray elements in figure 11.5, were connected in series and were used to generate an azimuthal B_1. This assembly was sandwiched between two large iron plates that formed the pole pieces of a large electromagnet. The field of this magnet is a downward-directed vertical B_0. An AC current flowing in the copper sector windings induces voltages in a small pickup coil suspended in approximately the center of the torus and the heater windings. Jameson measured the mutual inductance of those coils as a function of drive frequency. His results for the small coil are shown in figure 11.6.

The mutual inductance peaks just above 50 Hz, consistent with the predicted behavior. The phase of the inductance sweeps through 180° at the resonant frequency, confirming the existence of a resonance. The magnitude of the mutual inductance is nearly ten times the value measured without sodium in the torus, from which Jameson concluded that the fields within the torus were greatly amplified by the presence of a standing Alfvén wave.

---

[3]Jameson published "A demonstration of Alfvén waves" in the *Journal of Fluid Mechanics*.

FIGURE 11.6. Jameson measured the mutual inductance M between a coil suspended in liquid sodium and one exterior to the sodium. The magnitude (black dots) and phase (gray circles) of the inductance depict a clear resonance at about 53 Hz.

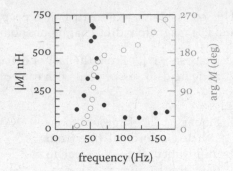

11.3. Astrophysical Plasmas

The sun has long dominated the interest of humanity but a detailed scientific description has only emerged in the relatively recent past. Due to the high intensity of light emerging from the sun, it has taken time to develop imaging technologies that are not overwhelmed by the radiation intensity. The picture that has emerged is even more complex than one might have imagined, even given that a roiling nuclear furnace in the interior is producing the energy emitted.

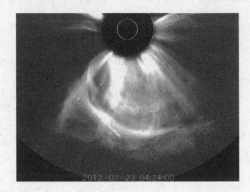

FIGURE 11.7. The solar corona imaged by the NASA Stereo A spacecraft on July 23, 2012 includes a large coronal mass ejection. The black disk is an artifact of the coronagraph. The solar size is indicated by the white circle. Image courtesy of NASA.

The sun does not have a surface, *per se*, as energy propagating outward from the interior creates a plasma that remains gaseous despite the gravitational force. (Actually, the gravitational acceleration at the nominal solar surface is only about 28 times that at the earth's surface.) Given the high temperatures, the solar atmosphere is ionized. Solar rotation gives rise to large electrical currents and, in turn, large magnetic fields.

The corona, of course, was known in antiquity, becoming visible during eclipses. Today we are not limited to waiting for the infrequent occurrences of planetary alignment to visualize the corona. The Stereo satellites, for example, launched by NASA in 2006 were placed in orbits at roughly 1 AU, but separated in phase from the earth, and provide simultaneous measurements of the sun from different aspects, thereby illuminating the three-dimensional nature of the observations. The coronagraphs in each satellite provide nearly continuous monitoring of the corona, revealing the dynamics of large scale events like the coronal mass ejection depicted in figure 11.7.

The corona is a manifestation of the particle flux emanating from the sun, that is given the name of solar wind. First anticipated in the late 1800s from the observations that the tails of comets always point away from the sun, details of the composition of the solar wind were not available until the late 1900s when spacecraft began to sample the particle distributions beyond our atmosphere. Today, we know that the wind is composed primarily of protons and electrons, with small admixtures of heavier elements, with energies in the 1–10 keV range. The existence of the solar wind confirmed the earlier speculation by German astronomer Ludwig Biermann and others who were looking to explain the behavior of comet tails.[4]

The energetics of the solar wind is still somewhat of a puzzle. We know that a blackbody radiates with a characteristic spectrum and that the solar spectrum is described reasonably well by a blackbody radiator with a surface temperature in the neighborhood of 5800 K, as can be seen from figure 11.8. Here, we used the Planck formula for the power spectral density:

$$(11.26) \qquad I(\lambda) = \frac{8\pi hc}{\lambda^5} \frac{1}{e^{hc/\lambda k_B T} - 1}.$$

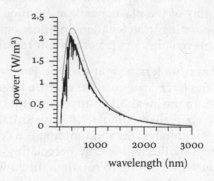

FIGURE 11.8. The power spectral density at the top of the earth's atmosphere (black) is compared to a blackbody radiator at 5800 K (gray). There are numerous absorption features in the experimental spectrum but the overall shape is well-described by the Planck formula.

---

[4]Biermann published "Kometenschweife und solare Korpuskularstrahlung" in the *Zeitschrift für Astrophysik* in 1951.

EXERCISE 11.5. The standard solar spectrum can be obtained from the National Renewable Energy Laboratory (nrel.gov) website. Use the Planck formula from equation 11.26 to study the temperature dependence. Note that there is an overall normalization not discussed in figure 11.8. What does a change of ±200 K to the nominal temperature do to the agreement? What about ±500 K?

The temperature of the corona is vastly higher: more than 10^6 K. So, one might well ask how the solar wind is heated to such astronomical temperatures? The answer is still not entirely resolved but it is in some way arising from the plasma interactions near the sun's surface. We do know that the simple MHD equations listed in equations 11.10–11.14 cannot explain this acceleration because they do not include processes that are physically relevant. More realistic results can be obtained with equations that include these terms. For example, the momentum conservation equation 11.11 can be expanded as follows:

$$(11.27) \qquad \rho\frac{\partial \mathbf{u}}{\partial t} + \nabla p - \mathbf{j}\times\mathbf{B} = \rho\mathbf{g} + 2\rho\Omega\times\mathbf{u} + \mathbf{F}.$$

Here the additional terms on the right-hand side of equation 11.27 take into account the gravitational acceleration near the sun's surface, the Coriolis force that arises from rotation of the sun and any other additional forces that might come to mind.

Similarly, the equation of state 11.14 can be expanded:

$$(11.28) \qquad \frac{\rho^\gamma}{\gamma-1}\frac{d}{dt}\left(\frac{p}{\rho^\gamma}\right) = \nabla\cdot(\kappa\nabla T) - \rho^2 Q(T) + \frac{j^2}{\sigma} + H.$$

Here, the additional terms represent thermal conduction, radiation, Ohmic heating of the plasma and unspecified other enthalpic contributions. Incorporating more challenging phenomena like the propagation of shock fronts remains an area of current research.

What is understood, at present, is that the plasma near the sun's surface carries large scale currents that are largely embedded with the plasma motion. In the ideal MHD approximation, the fields are fixed to the flow but in more realistic simulations, this situation is only approximately true. As the sun rotates, the fluid motion is a complicated three-dimensional flow forced by thermal energy percolating up from the interior and Coriolis-induced shear due to rotation. In addition to the surface currents, there is convective transport of material from the interior to the surface. These currents give rise to large magnetic fields that twist out of the surface and, through a still mysterious process, reconnect with significant energy release.

FIGURE 11.9. Stereo A image of the solar surface in the UV (171 Å). Image courtesy of NASA.

An image of this process is provided in figure 11.9 that was taken by the Stereo A satellite in December, 2006. The bright features in the right side of the figure are interpreted to be magnetic flux tubes emerging from the surface. These flux tubes are powered presumably by the solar dynamo responsible for the sun's magnetic field.[5] Solar physics is a challenging enterprise, involving fluid mechanics in a rotating frame of reference. Moreover, it remains challenging to provide laboratory-scale experiments of simplified systems.

As one might imagine, producing plasmas with very high temperatures is challenging. It is, of course, possible to do so by blasting solid surfaces with intense laser beams but those pulsed sources create plasmas that are parametrically far from those observed in the solar corona. More sustained plasma experiments require large magnetic fields, to ensure that 10^6 K plasmas do not touch vacuum chamber vessels whose walls melt near 10^3 K.

There are numerous astronomical plasmas outside our solar system that are not terribly well understood, due to the inability to incorporate all of the relevant physics at all relevant length scales. We have seen cosmic rays, presumably from extragalactic sources with energies above 10^{20} eV. It is quite difficult to imagine how particles could acquire such kinetic energies. Particle accelerators on earth can achieve 10^{13} eV through interactions with electromagnetic fields but finding another seven orders of magnitude is challenging. The mechanism that can produce such energies is not understood.

---

[5]Paul Charbonneau published "Dynamo models of the solar cycle" in *Living Reviews of Solar Physics* in 2009 and Yuhong Fan published "Magnetic fields in the solar convection zone" in the same journal, both in 2009.

EXERCISE 11.6.   What is the velocity of a 1 kg mass that has an energy of 10^{20} eV?

Astronomers have observed examples that can be interpreted as shock waves propagating through stellar nebulas. We can imagine that the sorts of magnetic reconnection events that drive coronal mass ejections on the solar surface are somehow related. The solar flares produce MeV-scale particles, so again there is some extrapolation required.

11.4. Fusion

Perhaps the most intriguing plasma experiments are those that have come from the pursuit of nuclear fusion as an energy source. It is well known now that the nuclear binding energy per nucleon peaks at iron. This explains why heavy elements like uranium can undergo fission reactions: the daughter products are more tightly bound and the process results in energy release. On the other end of the periodic table, lighter elements can fuse into heavier ones, also with a release of energy. As a practical matter, fission reactors are mechanically quite simple. If enough fissile material is present, a chain reaction can be sustained. Thus a fission reactor can be produced simply by stacking enough material in close proximity.

Indeed, there is at least one site on earth where natural reactors ran for several hundred thousand years. In 1972, French quality control engineers at the Pierrelatte uranium enrichment plant noticed that a batch of UF_6 contained somewhat less than the typical 0.72% ^{235}U. As they investigated the production pipeline to determine the source of the discrepancy, they eventually found that ore mined from the Oklo facility in Gabon had as little as 0.44% ^{235}U. Further investigations at the Oklo mine revealed the presence of daughter products, with the conclusion that, roughly two billion years ago, conditions at the site were right to permit sustained fission reactions to occur.

The practical downside of fission reactors is the production of long-lived radioactive isotopes. The daughter products tend to be neutron-rich isotopes that will β decay for potentially long times but there are also heavy, actinide isotopes produced through neutron capture that can have exceptionally long lifetimes.

EXERCISE 11.7.   The IAEA website (www-nds.iaea.org) maintains a portal to a variety of databases of nuclear data. What are the fission product yields for ^{235}U? What are the half-lives of the ten most commonly produced isotopes?

On the other hand, fusion of light isotopes generally produces stable isotopes plus neutrons. The neutrons, given time, will decay into protons but can also be captured into heavy nuclei surrounding the reaction volume. Mitigating the effects of the intense neutron bombardment produced by a fusion reactor is still a research project but the radioactivity problems associated with fusion reactors are not considered as distasteful as those of fission reactors.

At this writing, no commercial fusion reactors are operating and the possibility of ever attaining controlled fusion remains dubious. Such a situation is not due to lack of effort but controlling plasmas has proven extraordinarily difficult. Unlike fission reactors that are nominally simple to construct because the nuclear fission reactions are initiated by neutron capture, fusion reactions require two nuclei to be brought into close proximity. There is a sizable Coulomb barrier to surmount for fusion to occur. There was a spectacular flurry of interest provoked by the claims of cold nuclear fusion, catalyzed by palladium somehow, in 1989 by the publication of Martin Fleischmann and Stanley Pons.[6] While other groups have found the same "excess energy" in calorimetry experiments like those reported by Fleischmann and Pons, they have not done so consistently and there has never been the clear establishment of a nuclear signal. It is fair to say that the electrochemistry of deuterium/palladium systems is not fully understood but it does not appear likely that there is a shortcut to fusion.

EXERCISE 11.8.   What is the Coulomb energy associated with two deuterons separated by one deuteron diameter? The Boltzmann probability is proportional to $\exp[-\mathcal{E}/k_B T]$. What temperature is required to obtain a probability of 10^{-6}, when $\mathcal{E}$ is the Coulomb barrier?

In the sun, the interior reaches very high density due to the gravitational force and high temperatures exist both from the gravitational collapse and ongoing fusion reactions. In a reactor, it is necessary to provide adequate energy to overcome the Coulomb barrier and adequate density to ensure that reaction rates are large enough to sustain the reactions. There have been two principal pathways to achieve fusion in the laboratory. The first is known as inertially confined fusion (ICF). Here, a small target is compressed and heated by incident energy, predominantly from laser sources. The principal difficulties for laser-initiated fusion include (i) reflection of all laser power once a plasma exists around the target, (ii) the need to

---

[6]Fleischmann and Pons published "Electrochemically induced nuclear fusion of deuterium" in the *Journal of Electroanalytical Chemistry and Interfacial Electrochemistry*.

illuminate the target uniformly and simultaneously from several directions and (iii) construction of a suitable laser system that can achieve adequate cycling rates. Inertially confined fusion reactors would inevitably be pulsed, with one target imploded per laser pulse. How this process might be converted into a commercial reactor remains an unsolved challenge.

On the other hand, plasma-based reactors could, in principle, operate in a continuous fashion. As the plasmas burn, additional feedstocks could be introduced, although just how to do this in a stable fashion is problematic. Indeed, all of the years of serious work on plasma physics has revealed a host of instabilities that have had to be confronted one after another.

FIGURE 11.10. A plasma with circular cross-section carries a current j that generates a largely azimuthal B_ϕ magnetic field (loops).

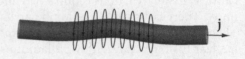

Consider that we have somehow managed to construct a plasma beam of nominal circular cross-section and that it is propagating in the z-direction, as indicated in figure 11.10. The current generates a nominally azimuthal magnetic field. Ideally, there is a Lorentz force acting on the plasma $q\mathbf{v} \times \mathbf{B}$ that is directed radially inward; this self-compression is known as a magnetic pinch. If the beam deviates from straight line motion, though, the current obtains a component in the transverse direction. This, in turn, leads to the generation of a field with a component in the z-direction and the inward-directed field is lessened. As a result, the kink will continue to increase its deviation from straight line motion. This is unfortunate, as realizable fields will always contain imperfections. To some extent, the kink problem can be mitigated by enclosing the plasma in a metal tube, where image currents in the metal counteract the instability to a large extent. Realistically, though, the plasma will almost certainly have more of a Gaussian density profile, not the simple circular cross-section pictured in figure 11.10. Keeping the outer plasma regions from impinging on the walls is challenging.

EXERCISE 11.9.   Consider a plasma particle at the outer radius of the tube, as indicated in figure 11.10. Draw the velocity and magnetic field vectors for points in the kink region. What is the direction of the Lorentz force on a positively charged particle?

A second instability that arises in cylindrical plasmas is the sausage instability sketched in figure 11.11. Here, the radius of the plasma tube

FIGURE 11.11. A second significant instability occurs if
the plasma diameter varies
along the length. Constric-
tions create an increase in the
local current density, further
increasing the confining magnetic field.

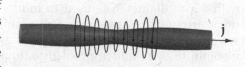

has decreased in a small region. Consequently, the current density in the
vicinity of the constriction will increase, increasing the azimuthal field.
This increases the magnetic pressure on the plasma and will continue to
decrease the diameter of the plasma tube at the constriction site. This
phenomenon is known as the sausage instability, as the plasma tube will
rapidly take on the appearance of a string of sausage links.

EXERCISE 11.10. Consider a plasma particle at the outer radius of
the tube, as indicated in figure 11.11. Draw the velocity and magnetic field vectors for points in the neck region. What is the direction
of the Lorentz force on a positively charged particle?

An early example of the kink behavior was obtained by Alan Sykes and
coworkers at the Culham Laboratory in the UK, as depicted in figure 11.12.
The device utilized a plasma pinch in an attempt to reach high densities
but plasma instabilities prevented further progress.

FIGURE 11.12. The plasma was contained in a circular tube with mirrors to constrain the plasma to a
fraction of 2π. Onset of the kink instability led to device failure.

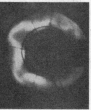

Today, the largest fusion program is the International Thermonuclear Experimental Reactor (ITER) currently under construction in St. Paul-lez-
Durance in southern France. The ITER is a large tokamak, a device invented by Soviet physicists Igor Tamm and Andrei Sakharov in 1951. The
plasma is contained within a toroidal region by magnetic fields provided
by 18 D-shaped toroidal field magnets, each of which is roughly 14 m tall
and 9 m in width, as illustrated in figure 11.13. The azimuthal field produced by the toroidal field magnets peaks at a value around 5.3 T in the
center of the rings. Six poloidal field magnets, two of which are identified
in the figure, form closed circles around the toroidal magnets and serve

to shape the field within the vacuum vessel and prevent the plasma from reaching the walls. A central solenoidal magnet, approximately 11 m tall with a 4 m diameter, is used to induce currents in the plasma. Additional trimming magnets for fine-tuning the fields within the tokamak are not shown. All of these magnets are wound with superconducting niobium-tin wire and are cooled to liquid helium temperatures. The internal plasma temperature will be over 150 million K, so it is critical that the plasma not touch the walls.

FIGURE 11.13. Schematically, the main magnets of ITER consist of 18 toroidal field (TF) magnets that generate an azimuthal field. Several are not displayed here to visualize the central solenoid (SC). The six poloidal field (PF) magnets form continuous rings around the toroidal magnets.

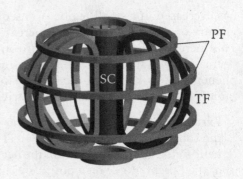

The central solenoid produces a field in the bore of approximately 11 T. As we have seen previously, the field of a solenoid is uniform in the interior and nearly zero outside the solenoid. One might well question the purpose of the solenoid if it does not affect the fields within the plasma region. The rôle of the central solenoid is to induce an azimuthal current in the plasma, which will attain a peak value of 15 MA. Ultimately, the plasma density will peak in the center of the D-shaped toroidal field magnets, as indicated in figure 11.14.

EXERCISE 11.11.   Use the expression for the magnetic field of a current loop developed previously, and the locations of the poloidal field magnets taken from figure 11.14 to calculate the poloidal magnetic field. Assume all magnets are carrying the same current. Use the StreamPlot or StreamDensityPlot functions to visualize the field. As a crude approximation, add another poloidal current at the center of the plasma density. How does the sign of the induced current affect the field patterns?

Everything about the ITER is extraordinarily large, although not exceptionally so given that the plant is expected to produce 500 MW of output power. Other, non-nuclear, power plants of similar power production capabilities have physical sizes that are crudely comparable but energy

FIGURE 11.14. The nominal plasma density within ITER is governed by the poloidal field (PF) coils and the six separate sections of the central solenoid (CS). The gaps gn are monitored to provide plasma control. The crosses (between g1 and g2 and adjacent to g5) mark points where the two magnetic separatrices intersect. The plasma is confined to the interior with the density indicated by the contours. Image courtesy of IAEA.

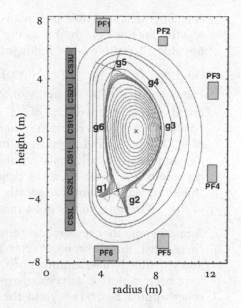

densities, of course, that are vastly smaller. ITER's size is based on the accumulated experimental evidence from a long series of tokamak designs and represents the best guess at a prototype power production facility. Currently, the general consensus among plasma physicists is that bigger is better.

There are, nonetheless, others who cling to hope that future reactors can be simpler. The energy from ITER will be generated from the burning of deuterium and tritium:

$$(11.29) \qquad {}^{2}H + {}^{3}H \rightarrow {}^{4}He \, (3.5 \text{ MeV}) + n \, (14.1 \text{ MeV}).$$

This process leads to the production of an extraordinary neutron flux. The neutrons are very penetrating and require shielding to avoid radiation damage to the machinery.

A drawback to this approach is that deuterium is common but tritium is not. Moreover, tritium is radioactive, which poses handling problems for fueling the reactor. One portion of the ITER research program is to study the possibility of breeding tritium from neutron capture reactions in a lithium blanket surrounding the reactor. Lithium is relatively plentiful and neutron capture on either isotope generally yields ${}^{4}He$ and a tritium nucleus (and another neutron in the case of ${}^{7}Li$). Designers believe that this process can be self-sustaining, reducing the need to obtain tritium from external sources.

It is, of course, possible to run a reactor on deuterium alone but the energetics are not as favorable as the notional deuterium/tritium design. A possible alternative is the following reaction:

$$(11.30) \qquad\qquad p + {}^{11}B \rightarrow 3\,{}^{4}He\ (8.7\ \text{MeV}).$$

Protons with a relative energy of 675 keV striking a boron target are resonant with an excited state of ^{12}C that can decay to $^{4}He + {}^{8}Be$ or three α particles. Using this reaction rather than the deuterium-tritium approach of ITER might provide greater simplicity in a reactor. In particular, the neutron flux will be greatly reduced and the charged α particles are not capable of the same depth of penetration as neutrons. As a result, the shielding requirements are greatly reduced. About 80% of boron consists of the ^{11}B isotope, so it is not a rare quantity.

A number of other researchers remain (often passionately) convinced that there must, somehow, be a better approach. As ITER struggles with cost and schedule issues, a number of alternative magnetic field configurations have been explored. Advocates exist for stellarators, mirror machines and other approaches but beyond the energetic feasibility of fusion power, there remain significant engineering issues that must be addressed before the promise of "burning the oceans" can become reality.

XII

On Stars

After our travails in the previous chapter, where we were only able to scratch at the surface of plasma physics, students may find it perplexing that we know more about the workings of the interior of the sun than we do the visible exterior. This rather surprising situation exists because we can perform laboratory experiments that mimic the reactions that take place within the sun and that statistical mechanics provides us with the framework to compute the macroscopic observables. Moreover, because there is always a logarithm involved in computing macroscopic properties, the sensitivity to details of the underlying microscopic models is reduced.

In trying to discuss the visible portions of the solar atmosphere, we were forced to construct highly simplified models that, at the outset, were not capable of explaining the majority of observations. Stunningly, we could describe a few phenomena, like Alfvén waves, but calculations that can approach a quantitative explanation of solar physics phenomena are still being developed. There are some hopes that the next generation of large computers will provide improvements but it is still very much a work in progress.

On the other hand, much of the detail underlying the nature of the solar furnace was summarized in a review article in 1957.[1] A significant amount of progress has been made since that time but it is telling that the ability to conduct experiments to understand the individual reaction rates led to stellar nucleosynthesis becoming a mature field much more rapidly than plasma physics.

---

[1]Margaret Burbidge, Geoffrey Burbidge, William Fowler and Fred Hoyle published "Synthesis of the elements in stars" in the *Reviews of Modern Physics* in 1957. Fowler was awarded the Nobel Prize in Physics in 1983 "for his theoretical and experimental studies of the nuclear reactions of importance in the formation of the chemical elements in the universe." Fowler shared the prize with Subramanyan Chandrasekhar, who was cited "for his theoretical studies of the physical processes of importance to the structure and evolution of the stars."

© Mark A. Cunningham 2018
M.A. Cunningham, *Beyond Classical Physics*,
Undergraduate Lecture Notes in Physics,
https://doi.org/10.1007/978-3-319-63160-8_12

12.1. Nuclear Fusion Cycles

In 1920, not long after Rutherford's discovery of the nuclear nature of
the atom, British astronomer Arthur Eddingtion noted that Francis As-
ton's precision measurements of atomic masses revealed that the mass of
four hydrogen atoms is greater than the mass of a ^{4}He atom.[2] Eddington's
suggestion contained the solution to a longstanding problem in astronomy. We know the luminosity of the sun and can compute the energy
flux, estimated to be of the order of 3.86×10^{26} W. The solar mass is about
1.99×10^{30} kg, and the solar radius is 1.39×10^9 m. The gravitational potential energy released by assembling this material is given by the following:

$$(12.1) \qquad\qquad \mathcal{U} = -\frac{3GM^2}{5R},$$

where G is the gravitational constant. From the values above, this is a
total gravitational energy of about 1.14×10^{41} J. At the observed rate of
energy release, this energy would be dissipated over about 9.5 million
years. As we believe that stars are much older, Eddington proposed that
fusion could be the source of stellar energy.

> EXERCISE 12.1.   Consider a sphere with a uniform density ρ. The
> gravitational potential energy of the sphere can be computed by in-
> tegrating over all the differential mass elements $dm = \rho \, dV$. Show
> that the following result holds:
>
> $$\mathcal{U} = -\frac{16}{15}\pi^2 \rho^2 G R^5,$$
>
> where G is the gravitational constant. Show that you can recover
> equation 12.1.

Hans Bethe provided a detailed model for the fusion engines in 1939.[3]
Depending upon the internal temperature of the star, Bethe found two
potential pathways for self-sustaining energy production. The first relies
on a chain of reactions built upon the nuclear weak interaction:

$$(12.2) \qquad\qquad p + p \rightarrow d + e^+ + \nu_e.$$

This reaction is quite slow, with estimates that the half-lives of protons
within the sun are of the order of a billion years. The vast majority of the
time, two protons encountering one another in the center of the sun will
simply scatter elastically; proton excited states (N and Δ) are generally
inaccessible. The internal temperature of the sun is presumed to reach

---

[2]Eddington published "The internal constitution of the stars" in *Nature* in 1920.

[3]Bethe published "Energy production in stars" in the *Physical Review* in 1939. Bethe was
awarded the Nobel Prize in Physics in 1967 "for his contributions to the theory of nuclear
reactions, especially his discoveries concerning the energy production in stars."

over 15 million Kelvin but this is only about 1.3 keV in terms of particle energies.

Whenever the deuterium production reaction succeeds, it is quickly followed by ^{3}He production:

$$(12.3) \qquad p + d \rightarrow {}^3\text{He}$$

where the half-lives of deuterons in the solar interior is measured in seconds due to the large proton capture cross-section of deuterium. From this point onward, there are three main competing reactions. The first is inelastic scattering of ^{3}He:

$$(12.4) \qquad {}^3\text{He} + {}^3\text{He} \rightarrow {}^4\text{He} + 2p,$$

which accounts for about 83% of solar energy production. The second involves intermediate production of lithium through electron capture:

$$(12.5) \qquad \begin{aligned} {}^3\text{He} + {}^4\text{He} &\rightarrow {}^7\text{Be} + \gamma \\ &\hookrightarrow {}^7\text{Be} + e^- \rightarrow {}^7\text{Li} + \nu_e \\ &\qquad \hookrightarrow {}^7\text{Li} + p \rightarrow {}^8\text{Be}^* \rightarrow 2\,{}^4\text{He}. \end{aligned}$$

Finally, a reaction pathway that is more important at higher temperatures than within our own sun involves the intermediate production of boron via proton capture:

$$(12.6) \qquad \begin{aligned} {}^3\text{He} + {}^4\text{He} &\rightarrow {}^7\text{Be} + \gamma \\ &\hookrightarrow {}^7\text{Be} + p \rightarrow {}^8\text{B} + \gamma \\ &\qquad \hookrightarrow {}^8\text{Be}^* + e^+ + \nu_e \\ &\qquad\quad \hookrightarrow 2\,{}^4\text{He}. \end{aligned}$$

One might wonder why we have excluded proton capture by ^{3}He as a possible reaction pathway. This is energetically feasible but the proton capture cross-section is small and this process has never been directly observed.

EXERCISE 12.2. From the masses of the constituents, what is the energy released in each of the steps of the p-p chain?

From his analysis, Bethe found that the temperature dependence of these reactions should be relatively modest, so he found it difficult to explain the vastly greater energy release in larger stars. Bethe found an alternative pathway in which ^{4}He is constructed through a catalytic process involving

^{12}C. The reactions involved can be summarized as follows:

$$p + {}^{12}\text{C} \rightarrow {}^{13}\text{N} + \gamma$$
$$\hookrightarrow {}^{13}\text{C} + e^+ + \nu_e$$
$$\hookrightarrow {}^{13}\text{C} + p \rightarrow {}^{14}\text{N} + \gamma$$
$$\text{(12.7)} \qquad\qquad \hookrightarrow {}^{14}\text{N} + p \rightarrow {}^{15}\text{O} + \gamma$$
$$e^+ + \nu_e + {}^{15}\text{N} \hookleftarrow$$
$$\hookrightarrow {}^{15}\text{N} + p \rightarrow {}^{4}\text{He} + {}^{12}\text{C}.$$

Of particular interest in this chain is that it begins and ends with ^{12}C, meaning that ^{12}C is not consumed in the process. This process is much more dependent upon the temperature than the p-p cycle and is expected to be the dominant energy source in massive stars. There are several variations on this pathway, where intermediates undergo proton capture reactions, for example, but the pathway illustrated in equation 12.7 is the most significant.

EXERCISE 12.3.   What are the half-lives of the positron emission steps in the CNO cycle?

A window into the nuclear physics underlying the stellar engine is provided by the neutrino production reactions. Despite the fact that the density of the solar interior is ten times that of gold, neutrinos have extraordinarily small capture cross-sections. We have

$$\text{(12.8)} \qquad \sigma(\nu_e + n \rightarrow p + e^-) = 5 \times 10^{-48}\ \text{m}^2\ \mathcal{E}_\nu^2,$$

where $\mathcal{E}$ is in units of MeV. The mean free path is given by $\lambda = 1/n\sigma$, where n is the number density. For neutrinos that have energies of about 1 MeV, this corresponds to a mean free path of fifty or sixty light years in water. Astonishingly, neutrinos generated within the solar interior leave immediately without even a single scattering interaction. Conversely, the photons emitted in the same processes possess mean free paths on the order of a centimeter and undergo crudely 10^{21} interactions during the several thousand years that it takes for them to reach the stellar surface.

It was confirmation of the relevant nuclear physics that Raymond Davis and his students sought in their neutrino experiments that we discussed in Chapter 5. By the time Davis began collecting data, there was widespread agreement between the nuclear physics, and the internal equation of state from the solar physicists that we understood the solar engine, at least to within 20% or so. There were always refinements to be made. The challenge in studying the cross-sections of the nuclear reactions is that one has to extrapolate to low energies. For example, one step in the CNO cycle is the production of ^{14}N by proton capture on ^{13}C. The cross section for this reaction is depicted in figure 12.1.

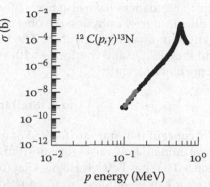

FIGURE 12.1. Experimental cross-section for proton capture in ^{13}C peaks at the 558 keV resonance. In the sun, the reaction proceeds with proton energies in the 1–2 keV range.

EXERCISE 12.4. Cross-section information can be obtained from the IAEA website, using their EXFOR utility. Plot the cross sections for other proton-capture reactions in the CNO chain.

In the sun, the gravitational pressure creates a very high density. In the lab, densities are more modest, so to overcome the Coulomb barrier, one must accelerate protons to at least 80 keV and then extrapolate down. Those extrapolations involve uncertainties that were the subject of debate until Davis found only a third of the expected neutrino flux. This called into question all aspects of the models until the unexpected discovery of neutrino mixing. Now that this exotic property of neutrinos has been confirmed through other means, it now appears that the models of the solar interior are in generally good agreement with the experimental observations.

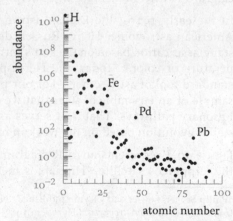

FIGURE 12.2. Elemental abundances established from solar spectral intensities for the light elements and meteor concentrations for the heavier elements are normalized to silicon (10^6). Abundances show distinct even-odd symmetries and enhancements for nuclei that involve multiple α particles, like ^{12}C and ^{16}O.

Others after Bethe extended the fusion chains into heavier and heavier elements. The observed elemental abundances are illustrated in figure 12.2.

These are normalized to an abundance of silicon of 10^6. The figure displays the total elemental abundances but more detailed results on isotopic abundances have also been collected. In one of the more gratifying results in nuclear physics, the observed abundances are in general agreement with model calculations. The began with efforts by William Fowler and Fred Hoyle and continued with work by Donald Clayton and his students more recently.[4]

12.2. Stellar Evolution

The concept that stars are formed of nuclear matter that burns in specified ways immediately raises the question of what happens when the fuel is exhausted? What Fowler, Hoyle, Clayton and others determined is that the interior can be defined by an equation of state that relates internal density, pressure and temperature to the release of thermal energy through fusion and gravitational collapse. These processes are in approximately equilibrium states until the fuel is spent and the system adjusts. Depending on the total mass of the star, it can be possible to continue nuclear fusion through heavier elements. Helium can be fused into carbon, oxygen, etc. This process can continue until the star is mostly iron. At that point, fusion becomes an endothermic process and can no longer compete against the gravitational collapse.

These assertions are, of course, purely speculative. One cannot conduct an experiment with a solar mass of material, partly because assembling it is inconceivable but also because the time scales that we are discussing are measured in at least millions of years, if not billions of years. So, we can turn to astronomers for their help.

In the early 1910s, the Danish astronomer Ejnar Hertzsprung and the American astronomer Henry Russell developed a graphical approach to star classification based on the star's intrinsic luminosity and surface temperature or color.[5] Today, the Hertzsprung-Russell (H-R) diagram is a standard tool of astronomers. For our purposes, it represents a statistical sample of an ensemble of stars that we can assume represent stellar evolutionary pathways. That is, it serves as a touchstone for our ideas about stellar evolution based on the work in nuclear physics.

Figure 12.3 represents an H-R diagram obtained from data acquired by the Hipparcos satellite. Hipparcos measured parallax for thousands of

---

[4]Philip Seeger, Fowler and Clayton published "Nucleosynthesis of heavy elements by neutron capture" in the *Astrophysical Journal* in 1965.
[5]The first publication of a such a plot is credited to Hans Rosenberg, whose "Über den Zusammenhang von Helligkeit und Spektraltypus in den Plejaden" in *Astronomische Nachrichten* in 1910.

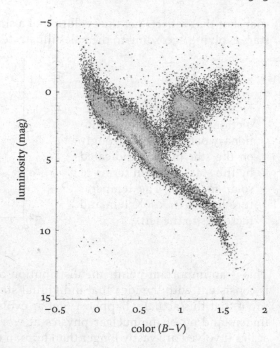

FIGURE 12.3. The Hipparcos satellite measured distances to nearby stars through parallax, which permitted determination of their absolute magnitudes (luminosity). Color index (B-V) is nominally independent of distance. Image courtesy of ESA.

stars in the immediate vicinity, providing an extraordinary opportunity to determine the absolute magnitudes (luminosity) of stars. Astronomers are beset by the difficulty of measuring distance. They can make exquisitely precise measurements of relative intensity but cannot determine absolute intensity without knowing the distance. The color index system UBV involves measuring light intensity integrated over the bands U = 300–400 nm, B = 350–550 nm and V = 480–650 nm. The color indices U-V and B-V should be independent of distance, provided that there is no absorption along the pathway. This permits the assignment of spectral class for even distant stars. Hipparcos provided orders of magnitude more data on stellar distances than were previously available, permitting, among other things, an independent check on the luminosities.

What can be seen from figure 12.3 is that stars are not uniformly scattered across the diagram; they are clustered into bands. The band that extends diagonally from lower right to upper left is known to astronomers as the main sequence. In figure 12.4, we depict the evolutionary pathways for the sun (black) and stars with masses that are multiples of one solar mass. These are distributed by mass along the luminosity/surface temperature plane and as the nuclear fuel is burned, the stars migrate along the indicated paths. Provided that the (B-V) color index is a reasonable representation of the surface temperature, we can infer that the H-R diagram is a snapshot that reflects not the evolution of a single star but an ensemble of stars evolving according to the rules illustrated in figure 12.4.

FIGURE 12.4. Theoretical predictions of the stellar evolutionary paths depend strongly on the star masses, indicated by the star's mass in terms of solar masses. The temperature here is in kiloKelvin and increases to the left.

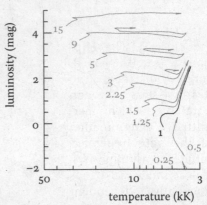

This is an important point: the distribution of stars along various bands is consistent with the idea that individual stars evolve over time but not proof that this actually happens. Stellar evolution is consistent with our understanding of the nuclear physics at work inside stars but the time scales involved are vastly longer than human existence.

The pathways drawn in figure 12.4 represent the evolution of stars of given mass going forward in time. We have not addressed how the stars arrived at their present condition and will defer that discussion until later. That challenge turns out to be one area in which physicists have achieved rather modest success, not surprisingly because star formation involves plasmas and these are exceptionally difficult to understand. Instead, we shall focus on the forward direction in time.

In 1931, the young Indian physicist Subramanyan Chandrasekhar examined the problem of the possible final state of stellar evolution.[6] Chandrasekhar examined the problem from the perspective of hydrostatic equilibrium. There are two main components to the interior pressure, he reasoned, one comes from the mechanical pressure due to gravity and the second comes from radiation pressure generated by fusion:

$$(12.9) \qquad p_{tot} = p_g + p_{rad} = \beta p_{tot} + (1 - \beta) p_{tot},$$

where β simply reflects the fraction of the pressure due to gravitation.

From the final MHD equation, we can deduce that the following is true:

$$(12.10) \qquad \frac{p_{tot}}{\rho^\gamma} = \kappa,$$

---

[6]Chandrasekhar published "The density of white dwarf stars" in the *Philosophical Magazine* and "The maximum mass of ideal white dwarfs" in the *Astrophysical Journal*.

where κ is a constant, independent of time. If we assume that the pressure is solely due to gravitation ($\beta = 1$), equilibrium is obtained when the following relation holds:

$$(12.11) \qquad \frac{dp_{tot}(r)}{dr} = -\frac{GM(r)}{r^2}\rho(r),$$

where ρ is the mass density and M is the mass contained within the radius r:

$$(12.12) \qquad M(r) = 4\pi \int_0^r d\zeta\, \zeta^2 \rho(\zeta).$$

Combining equations 12.11 and 12.12 yields an integro-differential equation, for which there is little mathematical support in generating solutions. We can convert this to a more tractable second-order differential equation by instead considering the derivative of equation 12.11. Then we obtain the defining equation for hydrodynamic equilibrium:

$$(12.13) \qquad \frac{1}{r^2}\frac{d}{dr}\frac{r^2}{\rho(r)}\frac{dp_{tot}}{dr} + 4\pi G\rho(r) = 0.$$

In cases where there is a simple relationship between the pressure and density, like equation 12.10, there is a (non-obvious) change of variables that permits the transformation of equation 12.13 to a previously solved equation. As we have seen, mathematicians can be quite ingenious.

Let us consider the following:

$$(12.14) \qquad \rho = \lambda\theta^n \quad \text{and} \quad p_{tot} = \kappa\rho^{1+1/n},$$

whereupon we note that $n = 1/(1-\gamma)$. Substituting this ansatz back into equation 12.13, we obtain the following:

$$(12.15) \qquad \left[\frac{(n+1)\kappa}{4\pi G}\lambda^{1/n-1}\right]\frac{1}{r^2}\frac{d}{dr}r^2\frac{d\theta}{dr} + \theta^n = 0.$$

Finally, we can redefine the radial coordinate to absorb the prefactor:

$$(12.16) \qquad \xi = \left[\frac{(n+1)\kappa}{4\pi G}\lambda^{1/n-1}\right]^{-1/2} r.$$

This allows us to finally obtain the Lane-Emden equation:

$$(12.17) \qquad \frac{1}{\xi^2}\frac{d}{d\xi}\xi^2\frac{d\theta(\xi)}{d\xi} + \theta^n(\xi) = 0,$$

that has been studied extensively.

Three Lane-Emden functions are depicted in figure 12.5. First of all, we notice that the functions are not positive-definite. Hence, their utility as a description of the density can be called into question. Certainly, density cannot be negative, so near the surface of the star equation 12.13 must

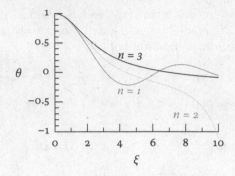

FIGURE 12.5. The Lane-Emden functions for $n = 1$ (gray), $n = 2$ (lightgray) and $n = 3$ (black) can be computed numerically. These satisfy the boundary conditions $\theta(0) = 0$ and $d\theta(0)/d\xi = 0$.

need modification. Next, because we have normalized the functions at the origin, we can assume that λ can be interpreted as the density at the center $\lambda = \rho_0$.

EXERCISE 12.5. We can use the function NDSolve to numerically solve equation 12.17. There is a technical issue that we will need to confront. If we rewrite the Lane-Emden equation in canonical form:

$$\frac{d^2\theta}{d\xi^2} + \frac{2}{\xi}\frac{d\theta}{d\xi} + \theta^n = 0,$$

we see that the term multiplying the first derivative is singular at the origin. This singularity will cause NDSolve to fail but it is something of a benign infinity, known as a regular singular point. As an expedient, let us choose initial conditions $\theta(\delta) = 0$ and $d\theta(\delta)/d\xi = 1$, where δ is a small number, like 10^{-5}.

Use NDSolve to compute solutions to the Lane-Emden equation for $n = 3$ and $\delta = 10^{-5}$ and $\delta = 10^{-6}$. Plot the difference between the two solutions. How sensitive is the result to the choice of δ?

EXERCISE 12.6. Use NDSolve to compute the Lane-Emden solution for $n = 3/2$, which corresponds to $\gamma = 5/3$. How does this differ from the $n = 3$ solution? Note for a fractional exponent, the solution will be complex if $\theta < 0$. To avoid this issue, one can replace $\theta(\xi)^n$ with sign$[\theta(\xi)]$abs$[\theta(\xi)]^n$.

Working back through the equations, Chandrasekhar found that he could provide values for the other constants that we have introduced in terms of physical parameters, like the mass of an hydrogen molecule. For the equations to be self-consistent, he found that there were limits to the size of the initial mass if the pressure arising from electron degeneracy were to balance the gravitational pressure. In 1931, Chandrasekhar calculated a value of 0.91 solar masses. Today, using three-dimensional codes and better equations of state, the commonly accepted value is about 1.4 solar masses. Beyond such values, the star would be unstable to gravitational collapse.

12.3. Black Holes

Chandrasekhar's results were not received well by Eddington, whose dominant personality dominated the field in those years. When confronted with Chandrasekhar's results, Eddington could not find a technical fault in their derivation but felt strongly that nature would find a way to prevent such collapses. As a result, Chandrasekhar wrote a monograph on his work and then set it aside to pursue other research interests. As a consequence, Chandrasekhar's results were not widely discussed until many years later.

Before continuing the discussion, we shall stop and review Einstein's theory of general relativity. One might anticipate that some relativistic correction or another could provide the missing mechanism that Eddington needed to prevent stellar collapse. As we shall see, it was Chandrasekhar's insights that proved to be ultimately correct.

As we mentioned in the introduction, Einstein's interests after the success of his special theory of relativity were directed to the study of the more challenging problem of motion in non-inertial frames. This includes situations for which the transformation between coordinate systems is not linear, i.e.,

$$(12.18) \qquad \frac{\partial^2 x'}{\partial x^2} \neq 0.$$

Fortunately, Einstein's friend from his university days, Marcel Grossmann was available to serve as interpreter of the mathematical tools that Einstein required: differential geometry. In any case, Einstein labored for a decade or so to grasp the nuances of the mathematics that we shall reveal here over the next few pages. In deference to Einstein, scientists have had a century to digest his efforts; it is always more difficult for pioneers.

As Grossmann suggested, the relevant mathematical technology for Einstein's program was provided by tensor analysis, an extension of the ideas of vector spaces. It is possible to construct theories without reference to specific coordinate systems, although we will not pursue that avenue here. It is possible to view physical processes as conserved flows in whatever number of dimensions you think necessary, three, four, ten, twenty-six or a million. The mathematical machinery exists to support at least the derivation of the equations of motion, if not the complete solution.

As we have seen, the relevant way to describe the evolution of a system in this methodology is to consider the evolution along the spacetime path defined by the following:

$$ds^2 = \sum_{i,k} g_{ik}dx^i dx^k,$$

where g is the metric tensor. Motion of a particle in a gravitational field is obtained through the geodesic equation, obtained by requiring the covariant derivative to vanish:

$$(12.19) \qquad \frac{d^2x^i}{ds^2} + \sum_{kl}\Gamma^i_{kl}\frac{dx^k}{ds}\frac{dx^l}{ds} = 0.$$

Here, Γ^i_{kl} is the Christoffel symbol and represents the parallel transport of the local Cartesian coordinate system in the (potentially) curved spacetime.

We have mentioned that it is the covariant derivative that specifies the coordinate-independent evolution of a system. So, what might happen if we need to obtain the second derivative of a vector, like computing an acceleration from a position, for example. The German mathematician Bernhard Riemann found that, in general, the order of differentiation matters. Recall that we defined the covariant derivative in chapter 9, in equation 9.26. Riemann determined that, in general, we must have the following:

$$(12.20) \qquad D_k D_l A_m - D_l D_k A_m = \sum_n R^n_{klm}A_n,$$

where A_m is some covariant vector. The fourth-rank tensor R is generally known as the curvature tensor but it is also called the Riemann tensor in honor of Riemann's work. It is defined in terms of the Christoffel symbols:

$$(12.21) \qquad R^n_{klm} = \frac{\partial \Gamma^n_{km}}{\partial x^l} - \frac{\partial \Gamma^n_{kl}}{\partial x^m} + \sum_q \left[\Gamma^n_{ql}\Gamma^q_{km} - \Gamma^n_{qm}\Gamma^q_{kl}\right].$$

Students may be aware of cases in which the order of integration matters but there are never examples presented in early calculus courses where exchanging the order of differentiation matters. In curved spaces, this is not always the case.

In figure 12.6, we illustrate the notion of parallel transport on a curved surface. Locally, one can define a Cartesian coordinate system that is aligned with one tangent to the surface, another tangent that is orthogonal to the first and the normal to the surface. A vector $\mathbf{v}$ at some point a can be decomposed into components in this coordinate system: $\mathbf{v} = v_1\mathbf{e}_1^a + v_2\mathbf{e}_2^a + v_3\mathbf{e}_3^a$. If we now move to an adjacent point b, there is another coordinate system defined by the tangent vectors and the normal. If we try to reconstitute the vector as $\mathbf{v} = v_1\mathbf{e}_1^b + v_2\mathbf{e}_2^b + v_3\mathbf{e}_3^b$, we obtain the

FIGURE 12.6. A vector at a point a can be decomposed into components in the local coordinate frame. Translation to neighboring points b or c is indicated by the black vectors. The gray vectors are obtained by using the components of the vector in the new coordinate system.

gray vector in the figure. If the vector field $\mathbf{v}$ is constant, then the field at the point b is depicted by the black vector. The gray and black vectors differ because the path is curved and our coordinate system can only be defined locally.

What Riemann demonstrated was that, if one conducted a parallel translation of a vector around a closed loop, the existence of curvature would yield a vector that was tilted with respect to the original. How tilted is quantified in the curvature tensor.

The curvature tensor has a number of symmetries that can be demonstrated from its definition:

(12.22) $$R^n_{klm} = -R^k_{nlm} = -R^n_{kml}.$$

That is, the Riemann curvature tensor is antisymmetric in its first two and second two indices,

(12.23) $$R^n_{klm} = R^l_{mnk}$$

but it is symmetric with respect to interchanging the first and second pairs of indices.

Contracting the Riemann tensor on its first and third indices produces another measure of curvature, the Ricci tensor:

(12.24) $$R_{km} = \sum_n R^n_{knm}.$$

Using equation 12.21, we see that the Ricci tensor is also defined in terms of the Christoffel symbols:

(12.25) $$R_{km} = \sum_n \frac{\partial \Gamma^n_{km}}{\partial x^n} - \frac{\partial \Gamma^n_{kn}}{\partial x^m} + \sum_{n,q} \left[\Gamma^n_{qn} \Gamma^q_{km} - \Gamma^n_{qm} \Gamma^q_{kn} \right].$$

It is a bit difficult to see the physical nature of the Ricci tensor. Mathematicians might argue that it represents an averaging of the Riemannian curvature and physicists might argue that it estimates the difference in a spherical volumetric element in a curved space from the volume that it would have had in a Cartesian space. We'll come back to this subsequently but it provides a measure of curvature.

EXERCISE 12.7. Show that the Ricci tensor is symmetric: $R_{ik} = R_{ki}$.

Einstein now had one necessary ingredient in his possession: a measure of the curvature. What he now wanted was to connect the curvature of spacetime with the presence of mass. Here, the earlier work we discussed on electromagnetics becomes quite useful. In order to demonstrate that Maxwell's equations were compatible with the concept of Lorentz invariance, we rewrote them in a tensor form. We didn't discuss it at the time but it is possible to derive Maxwell's equations through the principle of least action, where the action is defined as follows:

$$(12.26) \qquad S = \int d^4x \, \mathcal{L} = \int d^4x \left[-\frac{1}{4\mu_0} \sum_{ik} F^{ik} F_{ik} - \sum_k A_k J^k \right].$$

We have already seen that invariance of the Lagrangian density $\mathcal{L}$ with respect to gauge transformations leads to a conserved Noether current: the conservation of electric charge given in equation 3.3. The Lagrangian density is also invariant to translation and the resultant Noether current is given by the following:

$$(12.27) \qquad \sum_i D_i \left[\sum_m \frac{\mathcal{L}}{\partial(\partial A_m / \partial x^i)} \frac{\partial A_m}{\partial x^k} - g_{ik} \mathcal{L} \right] = 0,$$

where the term in brackets is known as the canonical stress-momentum-energy tensor T_{ik}. This form of the stress tensor has some deficiencies that can be rectified by producing a similar tensor that is symmetric and traceless, and maintains the property that the four-divergence vanishes. The advantage is that the symmetric tensor corresponds to the Maxwell tensor and the usual definitions of energy density in the fields. The point is that the fact that the four-divergence of the stress tensor vanishes is a manifestation of Noether's theorem, although Maxwell was unaware of this at the time of his development of the electromagnetic theory.

The symmetrized electromagnetic stress tensor is given by the following:

$$(12.28) \qquad T^{ik} = \epsilon_0 \sum_{lm} \left[g_{lm} F^{il} F^{mk} + \frac{1}{4} g^{ik} F_{lm} F^{lm} \right],$$

where we have reused the symbol T for the tensor. This discussion may seem a bit misplaced and awkward but the stress tensor is widely utilized in the mechanics of ponderous media. It provides the key link to Einstein's new gravitational theory.

We begin with the equation of the geodesic. Any object will follow a path, potentially a curved path, that minimizes the action. This is defined mathematically by the equation 12.19. Curvature is embodied within the Christoffel symbols which are defined within the Ricci tensor. The source of that curvature is simply the stress-energy tensor:

$$(12.29) \qquad\qquad T^{ik} = \kappa R^{ik},$$

where κ is some constant of proportionality. Einstein spent years developing this rather simple equation out of the need to ensure that it also incorporates Newtonian gravity as a limiting case. We know that Newton's theory of gravitation is quite good at predicting planetary motion. As a result, whatever theory you develop has to recover $F = -GM_1 M_2/r_{12}^2$, at least in the limit where the masses are relatively small.

A significant difficulty Einstein faced was that he was dealing with tensor equations and there are many of them. The symmetric stress tensor has 16 components but only ten of them can be independent, due to the symmetry. Similarly, the Riemann curvature tensor has 20 independent components, due to the antisymmetry properties that we have identified. Nonetheless, there are 20 equations for each of the components and, by hand, one must be quite diligent to avoid algebraic errors. Einstein adopted a number of quality control measures, as he gained facility with performing the calculations, not unlike introductory physics instructors' pleas to students to check their units. Progress was necessarily slow to avoid algebraic errors.

Ultimately, Einstein settled on the following statement for his general theory of relativity:

$$(12.30) \qquad\qquad R_{ik} - \frac{1}{2}g_{ik}\left[\sum_{jl} g^{jl} R_{jl}\right] + \Lambda g_{ik} = \frac{8\pi G}{c^4} T_{ik},$$

where G is the Newtonian gravitational constant and the term in brackets is the scalar curvature R. The second term has a technical purpose: it makes the four-divergence of the curvature vanish, in line with the vanishing four-divergence of the right-hand side. The third term has a cosmological purpose. Einstein left it out in his earliest work but recognized that the left-hand side of equation 12.30 could also contain a constant multiplier of the metric and still have a vanishing four-divergence. Subsequently, Einstein referred to the so-called cosmological constant as his greatest mistake, as he feared that it spoiled the ultimate symmetry of the equation. The existence of this term provides an independent scaling of the curvature of the universe, independent of the existence of local mass.

EXERCISE 12.8. Sometimes theorists will express equation 12.30 in its scalar form: $R = 8\pi GT$, where $c = 1$. Instead, let us expand the equation. Rewrite the left-hand side of equation 12.30 in terms of the metric tensor and its derivatives. Do not leave out any summation signs.

With his new equations in hand, Einstein sought experimental verification that he was on the right track. The one place in the solar system where there was a discrepancy between observation and Newton's theory was the precession of the perihelion of Mercury. The difference is 43 arc-seconds/century, noted by the French astronomer Urbain Le Verrier in 1843 from his analysis of the transits of Mercury dating from 1697 onward.[7] Le Verrier was engaged in the precise determination of planetary trajectories, deviations of the planet Uranus from Newton's predictions led to his discovery of the planet Neptune in 1846. His further studies on Mercury simply ended with no resolution until Einstein applied his new equations to the task. Remarkably, where two bodies in Newtonian gravity follow trajectories that are ellipses fixed in space, Einstein found that his theory predicted a small precession in the low field limit that was precisely that observed for Mercury. Resolving this small detail was the first significant victory for the new theory.

Einstein's successful construction of his general theory of relativity is an extraordinary accomplishment, unique in the history of theoretical physics. He was not driven by an unexplained experimental result that blocked further progress in understanding some physical phenomenon. He was not working in the most popular program; indeed, Einstein worked mostly in isolation, relying on a few colleagues for mathematical support. As a result, he created a theory based only on his own vision that physical theory should not depend upon the coordinate system. Along the way, he found that such a theory provides a framework for a new, geometrical approach to the understanding of gravitation but there was no particularly compelling reason to find a new theory of gravity when so little was known about quantum mechanics. Mathematicians sometimes wander off into the academic hinterlands to uncover new truths but physicists do not make such pilgrimages of discovery; they rely on experiment to guide their efforts. As a result, Einstein is a phenomenon that should not be emulated as a rôle model; his particular success has not been repeated.

In the small field approximation, the other notable success for Einstein's theory comes in providing a correction to the signal processing of the GPS satellites. Clocks on the ground measure the local time, which is slower than the clocks on satellites by about 45 μs/day. Without correcting for the effect of general relativity, the position error would accumulate at a

---

[7] Le Verrier's "Détermination nouvelle de lâĂŹorbite de Mercure et de ses perturbations" was published in the *Annales de l'Observatoire de Paris* in 1844.

rate of about 10 km/day. If Einstein had not bothered to finish his work, it is possible that some enterprising satellite engineer would have created a correction table that ultimately came to the attention of some physicist who constructed a modified version of Newton's equations that included this phenomenon. Possibly someone later would have then thought to make the new modified theory relativistically invariant and maybe then Einstein's field equations would have emerged.

In any case, Einstein's equations are generally quite difficult to solve in general. First, there are many components, despite the relatively tidy notation of equation 12.30. Second, the vast majority of physicists have never encountered tensors or the abstract concepts associated with differential geometry, so there is a significant mathematical barrier to surmount. Despite these hurdles, the German physicist Karl Schwarzschild obtained an exact solution for the metric tensor in the case of a spherical star with no charge and no angular momentum.[8] The geodesic equation becomes the following:

$$(12.31) \quad ds^2 = \left[1 - \frac{2GM}{c^2 r}\right] c^2 dt^2 - \left[1 - \frac{2GM}{c^2 r}\right]^{-1} dr^2 - r^2 (d\theta^2 + sin^2\theta \, d\varphi^2).$$

Here, M is the mass of the star and G is the gravitational constant. A puzzling feature of the Schwarzschild solution is that it is divergent at the origin $r = 0$ and when $r = 2GM/c^2 \equiv r_S$, known as the Schwarzschild radius.

EXERCISE 12.9. Plot the functions $1 - a/r$ and $(1 - a/r)^{-1}$. Let a be initially 1. What happens if you change the value of a?

At the Schwarzschild radius, the time coordinate vanishes and the radial coordinate diverges. This behavior is reversed at the coordinate origin. Hilbert, among others, investigated this problem of singularities and, ultimately, it has been found that there is no singular behavior at the Schwarzschild radius: this is a coordinate singularity, not unlike the coordinate singularity that arises at the north pole of a spherical coordinate system.

A more troubling point is that for $r < R_S$, the signs of the first two terms in the geodesic equation 12.31 are reversed. For points exterior to the Schwarzschild radius, we have seen that trajectories are conic sections; in particular, solutions with r constant are attainable. For points interior to the Schwarzschild radius, such solutions are not possible. In fact, the variable r must always decrease: trajectories fall into a real singularity at $r = 0$. This is the definition of a black hole. As it happens, Chandrasekhar

---

[8]Schwarzschild published "Über das Gravitationsfeld eines Masspunktes nach der Einsteinschen Theorie" in the *Sitzungsberichte der Königlich Preussische Akademie der Wissenschaften* in 1916.

was correct: stars of a certain size cannot ultimately sustain themselves against gravitational collapse.

EXERCISE 12.10. Download the Geodesics in Schwarzschild Space notebook from the Wolfram Demonstrations Project. The notebook makes use of the fact that the angular momentum and energy are still conserved quantities. By choosing different starting configurations, you can track the trajectories.

Start with beginning positions of 20 or larger. Demonstrate that elliptical orbits precess.

Looking at trajectories with initial positions of 5 or less, is it possible to stay outside the Schwarzschild radius?

These collapsed stars have been given the name black hole because no trajectories can leave the Schwarzschild radius. Light emanating outward from the vicinity of the black hole is increasingly red-shifted as it approaches r_S and reaches a frequency of 0 at r_S. Realistically, though, matter falling into the black hole will undergo tremendous acceleration and the resulting plasma will certainly radiate. So, black holes will not be black, exactly. Astronomers have subsequently searched for black holes in a number of surveys, without necessarily knowing quite what the signature of a black hole might be.

FIGURE 12.7. Adaptive optics and artificial guide stars have permitted high-resolution imaging of the galactic center. Observations over twenty years have identified Keplerian orbits of stellar objects. The mass of the unseen partner is approximately four million solar masses. This image was created by Prof. Andrea Ghez and her research team at UCLA and was constructed from data sets obtained with the W. M. Keck Telescopes.

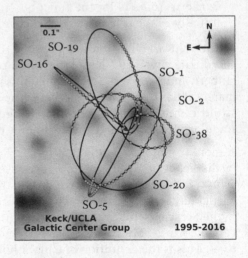

Perhaps the most compelling evidence for a black hole comes from observations of the center of the Milky Way galaxy by Andrea Ghez and her students. Ghez has utilized the Keck telescope and its adaptive optics to provide astonishing images in the infrared of the central core of the galaxy.[9] The Keck telescopes have 10-m, segmented mirrors, where the shape of individual segments is controlled by actuators on the back of the mirror. An artificial guide star is created by shining a laser up into the sky, where sodium atoms about 100 km above the observatory are excited and reradiate light back toward the telescope. Measuring the guide star image deviation from a point source permits the telescope to compensate for atmospheric refraction of light from distant sources. As a consequence, the Keck telescope can produce images of exceptional resolution, nearly free from atmospheric blurring.

Over the course of twenty years, the Ghez team has tracked the motions of stars within the central arc-seconds of the galaxy and have been able to produce the orbital trajectories of several, as illustrated in figure 12.7. These trajectories can all be described by ellipses, as one would expect from both Einstein's and Newton's theories and, when one extracts the mass of the unseen central object, the current best fit is 4.02×10^6 solar masses.

The center of the galaxy is opaque to optical wavelengths, due to the large amount of dust, so the precise nature of Sagittarius A* is still uncertain. Nevertheless, analysis of Keplerian orbits, even with Einsteinian modifications, is straightforward. There is something at the center of our galaxy with a mass of four million suns. Schwarzschild would argue that it must be a black hole and more sophisticated treatments that deviate from spherical symmetry do not alter that conclusion.

EXERCISE 12.11. From Kepler's third law, the period T of an orbit is proportional to the semimajor axis a:

$$T^2 = \frac{4\pi^2}{G(M_1 + M_2)} a^3.$$

Stellar objects SO-38 and SO-102 have orbital periods of 19 and 11.5 years, respectively. Assuming a circular orbit, and that $M_1 = 4 \times 10^6$ solar masses, what is the orbital radius and orbital velocity of these objects?

The case for smaller black holes is a bit more complex. The most widely accepted candidate is an x-ray source discovered in 1963 by US Air Force sounding rockets. The x-ray source in the constellation Cygnus was studied in more detail by the NASA Uhuru x-ray observatory, launched in

[9]Ghez and coworkers published "High proper-motion stars in the vicinity of Sagittarius A*: evidence for a supermassive black hole at the center of our galaxy" in *The Astrophysical Journal* in 1998 and many additional works subsequently.

1970, that demonstrated significant variation in x-ray flux over sub-second time scales.[10] Further Fourier analysis of the x-ray measurements, like those depicted in figure 12.8, did not find any sort of periodic behavior. The second-scale bursts appear to occur randomly.

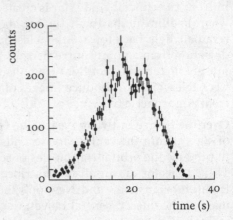

FIGURE 12.8. The Uhura satellite made many measurements of the x-ray flux from Cygnus X-1. The x-ray counts were accumulated for 0.384 s before reporting. The error bars represent statistical uncertainty.

From the duration of the bursts, one can estimate the source size: $d = ct$, where c is the velocity of light. If $t \approx 1$ s, then $d \approx 3 \times 10^8$ m, or roughly half the radius of the sun. This implies that Cygnus X-1 is a compact object. The only sensible mechanism for aperiodic x-ray production around small objects is matter infalling on a black hole. Further observations by NASA's Chandra x-ray observatory have lent further support to the identification of the x-ray source as a black hole: the x-ray bursts have millisecond-scale structure, for example.

Other strong-field tests of Einstein's theory include gravitational lensing. Einstein's theory provides that light also travels along geodesics, which was first studied by Eddington in 1919, who observed light from distant stars that was deflected by the sun.[11] Eddington's results were in agreement with Einstein's predictions and the observations were even reported in the popular press.

A more authoritative test of Einstein's theory in the strong field was discovered in 1979 by astronomers from the Kitt Peak observatory in Arizona. A survey identified two nearly identical quasars separated by 6 arc-seconds that possessed very similar spectra, subsequently identified

---

[10]Giacconi and coworkers published "X-ray pulsations from Cygnus X-1 observed from Uhuru" in *The Astrophysical Journal* in 1971.

[11]Eddington and colleagues published "A determination of the deflection of light by the sun's gravitational field, from observations made at the total eclipse of May 29, 1919" in *Philosophical Transactions of the Royal Society* in 1920.

to possess a light curve in which one lags the other by just over 400 days.[12] There have been several subsequent examples of gravitational lensing that have added to the general acceptance that Einstein's theory provides a good explanation of gravitation.

One of the most conclusive tests to date is the recent detection of gravity waves by the LIGO experiment.[13] Einstein had found wave-like solutions to his equations but was not convinced that they could ever be measured directly. LIGO is an interferometer that has 4 km long legs, schematically depicted in figure 12.9. Large test masses are vibration-isolated and suspended in a long vacuum tube. Laser power is accumulated into the long arms of the interferometer and a small amount is allowed to bleed back through the mirrors M_1 and M_3. A beam splitter routes the combined signal to a detector. Interference between the two beams causes the intensity on the detector to be nearly zero, unless a gravitational wave were to lengthen one of the legs of the detector. This would give rise to a distinct shift of the detector output. Two separate facilities are located in Hanford, Washington and Livingston, Louisiana.

FIGURE 12.9. The LIGO experiment consists of a laser interferometer that maintains a circulating power of 100 kW between distant (4 km) test masses (M_i). A low-intensity laser (20 W) supplies power to both legs of the system through a beam splitter. A change in length of either arm will produce a change in the interference pattern at the detector (D).

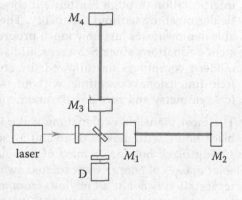

In October of 2015, shortly after the two facilities underwent a significant upgrade in sensitivity, the Hanford detector recorded the signal depicted in figure 12.10. Automatic processing software on the LIGO detectors routinely compares the outputs of the two detectors. A signal propagating at the velocity of light could take as long as 11 ms to traverse the distance

[12]Walsh, Carswell and Weymann published "0957+561 A,B: twin quasistellar objects or gravitational lens?" in Nature in 1979.
[13]Abbott et al. published "Observation of gravitational waves from a binary black hole merger" in the Physical Review Letters in 2016.

between the two facilities. So, the scanning software searches for similarities within the 11 ms window. Within a few minutes of the signal arrival, the scanning software had detected a matching signal from the Livingston detector within the time window. Gravitational waves were real.

FIGURE 12.10. The Hanford detector signal displays a frequency chirp in the interval between 0.35 and 0.45 s. Within the noise limits, the signal from Livingston is identical but shifted earlier by about 7 ms. Data courtesy of the LIGO Scientific Collaboration through the LIGO Open Science Center.

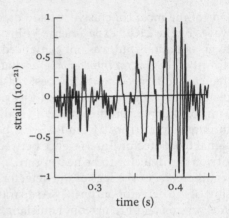

The fact is that the squiggly line in figure 12.9 is meaningless without interpretation through Einstein's theory. It is impossible to infer any particular meaning without the theory. The fact that an explanation is available demonstrates just how much progress has been made in solving Einstein's equations since Schwarzschild's earliest solution. The advent of modern computers has allowed the study of Einstein's theory fully in four-dimensional spacetime, without the restrictions imposed by spherical symmetry and zero angular momentum.

The most plausible explanation of the LIGO signal is the merger of two black holes, with masses of 36 and 29 solar masses, respectively. The resulting black hole has a mass of 62 solar masses, meaning that about 3 solar masses of energy was radiated away by gravitational waves. This interpretation is not at all obvious from an inspection of the data. It stems from computation of a number of different processes that might give rise to gravitational radiation. Each process has a characteristic radiation pattern and the LIGO researchers seek those patterns in their data through matched filters, a common signal-processing technique. After identifying the chirp expected from the rapidly inspiralling black holes, further calculations refined the masses to the values reported.

EXERCISE 12.12. Construct a signal vector from the chirped function $s(t) = e^{-3t}sin[2\pi(t + 10t^2)]$. Take 128 samples of s from $0 \le t \le 1$. The matched filter is the negative of the time-reversed signal. Create a noise vector with the RandomReal function of 1024 samples distributed from -0.2 to 0.2. The signal plus noise vector is the sum of the noise and (shifted by 150) signal vectors. Use the ListConvolve function to perform the matched filter. This should produce a peak at the position of the shifted signal. What happens if you increase the noise intensity to 0.5? Can you still detect the signal?

12.4. Open Issues

It is hard to imagine a higher-field limit than the collision of two black holes, so it appears that Einstein's theory of general relativity is successful over an extraordinarily large range of masses and distances. Increasing computational power and sophistication has permitted the expansion of modelling into the full four-dimensional spacetime. One can now calculate, with some reasonable certainty, the result of black hole collisions and the hydrodynamic models can incorporate convective processes within the stellar mass. This is a vastly more realistic model than was available a short time ago.

Nevertheless, there are difficulties when one asks the question how did we get to where we are now? The models that start with an existing star in basic equilibrium that then burns up its fuel and evolves into something else seem to do a pretty reasonable job of producing at least self-consistent results. When one tries to assemble a solar system or a star or a galaxy from a dust cloud, the models are not successful. This was a real concern for early researchers in computational astrophysics. If you start out with a thousand gravitationally interacting masses in a box and add a little angular momentum to the problem, you do not wind up with a big, rotating blob in the center, you end up with an empty box; the masses scatter out to infinity and are lost.

As a simple example, consider the following function:

```
VerletStep[x_, v_, a_, m_, delta_, first_] :=
Module[{x1 = x, v1 = v, a1 = a, m1 = m, del = delta, ifirst = first,
a2, aij, avec, dum, g, hdel, hdel2, Ntot, vmid, x2, v2},
Ntot = Length[x1]; hdel = 0.5*del; hdel2 = 0.5*del*del; g = 0.1;
(* First time through compute a(t) *)
If[ifirst == 1, aij = Table[{0, 0, 0}, {i, 1, Ntot}, {j, 1, Ntot}];
Do[avec = -g (x1[[i]] - x1[[j]])/Norm[x1[[i]] - x1[[j]]]^3;
aij[[i, j]] = m1[[j]] avec; aij[[j, i]] = -m1[[i]] avec,
{i, 1, Ntot}, {j, i + 1, Ntot}];
a1 = Table[Total[aij[[i, All]]], {i, 1, Ntot}], dum = 0];
```

```
(* x(t+del) = v(t)del + 0.5del^2a(t) *)
x2 = Table[x1[[i]] + del v1[[i]] + hdel2 a1[[i]], {i, 1, Ntot}];
(* v(t+del/2) = v(t) +0.5del a(t) *)
vmid = Table[v1[[i]] + hdel a1[[i]], {i, 1, Ntot}];
(* a(t+del) *)
aij = Table[{0, 0, 0}, {i, 1, Ntot}, {j, 1, Ntot}];
Do[avec = -g (x2[[i]] - x2[[j]])/Norm[x2[[i]] - x2[[j]]]^3;
aij[[i, j]] = m1[[j]] avec; aij[[j, i]] = -m1[[i]] avec,
{i, 1, Ntot}, {j, i + 1, Ntot}];
a2 = Table[Total[aij[[i, All]]], {i, 1, Ntot}];
(* v(t+del) = v(t+del/2) + 0.5del a(t+del) *)
v2 = Table[vmid[[i]] + hdel a2[[i]], {i, 1, Ntot}];
{x2, v2, a2, m1}]
```

This uses the velocity Verlet method to integrate one time step of the gravitational problem. It is not particularly efficient, as it includes a pairwise
sum over all of the particles. As cautionary note, setting the number of
particles to be a million will undoubtedly lock up any computer in existence; be realistic. To use the function, we can start with a (potentially
large) particle in the middle of a box and the remainder scattered around,
with an initial velocity and, potentially, a net angular momentum:

```
DustCloud[Motes_, Steps_, R_, V_, Mo_, L_: 0] :=
Module[{M = Motes, Nsteps = Steps, rmax = R, vmax = V, Mstar = Mo, L1=L,
a1, m1, v1, x1, xsave, z},
z = {0, 0, 1};
x1 = Join[{{0, 0, 0}}, Partition[RandomReal[{-rmax, rmax}, 3 M - 3], 3]];
v1 = Join[{{0, 0, 0}}, Partition[RandomReal[{-vmax, vmax}, 3 M - 3], 3]];
Do[v1[[i]] = v1[[i]] + L1 Cross[z, x1[[i]]];, {i, 1, M}]
a1 = Table[{0, 0, 0}, {i, 1, M}];
m1 = Join[{Mstar}, Table[1, {i, 2, M}]];
xsave = Flatten[x1];
{x1, v1, a1, m1} = VerletStep[x1, v1, a1, m1, 0.04, 1];
xsave = Join[xsave, Flatten[x1]];
Do[{x1, v1, a1, m1} = VerletStep[x1, v1, a1, m1, 0.04, 0];
xsave = Join[xsave, Flatten[x1]];, {i, 1, Nsteps}];
xsave = Partition[Partition[xsave, 3], M];
xsave]
```

To compute the results of fifty particles for 100 time steps, we just invoke
the function

```
pdata = DustCloud[50, 100, 10, 0.1, 10, 0.5];
rx=10;
Manipulate[
Show[ListPointPlot3D[pdata[[i]], PlotStyle -> PointSize[Large],
PlotRange -> {{-rx, rx}, {-rx, rx}, {-rx, rx}}, BoxRatios -> 1,
AxesLabel -> {x, y, z}], Graphics3D[Sphere[pdata[[i, 1]], 0.3]]],
{i,1, Length[pdata], 1}]
```

and use the ListPointPlot3D function to plot the data.

EXERCISE 12.13. Study the behavior of a 50-particle system for different values of the angular momentum L. What happens if the central mass is larger?

EXERCISE 12.14. The simulation in `VerletStep` could be improved if one incorporated collisions. The simplest mechanism is to assume that if the distance between two particles is less than some value, replace the velocity of the first particle with $(m_i v_i + m_j v_j)/(m_i + m_j)$ and the mass of the first particle with $m_i + m_j$ and set $m_j = 0$. To improve performance, one could then eliminate particles with zero mass from the lists but this will cause issues due to the list lengths changing throughout the simulations, so it is simpler to just do unnecessary work. What happens if particles stick together? Does it change things in a noticeable fashion?

A more realistic simulation would involve treating the initial nebula or gas cloud as a fluid but, as we have seen, those sorts of calculations can be quite involved, particularly if we are intending to include phase changes. If the gas can condense into liquid and then solid forms, as might be expected in star formation, this is obviously going to be a significantly more difficult problem.

Beyond just the phase problem, there is an additional problem with angular momentum. If star formation is triggered by some sort of density wave propagating through a gas cloud, it is hard to imagine how that can transpire without having some net angular momentum imparted to the cloud. Like an ice-skater folding in their arms to increase their rotational velocity, as the gas cloud condenses down towards the notional center of local mass, the rotational velocity of the gas will increase, thereby stopping the collapse.

In order to avoid this difficulty, it is necessary to provide a loss mechanism, whereby the kinetic energy within the gas is converted into something else. The most plausible alternative is magnetic fields: charged gas will support currents that generate magnetic fields. Precisely how this energy can be decoupled from the collapse is not yet well understood. An alternative suggestion is that turbulence within the gas might provide a mechanism for abating the angular momentum but incorporating turbulence in hydrodynamic simulations is also challenging.

With these caveats in mind, let us turn to two of the most contentious problems facing astronomers. In 1970, astronomers Vera Rubin and Kent

Ford published measurements of stellar velocities from the nearby Andromeda nebula.[14] The pair subsequently studied a variety of spiral galaxies, measuring the so-called light curves of many galaxies, like the one
depicted in figure 12.11. What is puzzling about this result, which is very
similar to that obtained in other galaxies, is that the light curve is flat far
from the galactic center.

FIGURE 12.11. Measuring the
red shift of stars from the spiral galaxy NGC 1325 leads
to an imputed velocity that
is roughly symmetric around
the center of the spiral.

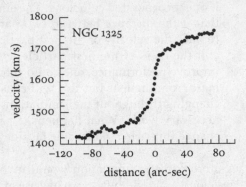

If one thinks of the orbits of the planets within our solar system, the gravitational acceleration due to the sun provides the centripetal acceleration
that maintains the planetary orbits. We have:

$$(12.32) \qquad \frac{GM}{r^2} = \frac{v^2}{r} \quad \text{or} \quad v = \left[\frac{GM}{r} \right]^{1/2}.$$

That is, a naïve estimation of velocity at the outer edges of the galaxy disk
suggests that the velocity should fall like $r^{-1/2}$ instead of becoming constant. Rubin and others have suggested that, in order to explain the light
curves, there must be a large halo of gravitational mass that surrounds
each galaxy. This dark matter is the subject of much current interest.

Let us try an alternative approach before joining the dark matter throng.
In Newtonian gravitation, the gravitational field **G** of a distributed source
can be obtained from the following equation:

$$(12.33) \qquad \mathbf{G}(\mathbf{r}_2) = -G \int d^3\mathbf{r}_1 \, \rho(\mathbf{r}_1) \frac{\mathbf{r}_2 - \mathbf{r}_1}{|\mathbf{r}_2 - \mathbf{r}_1|^3},$$

where G is the universal gravitational constant. This is analogous to the
distributed form of Coulomb's law from electromagnetics. It is generally
not possible to perform these integrations analytically, so one must resort
to numerical simulation. This is why equation 12.32 is most often used as
justification for the existence of dark matter.

---

[14]Rubin and Ford published "Rotation of the Andromeda nebula from a spectroscopic survey of emission regions" in *The Astrophysical Journal*.

Let us choose a very simple model for the galaxy: a disk with thickness h, radius R and a mass density ρ that is only a function of the radius. Students should be forewarned: it is possible to insert equation 12.33 into the Integrate function and press ⟨shift-enter⟩. Be prepared to wait a very long time, even if you choose ρ to be a constant. The problem is that equation 12.33 is singular when $r_2 = r_1$. This is not a problem if we restrict ourselves to the exterior of the mass distribution but it poses a problem when we want to understand the value of the field within the disk.

Unfortunately, we cannot simply use symmetry. In studying a charged sphere, we were able to compute the electric field inside the sphere by using the fact that the electric flux through the surface was equal to the charge enclosed. As the enclosing surface was a sphere, all points on the surface are equivalent, so we know that the field inside scales like r/R^3. A Gaussian surface on our disk has two flat surfaces and the cylindrical perimeter, the radial and longitudinal fields represent two unknowns and we only have one equation.

Instead of assaulting equation 12.33 directly, let us incorporate the astrophysical problem of light curves. That is, we want the centripetal acceleration v^2/r to equal the gravitational acceleration at each point along the radius. So, we can recast the problem into an integral equation for the velocity:

$$(12.34) \qquad -G \int d^3\mathbf{r}_1\, \rho(\mathbf{r}_1) K(\mathbf{r}_1,\mathbf{r}_2) = u(\mathbf{r}_2).$$

Here, K is known as the kernel, which we know from equation 12.33, and u is the centripetal acceleration, which we know from the light curve. In this representation, we are interested for determining the unknown ρ. By discretizing the problem, we can turn equation 12.34 into a matrix equation that we can solve readily.

If we utilize cylindrical coordinates centered on the galactic center and make life simple by restricting our observation points to the x-axis, i.e., $\mathbf{r}_2 = (r,0,0)$ then we can rewrite equation 12.33 as follows:

(12.35)
$$\mathbf{G}(r) = -G \int_0^R d\zeta_1\, \zeta_1 \int_0^{2\pi} d\varphi_1 \int_{-h/2}^{h/2} dz_1\, \rho(\zeta_1) \frac{(r - \zeta_1 \cos\varphi_1, -\zeta_1 \sin\varphi_1, z_1)}{[r^2 - 2r\zeta_1 \cos\varphi_1 + \zeta_1^2 + z_1^2]^{3/2}}.$$

Before panicking, we can readily notice that both the y- and z-components of the integrand are odd functions and, consequently, the integrals vanish.

EXERCISE 12.15. Plot the function $z/(1+z^2)^{3/2}$ over the domain $-3 \leq z \leq 3$. What will be the integral of this function over the domain?

So, we are left only with the x-component of the integral (which corresponds to the radial direction by symmetry). It is possible to perform the z_1 integration directly:

$$(12.36) \qquad \int_{h/2}^{h/2} dz_1 \frac{1}{[a^2 + z_1^2]^{3/2}} = \frac{h}{a^2[a^2 + (h/2)^2]^{1/2}}.$$

Using this result in equation 12.35 we obtain the following:

$$(12.37) \quad \mathbf{G}(r) = -\hat{\boldsymbol{\zeta}} Gh \int_0^R d\zeta_1\, \zeta_1 \int_0^{2\pi} d\varphi_1\, [r - \zeta_1 \cos\varphi_1]$$

$$\times \frac{1}{[r^2 - 2r\zeta_1 \cos\varphi_1 + \zeta_1^2][(h/2)^2 + r^2 - 2r\zeta_1 \cos\varphi_1 + \zeta_1^2]^{1/2}}.$$

If we were to make the usual assumption that h is negligibly small compared to the radial dimension, then we can actually compute the azimuthal integrals. If we pull out a factor of $r^2 + \zeta_1^2$ from the terms in the denominator, we can rewrite the integrand in a dimensionless way:

$$(12.38) \qquad\qquad \frac{r - \zeta_1 \cos\varphi_1}{[r^2 + \zeta_1^2]^{3/2}[1 - \xi \cos\varphi_1]^{3/2}},$$

where we have defined

$$(12.39) \qquad\qquad\qquad \xi = \frac{2r\zeta_1}{r^2 + \zeta_1^2}.$$

The azimuthal integrals produce elliptic integrals that are, unfortunately, singular at $\xi = 1$. As a result, we shall utilize the two-dimensional integral from equation 12.37 instead.

> EXERCISE 12.16. Plot the functions EllipticE and EllipticK over the domain $-1 \leq \xi \leq 1$.

To determine the source term, it has been noted that galactic light curves often follow a simple exponential behavior:

$$(12.40) \qquad\qquad v(r) = v_{\max}[1 - \exp(-r/R_c)],$$

where R_c sets the scale. For a galaxy like the one pictured in figure 12.11, it appears that a reasonable value for R_c is about $0.2R$.

> EXERCISE 12.17. Plot the function $f = 1 - \exp(-r/R_c)$ over the domain $0 \leq r \leq 1$. Use the Manipulate function to vary R_c from 0.1 to 1. Do any of the curves resemble figure 12.11?

To perform the integration, we will utilize Gaussian quadrature. This is a common numerical technique that utilizes the following estimate of the integral:

$$(12.41) \qquad \int_{-1}^{1} dx\, f(x) = \sum_{i}^{N} w_i f(x_i),$$

where the weights w_i and sample points x_i depend on the order N of the estimation. They can be obtained for an arbitrary interval $[a, b]$ instead of $[-1, 1]$ by scaling.

> EXERCISE 12.18. Obtain the quadrature weights from the Gaus-
> sianQuadratureWeights function. Compute the integral of $f(x) =$
> $\tan^{-1}[x/(1 + x^2)]$ from 0 to 1. Use the Integrate function to obtain
> the exact answer. Compute the integral using Gaussian quadratures
> for $N = 4, 6$ and 10. How does the Gaussian technique fare?

The following function will compute the density as a function of the radius. Note the Needs call loads the package that defines the quadrature weights. Here we utilize the fact that the φ_1 dependence is symmetric and double the value of the integral from 0 to π. The choice of Gaussian quadrature was motivated in part by the fact that the endpoints are not evaluated, thereby avoiding any numerical issues at the boundaries.

```
Needs["NumericalDifferentialEquationAnalysis`"]
GalaxyRho[Rc_,h_,Nr_,Nphi_]:=Module[{rc = Rc, hg = h, nr = Nr, np = Nphi,
cosp, Gr, h2, K, Kp, M, phi, r, rho, U},
h2 = (hg/2)^2;
phi = GaussianQuadratureWeights[np, 0, Pi];
cosp = Table[Cos[phi[[i, 1]]], {i, 1, np}];
r = GaussianQuadratureWeights[nr, 0, 1];
K = Table[0, {i, 1, nr}, {j, 1, nr}];
Do[Kp = Table[0, {i, 1, nr}];
Do[Kp[[k]] =
phi[[k, 2]] r[[i,1]]*(r[[j, 1]] - r[[i, 1]] cosp[[k]])/
(r[[j, 1]]^2 - 2 r[[j, 1]] r[[i, 1]] cosp[[k]] + r[[i, 1]]^2)/
Sqrt[h2 + r[[j, 1]]^2 - 2 r[[j, 1]] r[[i, 1]] cosp[[k]]+r[[i, 1]]^2];,
{k, 1, np}];
K[[i, j]] = 2 r[[i, 2]] Total[Kp], {i, 1, nr}, {j, 1, nr}];
Gr=Table[Total[K[[All,j]]],{j, 1, nr}];
U = Table[(1 - Exp[-r[[i, 1]]/rc])^/r[[i, 1]], {i, 1, nr}];
rho = LinearSolve[Transpose[K], U];
M = Total[Table[rho[[i]] r[[i,1]] r[[i, 2]],{i, 1, nr}]];
{M, Table[{r[[i,1]],rho[[i]]},{i,1,nr}],
Table[{r[[i, 1]], 1 - Exp[-r[[i, 1]]/rc]}, {i, 1, nr}],
Table[{r[[i,1]],Gr[[i]]},{i,1,nr}]}]
```

> EXERCISE 12.19. Use the GalaxyRho function to determine the mass
> density for $R_c = 0.1$ and 0.2. Try values of Nr=100 and Nphi=100.
> Are your results sensitive to these choices? Plot the velocities and
> densities.

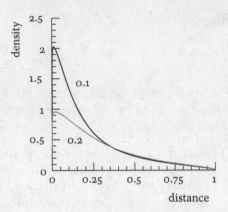

FIGURE 12.12. The density that generates the exponential velocity (equation 12.40) peaks at small values of the radial distance. The profile depends upon the value of R_c (0.1 black/0.2 gray).

In figure 12.12, we have illustrated the densities that provide exponential velocity curves with $R_c = 0.1$ and 0.2. Both densities are small at the origin, which seems inconsistent with the existence of a massive black hole. This is due to numerical issues with the evaluation of the kernel for small values of the radius that overestimate the gravitational field. Otherwise, the density profiles are consistent with experimental results on the mass distribution in galaxies. In any case, we can obtain flat light curves with nothing more than Newtonian gravitation.

EXERCISE 12.20. The GalaxyRho function uses a value of the disk thickness of $h = 0.01$. What happens if you alter the value of h?

EXERCISE 12.21. One can obtain the gravitational field of a disk by utilizing the NIntegrate function that can cope with the singular integrand in an explicit manner. The singular point can be included in the domain: {x1,0,x2,1} and the following directive included in the call to NIntegrate:

```
Method->{"GlobalAdaptive","SingularityDepth"->2,
"SingularityHandler"->"DuffyCoordinates"}
```

Compare the field obtained from the quadrature method and that obtained by the more sophisticated NIntegrate.

Consider the spiral galaxy NGC1309, pictured in figure 12.13. This image is the negated, grayscale conversion of the original Hubble Space Telescope image but will serve as a test case. We can find the pixel corresponding to the center of the galaxy. Load the image with the Import function and convert it to grayscale with the ColorConvert function. Clicking on the image will provide access to the Coordinates Tool. Find the coordinates of the bright point in the center.

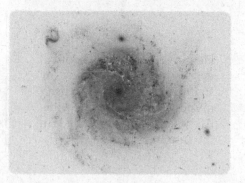

Figure 12.13. The spiral galaxy NGC1309 was imaged by the Hubble space telescope. Image courtesy of ESA/NASA

The following function will sum the intensity of the (not negated) image in bins that are radial to the center. The script assumes that an image file has been imported through the Import function.

```
GalDens[image_, center_] :=
Module[{vals = ImageData[image][[All, All]], ctr = center, ibin,
dens, Ni, Nj, Nd},
{Ni, Nj} = Dimensions[vals];
Nd = Floor[
Max[{Norm[{1, 1} - ctr], Norm[{1, Nj} - ctr], Norm[{Ni, 1} - ctr],
Norm[{Ni, Nj} - ctr]}]/10.];
dens = Table[0., {i, 1, Nd + 1}];
Do[ibin = Floor[Norm[{i, j} - ctr]/10.] + 1;
dens[[ibin]] = dens[[ibin]] + vals[[i, j]];, {i, 1, Ni}, {j, 1, Nj}];
dens]
```

The density obtained from the NGC1309 image is displayed in figure 12.14. Radial bins were ten pixels wide and no attempt was made to eliminate foreground objects from the sums. We see that the central density is small, owing to the fact that the bin areas increase as the radius increases. While these results are certainly influenced by the polar coordinate system and poor dynamic range of the image, they also reflect the difficulty that astronomers face when trying to equate luminosity with mass. Much of the light emanating from the highly populated galactic core is absorbed by dust and reradiated at much longer wavelengths. One can see something akin to this effect when driving through fog: the headlights of approaching vehicles are not apparent from a distance. They only appear dramatically at uncomfortably short range. As a practical matter, obtaining a meaningful correlation between observed galactic luminosity and galactic mass is a contentious problem. Consequently, our inferred galactic densities illustrated in figure 12.12 are not to be discounted.

Exercise 12.22. Use the GalDens function to compute the radial density of other galaxies. Many images are available from the Hubble Space Telescope site.

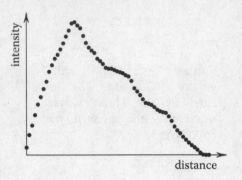

Figure 12.14. The optical density was obtained by summing the image intensity in radial bins 10 pixels wide. All pixels were included in the summations.

Real galaxies are, of course, not radially symmetric. Consequently, our simple disk model will need to be improved in order to establish a more direct comparison with experiment. More sophisticated analyses also yield results that do not require a large halo of unknown matter surrounding galaxies. So, one can wonder how dark matter has become ubiquitous in the discussions of modern science. In some sense, this is a sociology question. There are no experimental results available to physicists seeking grand unified theories; accelerators are many orders of magnitude too small. It is possible that many of these things might have been created during the initial big bang and, if they are weakly interacting with normal matter, might still be present. Hence, if astronomers say there is a halo of unknown stuff outside galaxies, there are numerous theorists with unseen particles who are grasping for any data. This is not to say that we understand the rotational curves in detail and we have all of the physics in hand. There is much that we do not know but we can state without question that the ubiquitous explanation that light curves demand dark matter is incorrect.

As a final topic, we come to the question that has intrigued astronomers for many years: is the universe open or closed? The answer to this question has been emphatically debated for many years. The discovery of the cosmic background radiation by Arno Penzias and Robert Wilson from Bell Laboratories in 1965 provided significant support for the Big Bang theory of the universe.[15] The 3 K microwave flux that permeates space can be interpreted as the remnants of the original energy density when the universe was created.

---

[15]Penzias and Wilson published "A measurement of excess antenna temperature at 4080 Mc/s" in the *Astrophysical Journal* in 1965. They were awarded the Nobel Prize in Physics in 1978 "or their discovery of cosmic microwave background radiation." They shared the award with Pyotr Kaptisa, who was cited "for his basic inventions and discoveries in the area of low-temperature physics."

The biggest astronomy question is what will be the ultimate fate of the universe? Will it collapse back upon itself or will it continue to expand for infinity? Of course, one might argue that the answer to this question is completely irrelevant, as it will have no impact whatsoever on the future of human existence. The question, though, is tied to our general interest in the nature of the universe and our rôle in it. As a result, there are some who are passionately interested in the answer.

If we look at the question a bit more dispassionately, we can observe that the fundamental problem facing astronomers is the lack of a distance measurement. There is no doubt that instruments have evolved that enable us to measure ever fainter sources with ever increasing angular resolution with broader spectral range. Yet most astronomical assumptions are keyed to the interpretation of the red-shift as a measure of distance. This reflects the widely held notion of an expanding universe after the big bang. Sources more distant will be red-shifted more. Unfortunately, there are other sources of red-shift than motion. The interaction of light with matter can produce frequency shifts, both red and blue, through quantum effects not typically discussed in astronomy. There may even be significant interactions with dilute gases that we have yet to uncover in laboratory-scale experiments. As we mentioned earlier, Ray Glauber won a Nobel Prize for work that indicated that low-intensity measurements require the electron/photon interaction to be treated quantum mechanically, not classically. As we have also discussed, the picture of the photon as a small blob of electromagnetic energy is incorrect; we do not really have a description for a photon that has travelled light years.

Recent debate on the issue has been provoked by a 1998 publication by Adam Riess and collaborators, who suggest that the measurement of a particular type of supernova provides evidence for a nonzero cosmological constant Λ.[16] Riess and coworkers rely on the observation of a particular type of supernova, known as Type Ia that are presumed to arise from the accumulation of mass on a white dwarf member of a binary system. Recall that Chandrasekhar found a maximum limit to the mass of so-called white dwarf stars but, if one is a member of a binary system, after the white dwarf formation more mass can be accumulated on the white dwarf. When enough accumulates, the star blows off its exterior layers in a brilliant flash.

Calculations suggest that there is a very small range of masses for which this can occur, meaning that one could use the known brightness of the

---

[16]Riess *et al.*, published "Observational evidence from supernovae for an accelerating universe and a cosmological constant" in *The Astronomical Journal*. The 2011 Nobel Prize in Physics was awarded to Saul Perlmutter, Brian Schmidt and Adam Riess "for the discovery of the accelerating expansion of the Universe through observations of distant supernovae."

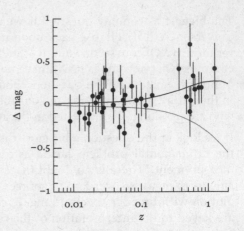

FIGURE 12.15. The difference between the predicted and observed magnitudes based on the MLCS method. The gray curve represents a universe in which there is no cosmological constant. The black curve represents a universe where 76% of the energy density arises from the cosmological constant.

flash with the measured intensity to establish the distance. Type Ia supernovae are characterized by a particular light curve, although recent evidence suggests that there may be more variability in the class than previously assumed. Using the light curve of the supernova to identify its validity as a Type Ia variety, one can then determine the distance independently from the redshift z. As can be seen in figure 12.15, there is significant scatter to the data but one can infer that roughly three quarters of the energy density of the universe can be attributed to the cosmological constant Λ, or something like it. This is an astonishing statement that we know very little of the universe.

This is an unsettling proposition because we generally feel that we know a lot about the universe. We have demonstrated within the pages of this text that we have developed models that are quite successful in explaining phenomena from the microscopic to the macroscopic. Only when we encounter the largest scales, do we seem to stumble. Of course, as aspiring physicists, students must decide for themselves whether or not to concern themselves with the problems of the universe. The difficulty, of course, is that one cannot travel great distances to discover the ground truth. Those experiments are completely impossible. Problems that can be addressed within laboratories are ones that can be solved in an individual lifetime, where it seems likely that the ultimate fate of the universe will continue to inspire debate until the end of the universe.

Mathematical Bits

We make extensive use of complex numbers throughout the text. They provide a succinct means of specifying a number of the concepts that we discuss. As many students are likely to be unfamiliar with the concept, we shall review a few of the more salient points from complex analysis.

Complex Numbers

Fundamentally, a complex number has two components called, somewhat unfortunately, the real and imaginary parts. We can write a complex number z as $z = x + iy$ or $z = (x, y)$, where $i = \sqrt{-1}$. In the second form, the complex number z looks very much like a two-dimensional vector but the space of complex numbers $\mathbb{C}$ is not equivalent to the space of two-dimensional vectors $\mathbb{R}^2$.

Addition of two complex numbers looks like addition of two two-dimensional vectors:

$$(A.1) \qquad z_1 + z_2 = (x_1 + iy_1) + (x_2 + iy_2) = (x_1 + x_2) + i(y_1 + y_2).$$

We define multiplication of complex numbers as follows:

$$(A.2) \qquad z_1 z_2 = (x_1 + iy_1)(x_2 + iy_2) = (x_1 x_2 - y_1 y_2) + i(x_1 y_2 + x_2 y_1).$$

This is not how we might have thought to multiply two two-dimensional vectors. In $\mathbb{R}^2$, we anticipate that the distance between two points would be obtained through the dot product:

$$(A.3) \qquad d = [z_1 \cdot z_1]^{1/2} = [x_1^2 + y_1^2]^{1/2},$$

but this is not what we obtain if we multiply z_1 by itself. Instead, we must multiply z_1 by the complex conjugate of itself:

$$(A.4) \qquad z_1 z_1^* = (x_1 + iy_1)(x_1 - iy_1) = x_1^2 + y_1^2.$$

Among the many advantages of complex numbers is the ability to easily deal with harmonic functions. Consider the exponential function:

$$(A.5) \qquad e^{ikx} = \cos kx + i \sin kx.$$

© Mark A. Cunningham 2018
M.A. Cunningham, *Beyond Classical Physics*,
Undergraduate Lecture Notes in Physics,
https://doi.org/10.1007/978-3-319-63160-8

Taking derivatives is straightforward:

(A.6)
$$\frac{d}{dx}e^{ikx} = ik\,e^{ikx}.$$

The x-dependence doesn't change. Moreover, if we allow k to become complex, the oscillatory sine and cosine functions become exponentially damped:

(A.7)
$$e^{i(k_r + ik_x)x} = e^{ik_r x}e^{-k_x x}.$$

These properties provide significant advantages when performing the algebra required to obtain solutions to differential equations.

Ultimately, the electric field is a real-valued triplet of numbers at each point in space, so students may not see the advantage of coping with a complex-valued triplet of numbers at each point in space but the advantages can be found *a posteriori*. Perform simpler algebraic steps along the path to solution and then take the real part of the result.

There are a number of powerful results in complex numbers, for example, that can be employed. We can define functions of complex numbers:

(A.8)
$$f(z) = u(x,y) + iv(x,y),$$

where u and v are real-valued functions of two real variables. The function f is differentiable if the limit exists:

(A.9)
$$\frac{df(z)}{dz} = \lim_{\Delta z \to 0}\frac{f(z + \Delta z) - f(z)}{\Delta z}.$$

At first glance, this is just the same definition that students were provided in their introductory calculus classes. Here though, the quantity Δz is a complex number, which means that we can take the limit as the real and imaginary parts independently go to zero. For this to make mathematical sense, the result cannot depend on the order of taking the limit. This condition on differentiability means that not all functions are differentiable. In particular, functions of z^* are not.

Analytic functions are those that are differentiable and the sufficient conditions for differentiability are that the Cauchy-Riemann conditions are met:

(A.10)
$$\frac{\partial u(x,y)}{\partial x} = \frac{\partial v(x,y)}{\partial y} \quad \text{and} \frac{\partial v(x,y)}{\partial x} = -\frac{\partial u(x,y)}{\partial y}.$$

Here, we also note that the partial derivatives must exist and be continuous.

Given the definition of analyticity, Cauchy was able to prove a remarkable theorem: the integral of an analytic function along a closed contour in the complex plane is identically zero, unless the function has singularities in the area enclosed by the contour. Cauchy's residue theorem can be stated as follows:

$$(A.11) \qquad \oint dz\, f(z) = 2\pi i \sum_i R_i,$$

where R_i is the residue of $f(z)$ at the singular point z_i. For simple poles, where $f(z) \propto (z - z_i)^{-1}$, then $R_i = (z - z_i)f(z_i)$. Recall that we have utilized a number of infinite integrals in the text. One way to solve them is to add a semi-circular contour at imaginary infinity. (I realize this sounds silly.) If there is an e^{ikz} in the integrand, such a term will be zero, so adding it to the integral does not alter the value of the integral. Now, though the integral is a closed loop and can be evaluated at points where the integral is singular. Consequently, a complicated integral can be converted to a sum over a small number of terms.

FIGURE A.1. An integral along the real axis (black) can be converted into a contour integral by adding a path at infinity (gray). The integral is obtained by computing the sum of the residues at poles of the integrand (dots).

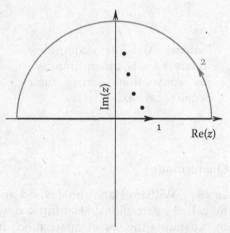

A sketch of the idea is presented in figure A.1. Poles of the function often follow a trajectory in the complex plane and can be found relatively easily numerically. As points for large values of the imaginary part of z are exponentially damped, often integrals can be approximated by a few (one) terms.

A potential issue that arises when attempting to convert real functions into complex ones is that some functions can become multi-valued. This is known technically as a branch point. Consider the function $f(z) = z^{1/2}$. If we use a polar representation of $z = r\,e^{i\varphi}$, then the square root can be seen to map the complex plane $0 \leq \varphi \leq 2\pi$ to a half plane $0 \leq \varphi \leq \pi$. The mathematician Bernhard Riemann envisioned the square root to be a continuous function, if you followed it across the branch cut onto another sheet. As a practical matter, one cannot integrate across a branch cut. The integral must be diverted around any such obstacle.

EXERCISE A.1. Visualize the result of the square root using the Plot3D function:

$$\mathrm{Plot3D[Im[Sqrt[x+Iy]],\{x,-4,4\},\{y,-4,4\}]}.$$

Plot the other half of the Riemann sheet by adding another function to the plot: the square root of $z^* = x - iy$.

We have plotted the imaginary part of the square root in figure A.2. There is, by convention, a discontinuity along the negative real axis. Not all computing languages support complex arithmetic but this choice of branch cut is not universal. This can create difficulties when porting code from one machine to another.

FIGURE A.2. The complex square root is discontinuous (by convention) across the negative real axis.

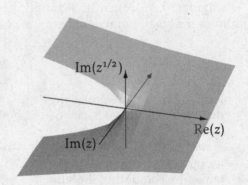

Quaternions

In 1843, William Hamilton devised an extension to complex algebra that he called quaternions. Hamilton devoted most of his remaining years to expanding the use of quaternions in mathematics and physics. Until Gibbs and Heaviside devised vector notation, Hamilton's quaternion approach was becoming more widely used. Today, it is a bit of an historical anecdote but it turns out to be quite useful in computer graphics.

A quaternion is a four-dimensional object (not $\mathbb{R}^4$) with three imaginary components:

$$(A.12) \qquad i^2 = -1 \quad \text{and} \quad j^2 = -1 \quad \text{and} \quad k^2 = -1.$$

These have the following relations:

$$(A.13) \qquad
\begin{array}{lll}
ij = k & jk = i & ki = j \\
ji = -k & kj = -i & ik = -j.
\end{array}$$

Addition follows the same rules as with complex numbers:

$$(A.14) \quad a + b = (a_r, a_i, a_j, a_k) + (b_r, b_i, b_j, b_k) = (a_r + b_r, a_i + b_i, a_j + b_j, a_k + b_k).$$

Multiplication becomes more complicated:

(A.15)
$$
\begin{aligned}
ab = {} & a_r b_r - a_i b_i - a_j b_j - a_k b_k \\
& + i(a_r b_i + a_i b_r + a_j b_k - a_k b_j) \\
& + j(a_r b_j - a_i b_k + a_j b_r + a_k b_i) \\
& + k(a_r b_k + a_i b_j - a_j b_i + a_k b_r).
\end{aligned}
$$

In fact, multiplication of quaternions is not commutative: $ab \neq ba$. Quaternions also require a conjugate to define their magnitude:

(A.16)
$$
a^* = (a_r, -a_i, -a_j, -a_k)
$$

and

(A.17)
$$
|a| = [aa^*]^{1/2} = [a^* a]^{1/2} = [a_r^2 + a_i^2 + a_j^2 + a_k^2]^{1/2}.
$$

The Quaternions package within *Mathematica* software supports quaternion use. Quaternions are most often used in computing rotations for computer graphics applications. This may seem like overkill but the usual approach to rotations involves the rotation matrices that we have encountered previously. These are straightforward to specify for rotations about any of the principal axes but become somewhat more challenging to compute for an arbitrary axis of rotation. Additionally, the strategy based on Euler angles has a coordinate singularity that causes a failure mode known as gimbal lock.

Consider that you are standing at the north pole. Which direction is south? Unfortunately, all directions are south, so codes using angular variables to define the orientation of objects will have to devise schemes to avoid such idiosynchratic behavior. With quaternions, this behavior does not occur.

In addition to multiplications, quaternions support the definition of a multiplicative inverse; quaternions form a division algebra; the space of all quaternions is usually identified as $\mathbb{H}$. Because multiplication is not commutative, to define the inverse of a, we really must mean one of the following:

(A.18)
$$
a_L^{-1} a = 1 \quad \text{or} \quad aa_R^{-1} = 1,
$$

where the left and right inverses need not be the same. If we multiply each equation by $a^*/|a|^2$, on the right and left, respectively, we obtain:

(A.19)
$$
a_L^{-1} = \frac{a^*}{|a|^2} \quad \text{and} \quad a_R^{-1} = \frac{a^*}{|a|^2}.
$$

So, the left and right inverses are the same and, for quaternions with unit magnitude, the inverse is the conjugate.

This last property is the one that makes quaternions useful for computer graphics calculations. We can specify a rotation by the similarity transform:

$$(A.20) \qquad p' = apa^{-1} = apa^*,$$

if a has unit magnitude.

Quaternions can be thought of geometrically, as a scalar a_r and a vector $\mathbf{v} = ia_i + ja_j + ka_k$. In this form, there are some curious properties:

$$(A.21) \qquad \mathbf{v}_1\mathbf{v}_2 = -\mathbf{v}_1 \cdot \mathbf{v}_2 + \mathbf{v}_1 \times \mathbf{v}_2.$$

The product of two quaternions with scalar components equal to zero has a scalar component that is the usual dot product of the two vectors (thought of as vectors in $\mathbb{R}^3$) and a quaternion vector that corresponds to the usual vector cross product. Maxwell utilized this in his original formulation of his theory of electromagnetics.

Octonions

Physicists have explored numerous mathematical pathways seeking a better representation for their ideas. Taking Hamilton's ideas one step further, John Graves found a means to again extend complex numbers, now into eight dimensions. These are called octonions and while one can still define a multiplicative inverse, multiplication is neither commutative nor associative:

$$(A.22) \qquad ab \neq ba \quad \text{and} \quad a(bc) \neq (ab)c.$$

The basis vectors $e_i, \{i = 1, \dots, 8\}$ satisfy the following relations:

$$(A.23) \qquad e_i e_j = -\delta_{ij} e_1 + \varepsilon_{ijk} e_k,$$

where ε_{ijk} is the antisymmetric tensor.

The octonions form a division algebra and the space of all octonions is denoted $\mathbb{O}$. There are no other division algebras beyond real numbers, complex numbers, quaternions and octonions. Physicists have explored the use of octonions to eliminate the need for spinors in the Dirac equation but this has not led to any particular simplifications or improvements in the ability to perform calculations.

Lie algebras

The German mathematician Wilhelm Killing classified all of the possible Lie algebras in a series of papers beginning in 1888.[1] There are three infinite families:

$$(A.24) \qquad \mathfrak{so}(n) \quad \text{and} \quad \mathfrak{su}(n) \quad \text{and} \quad \mathfrak{sp}(n).$$

[1] Killing published "Die Zusammensetzung der stetigen endlichen Transformationsgruppen" in four parts in the *Mathematische Annelen*.

Elements of $so(n)$ are $n \times n$ matrices, whose elements are real numbers ($\mathbb{R}[n]$). Similarly, elements of $su(n)$ and $sp(n)$ are taken from the complex numbers ($\mathbb{C}[n]$) and quaternions ($\mathbb{H}[n]$), respectively. These groups arise naturally as isometries: transformations that preserve a particular Riemannian metric.

Killing also found six other Lie algebras that did not fit the pattern and called them the exceptional Lie algebras. French mathematician Élie Cartan constructed all of the exceptional Lie algebras and realized that two 52-dimensional algebras of Killing's original six were actually the same. So, there are five exceptional Lie algebras: g_2, f_4, e_6, e_7 and e_8. The exceptional groups are all tied to the octonions. Details are to be found elsewhere but there are intriguing clues that these groups may prove useful in defining physical theories.

Of course, there are subtle mathematical connections amongst many of the tools in the physicists toolbox. Perhaps today's students will find the appropriate combination of tools that can blaze the trail towards better theoretical descriptions of nature.

Index

© Mark A. Cunningham 2018
M.A. Cunningham, *Beyond Classical Physics*,
Undergraduate Lecture Notes in Physics,
https://doi.org/10.1007/978-3-319-63160-8

Printed in the United States
By Bookmasters